Bioremediation of Agricultural Soils

Editor

Juan C. Sanchez-Hernandez

Laboratory of Ecotoxicology
Faculty of Environmental Sciences and Biochemistry
University of Castilla-La Mancha
Toledo, Spain

CRC Press
Taylor & Francis Group
Boca Raton London New York

CRC Press is an imprint of the
Taylor & Francis Group, an **informa** business

A SCIENCE PUBLISHERS BOOK

Cover illustrations reproduced by kind courtesy of the editor, Dr. Juan C. Sanchez-Hernandez

CRC Press
Taylor & Francis Group
6000 Broken Sound Parkway NW, Suite 300
Boca Raton, FL 33487-2742

First issued in paperback 2021

© 2019 by Taylor & Francis Group, LLC
CRC Press is an imprint of Taylor & Francis Group, an Informa business

No claim to original U.S. Government works

Version Date: 20181120

ISBN-13: 978-0-367-78017-3 (pbk)
ISBN-13: 978-1-138-65191-3 (hbk)

Library of Congress Cataloging-in-Publication Data
Names: Sanchez-Hernandez, Juan C., editor.
Title: Bioremediation of agricultural soils / editor: Juan C. Sanchez-Hernandez.
Description: Boca Raton, FL : CRC Press, Taylor & Francis Group, 2019. | Includes bibliographical references and index.
Identifiers: LCCN 2018044856 | ISBN 9781138651913 (hardback)
Subjects: LCSH: Soil remediation. | Bioremediation.
Classification: LCC TD878 .B55556 2019 | DDC 628.5/5--dc23
LC record available at https://lccn.loc.gov/2018044856

Visit the Taylor & Francis Web site at
http://www.taylorandfrancis.com

and the CRC Press Web site at
http://www.crcpress.com

Preface

Quality of agricultural soil is under permanent threat of degradation. Environmental phenomena such as erosion, landslides, flooding, decline of organic matter, loss of biodiversity, and pollution are considered the major threats to the quality of agricultural soils. Among them, chemical contamination is the most alarming and stealthy phenomenon because of its long-term adverse impact on soil biodiversity and functioning, ultimately affecting crop productivity.

Agricultural soil receives a wide variety of environmental contaminants through multiple input pathways. Pollutants such as polycyclic aromatic hydrocarbons, polychlorinated and polybrominated biphenyls, pesticides and fertilizers, metals, and more recently, pharmaceuticals and personal care products, reach the soil through irrigation with reclaimed wastewater, application of sewage sludges and agrochemicals to combat respectively nutrient deficit of soil and pests, or even by atmospheric deposition. Occasionally, land destined to agriculture is historically contaminated by toxic chemicals such as metals; this is a frequent challenge in countries with a high mining activity (e.g., Chile), or in urban farming. In addition, in recent years, compelling evidence in recent years show new families of contaminants in soil. This is the case of engineered nanomaterials and microplastics whose environmental fate and toxicity are nowadays topics of increasing concern in the scientific community. Plastic pollution of soil is particularly relevant if we consider that plastics may contain endogenous toxic chemicals that release during their degradation, or they may accumulate and transport exogenous contaminants bound on their surface.

Among the components of the chemical cocktail that contaminate the agricultural soil, pesticides and fertilizers are the main inducers of its deterioration. Decades of intensive research have led to a more sustainable use of agrochemicals that control pests and increase crop yield with a minimum impact in the environment. To keep this idea of sustainability, many methodologies for monitoring pesticide residues in the environment and a vast number of toxicity testing procedures and currently available to managing agricultural pesticides. However, new emerging issues strike this equilibrium in the coming years. For example, recent studies suggest that growing of biofuel crops and the climate change will be two global threats that will increase the agrochemical consumption with unpredictable side-effects in the agroecosystem.

Nowadays, environmental remediation technology needs innovative clean up strategies with a double scope, i.e., the removal of contaminants (and toxicity) from the soil and keeping soil quality as a preventive measure. In this context, bioremediation (i.e., the use of living organisms to remove or inactivate environmental contaminants) provides an attractive approach for removal of chemical stressors and, in turn, improving biological and chemical parameters of soil quality. Scientific literature is plenty of example of bioremediation methodologies to clean soil using microorganisms and plants. Although, most of the case studies of bioremediation have been performed in metal-contaminated sites, there is a growing concern in using *in situ* bioremediation strategies for cleaning up soil contaminated by pesticides. This is the main scope of this book, i.e., to provide the reader with a set of *in situ* bioremediation methodologies that, besides recovering contaminated soils, increase and maintain its quality. The book is set in three parts that collect 13 chapters written by experts in their field.

The first part (**Chemical Stressors in the Agroecosystem**) will introduce the main chemical stressors of current concern in agriculture, offering a cutting-edge knowledge on their environmental fate and toxicity. Four chapters will deal with the most common pesticides and fertilizers used in conventional agriculture, providing a general vision about sources of contamination and potential environmental risks. Particular attention will be put on plastic debris and microplastics as emerging pollutants in soil, and to urban agriculture as an increasing option of sustainable agriculture although not exempt from contamination and risk to human health. The second part (***In situ* Bioremediation**) will provide an up-to-date knowledge on the major *in situ* bioremediation approaches to clean up polluted soils. Readers will find seven chapters that describe the most common bioremediation strategies (e.g., phytoremediation, biostimulation and bioaugmentation) and their principal achievements and limitations. In this part of the book, we describe emerging in situ bioremediation methodologies compatible with sustainable agriculture and the concept of bioeconomy. Among them, vermicompost, biochar and earthworms appear as promising and complementary remediation strategies. Finally, the third part (**Biological Methodologies for Monitoring Bioremediation**) deals with methodologies to be used in the evaluation of bioremediation effectiveness. The major goal of any remediation action is the decrease of concentrations of environmental contaminants below regulatory limits. However, this end does not mean necessarily that soil deterioration disappears. Indeed, adverse effects on the biological components of soil still may persist because of long-term exposure to low-level contamination. The last section will cover this important issue, occasionally forgotten in the remediation programs, describing the most innovative methodologies for monitoring soil degradation through toxicity testing, biomarkers, and soil enzyme activities.

It is expected that the reader finds in this book inspiration for developing novel ways and strategies for the *in situ* bioremediation of agricultural contaminated soils. The interesting point around the multiple strategies discussed in this book coming from the ecological interactions and synergistic effects that can be developed between the biological actors involved in bioremediation, from microorganisms and plants to soil fauna (e.g., earthworms).

Finally, I am very grateful to acknowledge the efforts of all contributors that made this book comes the light, several colleagues from the academic and business sectors for their criticism and suggestions. I also extend my appreciation to the Editorial Department of CRC Press for continuous assistance.

Toledo, November 2018   **Juan C. Sanchez-Hernandez**

Contents

Part 3: Biological Methodologies for Monitoring Bioremediation

Part 1
Chemical Stressors in the Agroecosystem

Chapter 1

Current-use Pesticides
A Historical Overview and Future Perspectives

Francisco Sánchez-Bayo

INTRODUCTION

The use of poisonous substances to control pests is ancient, probably dating to Neolithic times, when agriculture began. Typical pests include not just insects specific to the various crops but also more generic animals like rodents and birds. Plants also interfere with the productivity of crops, and the hampering effects of weeds in food production are mentioned in the Bible.

The vast majority of poisons used for pest control (pesticides) in the past were obtained from plants, with only a few being made from inorganic materials. Among them, volatile substances were used as fumigants to preserve grain and fruit in storage as well as clothing. The use of chemicals to treat plant diseases, mainly fungal infections, developed in the early 20th century on a par with scientific advances in microbiology and biochemistry. Chemicals for weed control, however, were not used on a large scale until the advent of the Green Revolution in the 1950s, as traditionally this type of control relied on human power alone.

The development of the chemical industry was boosted by the Green Revolution, when an enormous diversity of synthetic pesticides was produced to combat not just crop pests and plant diseases but all kinds of living organisms. From primary producers (algae, plants) to animal consumers (insects, snails, rodents and birds), parasites (worms, mites and insects) and saprophytic organisms (bacteria, fungi), pesticides are named according to their target group: algicides, insecticides, rodenticides, fungicides, nematicides, etc.

School of Life and Environmental Sciences, The University of Sydney, Australia.
Email: francisco.sanchez-bayo@sydney.edu.au

Chemical industries that manufacture pesticides have grown huge, consolidating in only five major companies (Jeschke 2016). Together with the fertilizer, pharmaceutical and plastic industries they are having such a vast influence on the biosphere that some authors recognize the modern times as a new geologic epoch called Anthropocene (Waters et al. 2016). The global output of pesticides is currently estimated as 6 million tons per year (Bernhardt et al. 2017), with Europe using one third and the USA one quarter of this production. About 74% of pesticides are used in agriculture for either crop or livestock protection. Other uses include gardening, golf courses, industry and urban pest control (25% of the total consumption), with only 1% being used in forestry (Sánchez-Bayo 2011). Also, the economic value of both the pesticide and pharmaceutical industries "is increasing at a rate more than double that of any other global-change factor", with the pesticides market alone being valued at US$ 29 billion (Bernhardt et al. 2017).

Whether pesticides increase significantly the productivity of the crops remains controversial (Lechenet et al. 2017), although there is evidence that herbicides have a significant increase in yields (Gianessi 2013). However, it is accepted that their continuous use for several decades have created a widespread contamination of the environment that threatens the function and integrity of ecosystems (Vörösmarty et al. 2010, Beketov et al. 2013, Stehle and Schulz 2015, Sánchez-Bayo et al. 2016). Despite these threats, the development of new pesticides continues unabated, driven by a growing agrochemical industry and facilitated by government subsidies, creating more synthetic chemicals that are not less damaging to the environment than the previous ones they intend to replace (Sánchez-Bayo 2018).

Application of Pesticides to Agricultural Crops

Traditionally, pesticides in powdery or liquid formulations have been sprayed over the growing plants using either hand-held sprays or machinery, i.e., ground-rigs attached to tractors or aircraft. The pests and weeds are covered or come in contact with the tiny poisonous droplets thus sprayed. Unfortunately, spraying operations always produce drift, which results in contamination of the landscape at the edges, including irrigation ditches and creeks (Arts et al. 2006, Langhof et al. 2015).

Fumigants, fungicides and herbicides can also be applied directly onto the soil, either as liquid drenches or using granular formulations that dissolve in rain water and penetrate the soil profile. Granular formulations of systemic insecticides can be used also in nursery pots (Phong et al. 2009). Some herbicides are also applied directly into irrigation ditches. These practices contaminate the soil in the treated fields, with residues subsequently moving into ground water or in surface runoff, thus contaminating the nearby creeks and land.

Systemic insecticides and fungicides can be applied in any of the above ways, but recently it has become a common practice to coat the seeds with a formulation containing the two pesticides, so that they protect the seedlings as they grow against fungal and pest attacks—effectively, the plant becomes poisonous. Although this avoids the risk of drift, most of the coating pesticide (> 80%) remains in the soil and eventually ends up in the ground water and surface runoff, contaminating the environment as well (Hladik et al. 2014).

Rodenticides are typically applied as baits so as to attract only the target vermin, e.g., oily meals for rodents, meat for foxes, carrots for rabbits, etc. Powdery or liquid formulations of insecticides are also used for controlling parasites in livestock and animals.

Natural Pesticides

A look at the discovery of the natural compounds that prompted the development of this industry in the 20th century seems necessary. Most of these substances are still in use and registered as pesticides. However, there are hundreds of toxic compounds found in plants, fungi and even animals that have not been used commercially. For a comprehensive account on these substances please refer to Harborne and Baxter (1993).

Pyrethrum is a natural insecticide made from the flowers of *Chrysanthemum cinerariifolium* and *Chrysanthemum coccineum* (Asteraceae). The insecticidal powder (a mixture of six pyrethrins) was used in Persia since 400 BC to kill flies and other nuisance insects, and was introduced in Europe in the early 19th century by "an Armenian merchant, who had discovered the secret of how the powder was prepared while traveling around the Caucasus" (Aromatica, http://web.archive.org/web/20100324061424/http://www.aromatica.hr/eng/page.asp?id=buhac&sub=buhac3). The insecticide also reached America by 1860, but it was not widely used until the end of World War I, when purified insecticidal extracts replaced the powder. The identification of the active compound structure by Lavoslav Ruička and Hermann Staudinger in 1924 enabled later the synthesis of artificial pyrethrum and its derivatives, the pyrethroids, which overcome the photo-instability of the natural pyrethrins. Pyrethrum is still used today all over the world, with Kenya and other African countries being the main producers.

Camphor is a terpenoid found in the wood of the camphor tree (*Cinnamomum camphora*, Lauraceae) and some other plants, e.g., oil from rosemary leaves (*Rosmarinus officinalis*, Lamiaceae) contains up to 10% camphor. It can also be made from turpentine, a distillation product of the resin of pine trees. Camphor was used as insect repellent in eastern Asia for centuries, and was a valuable trade item of the Egyptians and Babylonians, but its main uses were medicinal due to its anesthetic and antimicrobial properties. It was used as a fumigant in Europe during the outbreaks of the plague (Chen et al. 2013) and is still used in moth balls to preserve clothing.

Leaves and seeds of the neem tree (*Azadirachta indica*, Meliaceae) have been used as medicine in India and Southeast Asia over two millennia. The antifungal, antibacterial, antiviral and anthelmintic properties of crude extracts are still used for various purposes in pest control. Their main active compound, azadirachtin, is anti-feedant, repellent and egg-laying deterrent in insects, thus protecting the crop from pest development as well as protecting trees from borer infestations (Immaraju 1998, Kreutzweiser et al. 2011). Azadirachtin is considered a safe pesticide for use in integrated pest management programs.

Strychnine is a highly poisonous alkaloid found in the seeds of the poison nut tree (*Strychnos nux-vomica*, Loganiaceae) that grows in India and Southeast Asia. Powdery preparations of the seeds have been used in baits to kill rats, foxes and other animals over the centuries. Given its poisonous nature to mammals and birds, health regulations in many countries prohibit its use without special permits.

Nicotine is an insecticide found mainly in tobacco (*Nicotiana tabacum*) and other plants of the Solanaceae family. Tobacco plants from South America were introduced in Europe in 1559 as a medicinal remedy against the plague. The insecticidal properties of tobacco were discovered in the 18th century, but the identification of the active substance, nicotine, was first done in 1828 by Wilhelm Heinrich Posselt and Karl Ludwig Reimann in Germany. Nicotine sulfate (40% free nicotine) became one of the main insecticides used during the 19th and early 20th centuries until the introduction of DDT and other synthetic chemicals; the novel neonicotinoids are derived from nicotine (Yamamoto 1998). Nicotine is still used as an insecticide, despite the resistance that some insect pests have developed after many years of usage, but it is no longer registered in the USA (Nicotine 2009). The pyridine anabasine, found in the tree tobacco (*Nicotiana glauca*), resembles nicotine and has the same mode of action (MoA) and insecticidal properties.

Paris green (copper acetoarsenate) is a pigment with broad spectrum toxicity that was used in the French capital to kill sewer rats (*Rattus norvegicus*) during the 18th century expansion of this rodent. The toxic mixture of lead arsenate with this compound was used to control insect pests in apple orchards of America and many European countries around 1900, but it was noticed that the poison also affected trees and other plants. Before the introduction of modern organic insecticides, Paris green was sprayed by airplane to control malaria in Italy and some Mediterranean islands. Due to its non-degradable nature, which leads to persistence and accumulation of arsenic residues in the environment, it is now banned for use in agriculture.

Rotenone is a complex flavonoid found in several plants of the Fabaceae family, such as tuba root (*Derris ellyptica*) from Southeast Asia and New Guinea, hoarypeas (*Tephrosia* spp.), South American robinia (*Lonchocarpus nicou*) and the seeds of the Mexican jicama (*Pachyrhizus erosus*). The powder from the tuba root or seeds from the other plants was traditionally used for fishing by indigenous peoples. The insecticidal properties of rotenone were discovered in 1848, and since then the root powder has been used to kill caterpillars and other insects. The active constituent was purified by Nagai Nagayoshi in 1902, although it was already isolated in 1895 from the South American robinia and named nicouline by Emmanuel Geoffroy (Ambrose and Haag 1936).

Veratridine is a neurotoxic steroid-alkaloid found in some Melanthiaceae plants such as the white hellebore (*Veratrum album*) and the seeds of sabadilla (*Schoenocaulon officinale*). Preparations of the latter seeds were used by the indigenous Americans as insecticides, since the active compound acts in the same way as pyrethrum. Its molecular structure was discovered in the late 1980s (Codding 1983).

Naphthalene, the major component of moth balls, was first extracted from coal tar in the 1820s, but it occurs naturally in the bark of certain species of magnolia. Its pungent odour repels many animals, but naphthalene also kills moth larvae and other insects. Due to its volatility, it is still used as fumigant in buildings and grain storages. The simple structure of two benzene rings makes naphthalene a suitable precursor of many chemicals, one of the best known being the carbamate insecticide carbaryl, produced in 1958 by the Union Carbide (now Bayer).

Coumarin is a fragrant benzopyrone found in tonka beans (*Dipteryx odorata*), sweet clover (*Melilotus officinalis*) and other Fabaceae plants. The active substance was isolated in 1820 and used in European perfumes since the 1880s. In 1920, after cattle that fed on hay containing sweet clover suffered haemorrhaging, it was found that coumarin is converted by certain fungi to dicoumarol, a potent vitamin K antagonist. This discovery led to the production of the so-called coumarin rodenticides, anticoagulants that are widely used nowadays (Watt et al. 2005).

Potassium fluoroacetate occurs naturally in plants of Western Australia, South Africa and Brazil. This poisonous substance was first identified in the leaves of South African gifblaar (*Dichapetalum cymosum*, Dichapetalaceae) by Marais in 1944, but extracts of the poison-leaf *Chailletia toxicaria*, which also contain fluoroacetic acid or its salts, were used by colonists in Sierra Leone in the early 1900s to kill rats. Fluoroacetamide (1081) and sodium fluoroacetate (1080) are the current commercial products derived from this natural poison, and were first used to control rabbits in Australia in the 1950s (Fenner et al. 2009).

The toxic crystals produced by the soil bacterium *Bacillus thuringiensis* (Bt) have been sprayed as insecticides since the 1920s. Only larvae from the Lepidoptera, Diptera, Coleoptera and Hymenoptera orders (e.g., insects with complete metamorphosis) are affected by these toxins: when the caterpillars or grubs ingest the Bt spores, their alkaline digestive tracts dissolve the crystals and liberate the Cry toxin, thus destroying the insect gut cell membrane, paralyzing the digestive tract and starving the insect to death (Boisvert and Boisvert 2000). In the 1990s, the gene that codes the Cry toxic protein was isolated and transferred to maize and cotton plants, starting the so-called transgenic Bt-maize and Bt-cotton so the plants become toxic to their target pests.

In recent decades, many natural toxins have been discovered in fungi, plants and animals and marketed later as insecticides. Among them are the nereistoxins produced by the marine polychaete worm *Lumbriconereis heteropoda*; their structure was established in the early 1960 by Japanese researchers, paving the way for the development of a new class of insecticides that act in a similar way to nicotine (Lee et al. 2003). In the mid-1970, a mixture of two neurotoxins (avermectins B1a and B1b), produced by the soil bacterium *Streptomyces avermitilis* (Actinomycetales), was discovered by William Campbell and Satoshi Omura; it was later marketed as abamectin, the first of a new class of avermectin insecticides (Ishaaya et al. 2002). A decade later, fermentation products from another bacterium, *Saccharopolyspora spinosa* (Actinomycetales), the so-called spinosyns, were found to have insecticidal properties; spinosad is a mixture of two spinosyns (A and D) (Crouse and Sparks 1998). Secretions from the fungus *Strobilurus tenacellus* (Agaricales) contain strobilurins (strobilurin A and oudemansin A), potent substances that help the mushrooms defend against other fungi; their discovery in the mid-1990s has led to the production of this major group of fungicides (Bartlett et al. 2002). About the same time, the alkaloid ryanodine was found in the South American plant *Ryania speciosa* (Salicaceae), giving rise to a novel class of ryanoid insecticides (Usherwood and Vais 1995).

Synthetic Pesticides

Insecticides

Chlorinated aromatic compounds such as dichloro diphenyl trichloroethane (DDT) and hexachlorohexane (HCH) were the first synthetic insecticides introduced to the market in the 1940s, followed by the cyclodienes in the mid-1950s and toxaphene in the late 1960s. All these organochlorines are lipophilic, volatile, recalcitrant to degradation in animal tissues and microorganisms, and many of their metabolites are also toxic and more persistent than the parent compounds. All of them are now banned for use in agriculture and forestry due to their persistence in the environment and bioaccumulation in organisms, although DDT continues to be used for controlling insect-borne infectious diseases (e.g., malaria) in tropical countries.

DDT was originally synthesized by Othmar Zeidler in 1874, well before Hermann Müller discovered its insecticidal properties in 1939. DDT acts upon the sodium channels located in neuronal membranes of all organisms, causing the continuous discharge of sodium ions and effectively inactivating the nervous system, so paralysis ensues (Eldefrawi and Eldefrawi 1990). Sprayed profusely during World War II to control insect vectors of malaria (*Anopheles* mosquitoes) and typhus (lice, fleas), it was first applied to agriculture and forestry in 1946. Hand sprayers for household uses and powder preparations for ectoparasite control in livestock and pet animals were also available. Other compounds with the same mode of action, DDD and methoxychlor were also used but to a much limited extent. Two decades after its introduction on a large scale, its persistent residues were found to accumulate and magnify through the food web, a finding that was linked to population declines of a number of predatory birds (Ratcliffe 1970, Peakall 1993). Further harmful side-effects on rats led to a ban of DDT for agricultural uses, first in the USA (1974) and subsequently in all countries.

All other organochlorines act upon the γ-aminobutyric acid (GABA) receptors located in the post-synaptic membranes of neurons, causing a generalized hyper-stimulation and neuronal activity that results in convulsions (Gant et al. 1987). Since GABA receptors are found in the brain of vertebrates and invertebrates as well as in the neuromuscular junctions and ganglia of arthropods, organochlorines are broad-spectrum poisons. They include gamma-HCH (lindane) and technical HCH (a mixture of alpha-, beta- and gamma-HCH) that were used until recently as seed dressings and foliar sprays over crops, for ectoparasite control in livestock and for treating timber against borers and termites. The water solubility of HCH is 1–2 orders of magnitude higher than that of other organochlorines, allowing its metabolism and excretion by vertebrates as well as its readily dispersion through environmental waters. The cyclodienes are derivatives of cyclopentadiene, and include aldrin, chlordane, chlordecone, dieldrin, endosulfan, endrin, heltachlor, mirex and telodrin. All cyclodienes are very toxic to aquatic organisms, including fish, and more toxic to birds and mammals than DDT (Brown 1978). Residues in prey can produce secondary poisoning in predators, and persistent residues of aldrin, dieldrin, heptachlor and endosulfan, still present in sediments of many aquatic environments, enter the food web and accumulate at the apex of the food pyramid, having similar impacts on populations of birds as DDT (Sibly et al. 2000, Walker 2001). Toxaphene is a mixture of compounds derived from the chlorination of camphene, an aromatic terpene found in many essential oils (e.g., camphor, turpentine, citronella, valerian). It causes a general disruption of the nervous system, including inhibition of GABA receptors and neuronal ATPases (Rao et al. 1986). It was mainly used on cotton and some broad-acre crops in the USA and other countries until it was banned in 1990.

Organophosphorus (OP) insecticides are triesters of orthophosphoric acid ('oxon' form) that were developed in the early 20th century for warfare (i.e., nerve gas). Parathion was first synthesized by G. Schrader in Germany during World War II, opening the way for a new class of highly neurotoxic but degradable insecticides that would eventually replace the recalcitrant organochlorines. About 150 commercial compounds have been manufactured to date, although most of them are phased out. They include well-known compounds like acephate, chlorpyrifos, coumaphos, diazinon, ethion, malathion, naled, omethoate, phorate and terbufos. The 'thion' forms have the oxygen in the acid moiety substituted by sulfur, and only become active after oxidation to the 'oxon' form inside the organisms (Matsumura 1985). Carbamate insecticides are derivatives of carbamic acid and were developed in the early 1960s, also in substitution for the persistent organochlorines. Carbaryl was developed from naphthalene, and carbofuran resembles the natural alkaloid physostigmine found in Calabar beans (*Physostigma venenosum*, Fabaceae), a toxin that shares the same mode of action. About 40 carbamates are still in use, including bendiocarb, fenobucarb, methomyl and propoxur. OP and carbamate insecticides are inhibitors of the enzyme acetylcholine esterase (AChE), which is present in the synaptic cleft of neurons of all organisms, both in the central nervous system and muscular junctions (Matsumura 1985); so, they all are broad-spectrum biocides (Table 1). This inhibition leads to an excess of the neurotransmitter acetylcholine in the cleft, causing a hyper-excitation of the receptors on the post-synaptic membrane that results in spasms and convulsions. The strength of the binding depends on the molecular configuration of each compound: the binding of carbamates lasts only a few hours (i.e., reversible), whereas that of organophosphates is irreversible. All of them are very hazardous to operators, whether farmers or professional applicators (Forget 1991), but after withdrawal from exposure recovery from poisoning can occur by regeneration of the AChE enzyme. Although the majority of OP insecticides sprayed on crops are lipophilic, they usually do not accumulate in the environment because are hydrolysed or metabolised by soil microorganisms (Freed et al. 1979). Most carbamates and some OPs are highly soluble in water and have systemic properties, thus posing serious hazards to consumers: aldicarb has caused human poisoning through contamination of water and fruits (Ragoucy-Sengler et al. 2000).

Synthetic pyrethroids are derivatives of the natural pyrethrins, esters of either pyrethric or chrysanthemic acids with several alkaloids. They became the best choice for replacing organochlorines during the 1970s due mainly to two features: (i) their biodegradability in vertebrates, which does not lead to bioaccumulation and biomagnification through the food chain; and (ii) their low toxicity to terrestrial vertebrates. The selective toxicity also allowed them to replace the broad-spectrum and hazardous OP and carbamates. Some 60 compounds are used nowadays, including bifenthrin, cypermethrin, cyfluthrin, deltamethrin, fenvalerate, permethrin

Table 1. Insecticides and other pesticides used for controlling animal pests.

Mode of action (MoA)	Chemical groups	Target	Toxicity, selectivity	Environmental features
Neurotoxic				
Acetylcholine esterase (AChE) inhibitors	Carbamates	Insects, mites, nematodes, snails	High, broad-spectrum, systemic	Biodegradable, hydrolysis, toxic metabolites
	Organophosphates	Insects, mites, nematodes, snails	High, broad-spectrum	Biodegradable, hydrolysis, some persistent
Glutamate gated chloride channel agonists	Avermectins*	Insects, mites, nematodes	Very high, invertebrates, selective	Biodegradable
γ-aminobutyric acid receptor (GABA) inhibitors	Cyclodienes	Insects, mites, nematodes	High, invertebrates	Volatile, recalcitrant, persistent, bioaccumulation, toxic metabolites
	Phenyl-pyrazoles	Insects, mites	Very high, arthropods	Persistent, toxic metabolites
Nicotinic acetylcholine receptor (nAChR) agonists	Neonicotinoids, nicotine*	Insects	Very high, insects, systemic	Persistent, leaching, toxic metabolites
	Spinosyns*	Insects, mites, nematodes	High, invertebrates	Persistent, possible accumulation
Nicotinic acetylcholine receptor (nAChR) blockers	Dithiols (thiocarbamates), nereistoxin*	Insects, mites	High, broad-spectrum	Biodegradable, hydrolysis
Glycine receptor inhibitors	Strychnine*	Rodents	High, vertebrates	Biodegradable
Octopamine receptor inhibitors	Amitraz	Mites	Low, selective	Biodegradable, hydrolysis
Sodium channel (neurons)	DDT, DDD, dicofol	Insects, mites, nematodes	High, invertebrates	Recalcitrant, persistent, bioaccumulation, toxic metabolites
	Pyrethroids, pyrethrum*	Insects	Very high, arthropods, fish	Biodegradable, photolysis, some persistent
	Indoxacarb, metaflumizon, veratridin*	Insects	Moderate, broad-spectrum	Biodegradable, hydrolysis

Respiration inhibitors

ATPase inhibitors	Diphenyls, sulfluramid	Insects, mites	Moderate, invertebrates, fish	Volatile
	Organotins	Insects, mites, nematodes, snails	High, invertebrates, fish	Persistent, bioaccumulation
Mitochondrial phosphorylation	Chlorfenapyr, diafenthiuron, rotenone*	Insects, mites, nematodes, snails	High, broad-spectrum	Some persistent, bioaccumulation
Hormone mimics				
Ecdysone agonists	Diacylhydrazines, azadirachtin*	Insects	Low, insects	Moderately persistent, leaching
Juvenile hormones	Fenoxicarb, methoprene, etc.	Insects, mites	Low, arthropods	Stable in water, non persistent
Other				
Acetyl CoA carboxylase (ACCase) inhibitors	Ketoenoles (tetronic acid)	Mites	Moderate, broad-spectrum, systemic	Stable in water, non persistent
Chitin inhibitors	Benzoylureas	Insects, mites	Moderate, arthropods	Persistent, bioaccumulation
Citric acid (Krebs cycle) inhibitors	Fluoroacetamides	Rodents	High, vertebrates	Biodegradable
Phosphine (nerve) gas	Metal phosphides	Rodents	Very high, broad-spectrum	Volatile
Ryanodin receptor (RyR) inhibitors	Diamides, ryanodine*	Insects, mites, nematodes	Moderate, broad-spectrum, systemic	Persistent, leaching
Stomach poisons	Bacillus thuringiensis*	Insects	Moderate, insects (selective)	Biodegradable
Vitamin K inhibitors (anticoagulants)	Coumarins	Rodents	High, vertebrates	Persistent, bioaccumulation
	Indandiones	Rodents	High, vertebrates	Moderately persistent, bioaccumulation

* natural compounds

and tau-fluvalinate. Most compounds consist of a mixture of enantiomers. All pyrethroids act upon the voltage-dependent sodium channel of the nerve axon in the same way as DDT, with a characteristic 'knock-down' effect that eventually leads to the death of the insect. They are more toxic to insects and aquatic invertebrates than the organochlorines, but are also toxic to fish (Edwards et al. 1986), so spray drift over water bodies and run off from treated fields are often a problem (Yaméogo et al. 1993, Davies and Cook 1993). Synergistic chemicals such as piperonyl butoxide and azole fungicides, which inhibit the detoxification by cytochrome P450, can increase their toxicity to insects several-fold (Pilling and Jepson 1993). Because of their high hydrophobicity, residues of pyrethroids in aquatic environments tend to adsorb strongly onto sediments, where they can remain and cause toxicity to benthic organisms and bottom fish feeders (Kellar et al. 2014, Weston et al. 2013). More modern insecticides that are antagonist of the sodium channels include indoxacarb and metaflumizone, which block the nervous transmission. The former is degradable by hydrolysis whereas the latter is quite persistent in the environment.

Amitraz is a synthetic, lipophilic compound with insecticidal activity first produced by the Boots Corporation in 1969. It is antagonist of the octopamine receptors in the central nervous system of invertebrates, also inhibiting energy demanding activities such as jumping, flying, or emitting light. Terrestrial vertebrates are not affected by this mode of action, so amitraz is selectively used for controlling mites and ticks in animals.

Benzoylurea insecticides are considered insect growth regulators because they inhibit the biosynthesis of chitin, the main component of the exoskeleton of all crustaceans, therefore stopping their growth. They are not toxic to vertebrates, but their selectivity towards all arthropods causes many problems to aquatic crustaceans (Brock et al. 2009). Some 15 compounds have been developed since the 1980s, including diflubenzuron, flufenoxuron and lufenuron. These compounds are very hydrophobic and recalcitrant to biodegradation, so they bioaccumulate in organisms and biomagnify through the food chain. For this reason, some of them have been banned for use in agriculture. Lufenuron is mainly used for flea control in pet dogs and cats.

As mentioned earlier, avermectins are fermentation products of the bacterium *Streptomyces avermitilis*, which has potent anthelmintic and insecticidal properties. Abamectin and emamectin are insecticides, whereas doramectin, ivermectin, milbemycin and selamectin are used to treat parasites and nematode diseases in livestock and pets. They are agonists of the glutamate-gated chloride channel (GluCl), enhancing the function of these receptors to stop the nervous transmission, thus causing paralysis. Since vertebrates do not have GluCl receptors, these insecticides are very selective towards all invertebrates, including worms, but quite safe to vertebrates. They are biodegradable.

Fipronil is a phenyl-pyrazole antagonist of the GABA-gated chloride channel and also the GluCl channel, binding irreversibly to these receptors and impeding the nervous transmission in arthropods (Cole et al. 1993). This mode of action, shared with the avermectins and cyclodiene organochlorines, provides some selectivity for insect pests, but fipronil is mostly systemic, very toxic to insects and crustaceans, although not so much to fish and vertebrates. Because of its high toxicity to bees, its persistence in the environment and the fact that its metabolites are as toxic as the parent compounds, fipronil was banned in France and is undergoing review for de-registration in other countries.

The spinosyns are agonists of the nicotinic acetylcholine receptors (nAChR) located in the post-synaptic membranes of all neurons, thus competing with acetylcholine for the same receptor and producing hyper-excitation that leads to neuronal death. It appears they are also agonists of the GABA receptors. Due to its lipophilic properties, spinosad may accumulate in organisms although it appears to be degradable (Crouse et al. 2012). The dithiols (thiocarbamates) bensultap, cartap and thiosultap, which are based on the natural nereistoxins, appear to be antagonists of the nAChRs, with strong blocking of these receptors (Lee et al. 2003). They are broad-spectrum but not persistent in soil and easily hydrolysed in the environment.

Neonicotinoid insecticides are derived from nicotine, having the same mode of action as this natural insecticide, which is also shared with spinosyns. However, neonicotinoids bind specifically to the $\alpha 4\beta 2$ subunit of nAChR in insects. The selective toxicity of neonicotinoids towards insects is due to the differential affinity for the subunits that make up the nAChR in insects and vertebrates (Tomizawa and Casida 2005). First launched to the market in the early 1990s, imidacloprid is also antagonist of the nAChR in the neuromuscular junctions of insects. Neonicotinoids are as toxic to insects as pyrethroids (Douglas and Tooker 2016), but less toxic than the latter compounds to crustaceans; surprisingly, waterfleas are as tolerant as fish and most vertebrates (Hayasaka et al. 2012). This feature has given them an advantage over the previous classes of insecticides, so new compounds are constantly being developed. Two main groups can be distinguished: chloropyridinyls (acetamiprid, clothianidin, cycloxaprid, guadipyr, imidacloprid, imidaclothiz, nitenpyram, paichongding, thiacloprid and thiamethoxam) and fluropyridinyls (flupyradifurone and sulfoxaflor). Their systemic properties allow the coating of seeds, granular applications and soil drenching in addition to sprays, so they have become the largest group of insecticides in the market (Jeschke and Nauen 2008). Neonicotinoids are stable in plant tissues, so their residues appear in fruits, seeds, pollen and nectar, posing a high risk to insect pollinators (Pisa et al. 2015). Many of their metabolites are also toxic (Simon-Delso et al. 2015). Most neonicotinoids are persistent in soil and very water soluble, thus leaching through the soil profile (Wettstein et al. 2016). Because of this and their resistance to degradation in water treatment plants, neonicotinoids are contaminating surface and ground waters worldwide (Sánchez-Bayo et al. 2016).

Diamides are a new class of compounds based on the natural ryanodine. They interact with the ryanoid receptors (RyR), which controls the release of calcium in the endoplasmic reticulum of heart and muscle cells to produce muscle contractions (Lahm et al. 2012). Included here are chlorantraniliprole, cyantraniliprole, cyclaniliprole and flubendiamide, which was the first of this group launched in 2007. They have low toxicity to vertebrates, but are very persistent in soil and stable in water, making them very prone to leaching.

Ketoenoles are inhibitors of the acetyl CoA carboxylase (ACCase), involved in the metabolism of fatty acids in all organisms. Spirodiclofen is non-systemic with long-lasting activity, whereas spiromesifen and spirotetramat are systemic. It is not well understood why this group is effective against mites and other animals but not upon fungi or plants (Wenger et al. 2012).

Other insecticides and acaricides with broad-spectrum toxicity include ATPase inhibitors such as the organotins and propargite, which are particularly toxic to aquatic organisms (e.g., crustaceans, fish and amphibians), but are relatively harmless to terrestrial vertebrates. Chlorfenapyr, diafenthiuron, DNOC, dicofol, rotenone and many others disrupt the electron transport mechanism in the mitochondria, each compound targeting a particular complex. All of them are highly toxic to fish, amphibians, zooplankton, worms, mammals and birds (Albers et al. 2006).

The production of safer and more selective insecticides has prompted the development in recent years of insect growth regulators (IGRs), i.e., substances targeting physiological mechanisms that are specific to insects, such as the metamorphosis. Two main groups can be distinguished: (i) ecdysone agonists mimic the steroidal hormone ecdysone, which induces moulting in the larval stages of some insect taxa such as Lepidoptera. Thus, azadirachtin, tebufenozide and other diacylhydrazines cause a premature aging of larvae before reaching the adult stage; they are non-hazardous chemicals; (ii) juvenile hormone analogues like methoprene, hydroprene, kinoprene, pyriproxyfen and the carbamate fenoxycarb are also agonists of the hormonal system, but they prevent the pupae to moult into adult insects (Ishaaya and Horowitz 1995); however, some compounds can also affect zooplankton cladocerans (Chu et al. 1997). All IGRs are degradable and pose little risk to organisms other than arthropods.

Rodenticides

Named after their main target, these poisons are usually applied in baits to attract the target animals. Despite their name, they are poisonous to all vertebrates, not just rodents, and some of them are broad-spectrum biocides.

Indandiones (chlorophacinone and diphacinone) and coumarins (brodifacoum, bromadiolone, coumatetralyl, difenacoum, difethialone, flocoumafen and warfarin) are anticoagulants that effectively block the vitamin K cycle, resulting in the inability to produce essential blood-clotting factors. Hemorrhaging ensues until eventually the animal dies. Calciferols (vitamins D) are synergists of these poisons. Because of their lipophilic properties and stability, most coumarins bioaccumulate and produce secondary poisoning in mammal and bird predators (Stone et al. 1999, Rattner et al. 2011).

Other organic compounds that have been used as rodenticides include 2,4-dinitrophenol (DNP) and bromethalin, both of which inhibit the production of ATP in the mitochondria, and chloralose. DNP is an environmental contaminant, although can be degraded by soil microbes. The fluoroacetate salts 1080 and 1081 mentioned above disrupt the citric acid cycle (Krebs cycle) by combining with coenzyme A, and this results in an accumulation of citrate and fluorocitrate in the blood that impedes the metabolism of sugars. Although they are broad-spectrum poisons, their use in specific baits avoids the exposure of non-target animals. They are water soluble and biodegradable. Australian native animals have developed resistance to these poisons because they occur naturally in many Australian plants.

Among the broad-spectrum rodenticides are metal phosphides, which are used when pest animals have developed some resistance to the previous chemicals. The acid in the digestive system reacts with the phosphide to generate the neurotoxic phosphine gas that kills the animal. The odor of the gas repels most other animals, but not rodents or birds, which become the unfortunate targets. Apart from these, many inorganic salts containing arsenic, barium, calcium and phosphorus are used to kill vermin.

Fungicides, Fumigants and Biocides

As with the previous pesticides, several fungicide classes can be distinguished based on their mode of action (Table 2). Most fungicides are broad-spectrum biocides that act upon biochemical mechanisms common to all fungi, with very few being selective towards one or several fungi classes. Many fungicides and biocides are water soluble and therefore have systemic properties; however, some chemical classes act upon contact and require foliar sprays.

Inorganic compounds have long been used as broad-spectrum fungicides and fumigants. Included in this group are boric acid (borax) and the salts of calcium (e.g., lime sulfur, for treating tree bark), copper (e.g., cuprous oxide, copper acetate, carbonate, hydroxide and sulfate), copper and sulfur (Bordeaux mixture) and ammonia (ammonium acetate, carbonate, hydroxide, sulphamate and thiocyanate). Fungicides containing mercury (e.g., mercuric oxide and mercurous chloride) are no longer used in agriculture because of their risks to humans and other organisms (Ackefors 1971). Common fumigants used in the past were methyl bromide, naphthalene and hexachlorobenzene (currently banned); metam-sodium, sulforyl fluoride, 1,3-dichloropropene and the newly developed iodomethane are used nowadays.

Phenolic compounds are still used as fungicides and fumigants because of their disinfectant properties and generic toxicity, which prevents germination of spores. Included in this group are 2-allyphenol, 8-hydroxiquinoline and pentachlorophenol. The latter compound results from the chlorination of phenol and is an all-purpose biocide that was also registered as insecticide in the past; it is very persistent and prone to leaching. Chlorophenyls (e.g., dicloran, quintozene), biphenyl and dinitrophenyls (e.g., meptyldinocap) inhibit lipid peroxidation and respiration during the germination of spores. Many of them are volatile compounds used as fumigants and are usually persistent in the environment.

Organometallic compounds were discovered in the mid-1800s and were the first group among a large array of chemicals better known as respiration inhibitors. Specifically, organometals inhibit the production of ATP, hence their poisonous nature to all organisms. Among the compounds still used in agriculture are fentin acetate (an organotin) and dithiocarbamates containing manganese (mancozeb, maneb) or zinc (zineb, ziram, metiram).

The discovery of antibiotics at the turn of the 20th century led to the production of many products, some natural and others synthetic, several of which are used as fungicides in agriculture. Later, sulfonamides (e.g., amisulbrom) were discovered in Germany in 1932, which together with sulfamides (e.g., tolylfluanid) are inhibitors

Table 2. Main groups of fungicides and biocides used in agriculture.

Mode of Action (MoA)	Chemical Groups	Selectivity and Application	Environmental Features
Sterol biosynthesis			
Ergosterol biosynthesis inhibitors (SBI)	Conazoles	Broad-spectrum, foliar fungicides	Persistent
	Imidazoles	Broad-spectrum, systemic fungicides	Stable in water, moderately persistent
	Morpholines (piperazines)	Broad-spectrum, systemic fungicides	Biodegradable
	Pyrimidines	Broad-spectrum, systemic fungicides	Moderately persistent, prone to leaching
	Triazoles	Broad-spectrum, systemic fungicides	Stable in water, persistent
Respiration inhibitors			
ATPase inhibitors	Organometallic	Broad-spectrum, fumigants and foliar fungicides	Persistent in soil and organisms, bioaccumulation, biomagnification
Respiration blockers – multisite	Dithiocarbamates	Foliar fumigants	Biodegradable
Respiration and lipid peroxidation inhibitors	Chlorophenyls, dinitrophenyls, biphenyl	Broad-spectrum, foliar biocides, fumigants and fungicides	Volatile, moderately to very persistent
Respiratory chain (Qo-I)-complex III	Strobilurins	Broad-spectrum, foliar fungicides	Moderately persistent, prone to leaching
Respiratory chain (Qi-I)-complex III	Sulfamides, sulfonamides	Broad-spectrum, foliar or systemic fungicides	Moderately persistent
Succinate dehydrogenase (SDHI) inhibitors-complex II	Pyridinyl-benzamides , pyrazole-carboxamides	Selective, foliar or systemic fungicides	Stable in water, persistent
	Carboxamides, benzamides	Broad-spectrum, systemic fungicides	Persistent, prone to leaching
Cell division			
Adenosin deaminase inhibitors – DNA biosynthesis	Pyrimidines	Systemic fungicides	Moderately persistent
β-tubulin assembly-mitosis inhibitors	Benzimidazoles	Systemic fungicides	Stable in water, moderately persistent
	Benzamides	Selective, foliar or systemic fungicides	Persistent, prone to leaching

Cellulose biosynthesis inhibitors	Amides, carbamates	Systemic fungicides, selective	Biodegradable
Melanin inhibitors – multisite activity	Amines, amides	Systemic fungicides	Stable in water, non persistent
	Phthalimides	Foliar biocides and fungicides	Non persistent
RNA polymerase inhibitors	Phenylamide, furalaxyl	Systemic fungicides	Moderately persistent, prone to leaching
Other			
Corrosive, chelating	Inorganic	Broad-spectrum, fumigants and foliar fungicides	Volatile or persistent in all media
	Phenolic	Broad-spectrum, fumigants and foliar fungicides	Mobile
Inducers of plant defence mechanisms	Thiadiazoles, benzothiazoles	Selective, foliar or systemic fungicides	Moderately persistent
Lipid inhibitors	Guanidines	Foliar fungicides	Non persistent
	Dithiocarbamates	Foliar fungicides	Biodegradable
Membrane proteins	Antibiotics	Selective, systemic biocides	Biodegradable
Phospholipid biosynthesis inhibitors	Carbamates, organophosphates	Systemic fungicides	Biodegradable, hydrolysis
Triglyceride biosynthesis inhibitors	Dicarboximides	Systemic fungicides	Non persistent

of sporulation, acting upon the complex III of the electron transport chain (quinone inside inhibitors or Qi-I). They are moderately persistent.

Carboxamides are broad-spectrum fungicides that inhibit the succinate dihydrogenase (SDHI) in the complex II of bacteria and mitochondrial electron transport systems (Gisi et al. 2012). They include systemic but persistent compounds such as boscalid and penthiopyrad, among many others that are now obsolete, but carboxin is less persistent. Pyrazole-carboxamides (e.g., fluxapyroxad, penfuflen) have the same mode of action and persistence but are not systemic; new fungicides in this group include bixafen, isopyrazam and sedaxane (Walker 2012). Pyridinil-benzamides (e.g., flutolanil) are also SDHIs, with new compounds in this class including the foliar fungicide benzovindiflupyr and the systemic fluopyram. Together, SDHI fungicides comprise about 6% of the current fungicide market (Jeschke 2016).

Strobilurins are derived from the fungus *Strobilurus tenacellus* and comprise about 20% of all fungicides used in agriculture nowadays. The first synthetic derivative, azoxystrobin was introduced in the market in 1996 (Bartlett et al. 2002). These compounds inhibit the respiratory system, acting upon the cytochrome bc_1 at complex III (quinone outside inhibitors or Qo-I), located in the inner mitochondrial membrane of fungi and therefore are broad-spectrum fungicides towards all four major classes of fungi (Sauter 2012). Most of them are foliar fungicides and moderately persistent, but some are systemic and prone to leaching. Newly developed compounds in this group are fluoxastrobin, orysastrobin and pyriminostrobin.

The largest chemical group of fungicides used in agriculture is the azoles, comprising about 30% of all compounds. Triazoles inhibit the enzyme lanosterol 14α-demethylase (CYP51A1), which synthesizes a precursor of ergosterol (SBI). Since ergosterol in an essential component of the fungal cell wall, these fungicides are very selective inhibitors of fungal growth (Kuck et al. 2012). Widely used triazole fungicides include difenoconazole, myclobutanil, propiconazole, tebuconazole and triadimenol among 40 others. Newly developed conazoles (e.g., mefentrifluconazole and prothioconazole) and the imidazoles imazalil and prochloraz also inhibit ergosterol synthesis. The majority of azoles are systemic, very stable in water and soil, and resistant to degradation. Because the mechanism of azoles involves inhibition of the cytochrome P450 detoxification system in animals they also act as synergists of pyrethroids and neonicotinoids insecticides (Pilling and Jepson 1993, Iwasa et al. 2004).

Morpholines (e.g., dodemorph, piperalin) are modern systemic fungicides that also inhibit the biosynthesis of ergosterol (SBI) by acting upon the sterol-Δ^{14}-reductase. They are systemic and degradable in the environment. Some pyrimidine fungicides (e.g., fenarimol, pyrifenox) are also systemic SBIs but prone to leaching due to their persistence.

Apart from SBIs, many other fungicides target fungal cell division by a variety of processes. Benzimidazoles (e.g., benomyl, carbendazim) are systemic fungicides that inhibit the assembly of beta-tubulin during mitosis, thus preventing cell division, sporulation and growth of fungi (Young 2012). The newly developed flupicolide (benzamide) also prevents cell division by a different mechanism, i.e., inducing a restructure of the spectrin-like proteins between the cytoskeleton and the plasma membrane; it appears to be selective towards oomycetes. Phthalimides (e.g., captafol, captan and folpet) also inhibit the germination of fungal spores by a variety of mechanisms. Many pyrimidine fungicides (e.g., bupirimate) and debacarb inhibit adenosine deaminase, which synthesizes DNA; mandipropamide inhibits the production of cellulose, one of the components of the cell wall in certain classes of fungicides. Most carbamate and OP fungicides disrupt membrane functions by blocking the synthesis of phospholipids, whereas guanidines and dithiocarbamates inhibit lipid production in multiple ways. Furalaxyl and the phenylamides (e.g., benalaxyl, metalaxyl) inhibit the polymerase that synthesizes ribosomal RNA, stopping protein production, also necessary for cell division and growth. Recent developments include novel modes of action such as inducers of the host plant defense mechanisms (Toquin et al. 2012), as in the case of the systemic fungicide isotianil (thiadiazole).

Herbicides

Despite the complexity of plant biochemistry, only a few mechanisms specific to plant biochemistry and physiology are targeted: about 80% of the current herbicide market consists of chemicals acting upon six MoAs (Jeschke 2016). Some chemical groups are selective to either grasses (monocotyledon) or broad-leaved plants (dicotyledon), but most of them are broad-spectrum and act upon common biochemical pathways of plants and algae. Because of their mode of action, the vast majority of herbicides are non-toxic to animals, but there is always the possibility of certain side-effects such as endocrine disruption (Hayes et al. 2010). As with fungicides, most herbicides are soluble in water and therefore are systemic (Table 3).

Certain organometallic compounds are phytotoxic and have been used since the 19th century as herbicides. While most of them are not used anymore due to their persistent contamination of soil, arsenic compounds like monosodium methylarsonate (MSMA) are still in use in some countries.

Dinitrophenols are also an old class of broad-spectrum herbicides that uncouple the phosphorilation process in the mitochondria, thus preventing the germination of seeds and growth. Due to their persistence, most of them have been phased out and only one, dinoterb, is used nowadays.

Dinitroanilines (e.g., pendimethalin, trifluralin) and benzamides (e.g., isoxaben, propyzamide) disrupt the assembly of beta-tubulin during mitosis, thus stopping cell division and germination in the same way as the fungicides that share this mode of action. They are all broad-spectrum herbicides, but while the former group comprises persistent and recalcitrant herbicides with residual activity, many of which are now obsolete, the second group contains systemic compounds.

Bipyridilium herbicides (diquat and paraquat) inhibit the electron transport in the photosystem I, whereas benzonitriles (e.g., bromoxynil, dichlobenil), phenylureas (e.g., diuron, thidiazuron), triazines (e.g., atrazine, simazine), triazinones (e.g., hexazinone, metribuzin) and urea (e.g., chlorotoluron, linuron) herbicides do the same at the photosystem II in the chloroplasts. Many chemicals in the latter four groups have also phased out due to their persistence in the environment and leaching properties, thus causing contamination of soil and water which can have a negative impact on the primary productivity of many agricultural lands (Walker 2001) due

Table 3. Main groups of herbicides used in agriculture.

Mode of action (MoA)	Chemical groups	Selectivity and application	Environmental features
Auxins			
Growth regulation	Synthetic auxins	Selective (dicots), systemic, post-emergent	Non persistent
	Phenoxy		Non persistent
	Pyridine		Moderately persistent, prone to leaching
Cell division			
β-tubulin assembly—mitosis inhibitors	Benzamide	Selective, pre-emergent, systemic	Moderately persistent, prone to leaching
	Dinitroaniline	Broad-spectrum, pre-emergent	Persistent, volatilisation
Cellulose biosynthesis inhibitors	Alkylazine	Broad-spectrum	Persistent
	Benzamide	Selective, pre-emergent	Persistent, prone to leaching
Very long chain fatty acid (VLCFA) biosynthesis inhibitors	Chloroacetamide, oxyacetamide, Phosphorodithioate	Broad-spectrum, systemic, pre-emergent	Moderately persistent
Chloroplasts—electron transport inhibitors			
Photosystem I	Bipyridilium	Broad-spectrum, foliar	Persistent
Photosystem II	Benzonitrile	Selective, systemic or foliar	Non persistent
	Phenylurea	Broad-spectrum, systemic	Moderately persistent
	Triazine	Selective, systemic or foliar	Persistent, prone to leaching
	Triazinone	Selective, systemic	Non persistent
	Urea	Selective, systemic or foliar	Persistent, prone to leaching
Plant growth			
4-HPPD inhibitors—carotenoid biosynthesis	Diketonitrile	Systemic	Non persistent
	Pyrazolone	Selective, systemic	Moderately persistent
	Triketone	Broad-spectrum, foliar	Non persistent

PPO inhibitors—chlorophyll biosynthesis	Carboxamide	Broad-spectrum, foliar	Moderately persistent
	Diphenyl ethers	Selective, foliar	Moderately persistent, prone to leaching
	Phenylpyrazole	Selective, foliar	Moderately persistent
	Triazolinone	Broad-spectrum, foliar or systemic	Moderately persistent
	Uracil	Selective, systemic	Persistent, prone to leaching
ACCase inhibitors—fatty acid biosynthesis	Cyclohexanedione	Selective, systemic	Non persistent
	Phenoxypropionate	Selective, systemic or foliar	Moderately persistent, prone to leaching
	Phenylpyrazolinone	Selective, systemic	Non persistent
ALS inhibitors—branched amino acid biosynthesis	Imidazolinone	Mostly selective, systemic or foliar	Persistent, prone to leaching
	Pyrimidine	Selective, systemic	Moderately persistent
	Sulfoanailide	Mostly selective, systemic or foliar	Moderately persistent
	Sulfonylurea	Selective, systemic	Persistent, prone to leaching
	Triazolone	Selective, systemic	Moderately persistent, some leaching
	Triazolopyrimidine	Broad-spectrum, systemic	Moderately persistent, some leaching
EPSP synthase inhibitors—aromatic amino acid biosynthesis	Glycine	Broad-spectrum, foliar	Non persistent
Glutamine synthase inhibitor—photosynthesis	Phosphonates	Broad-spectrum, foliar	Non persistent
Lipid biosynthesis inhibitors	Thiocarbamates	Selective, systemic	Non persistent
Other			
Crassulacean acid metabolism	Organometallic	Broad-spectrum, foliar	Persistent
Phosphorylation	Dinitrophenol	Broad-spectrum, foliar	Non persistent
Unknown	Inorganic salts	Broad-spectrum, systemic	Persistent, prone to leaching

to their residual activity—they are called residual herbicides. Some carbamate herbicides (e.g., desmedipham) act the same way but are systemic and degradable.

Auxins are plant hormones that control cell division and plant growth. Apart from the natural indole-3-acetic acid (IAA), since the 1940s a number of synthetic auxins derived from benzoic acid (e.g., dicamba, 2,3,6-TBA) were marketed as herbicides that promote uncontrolled plant growth beyond its nutrient supply until the plant dies. Phenoxy acids, esters and salts (e.g., 2,4-D, halauxifen-methyl, mecoprop-P-potassium) and pyridine herbicides (e.g., chlopyralid, picloram) comprise two large groups of herbicides that behave like auxins, causing senescence and/or uncontrolled cell growth. All of them are systemic and quite selective towards broad-leaved plants, degradable and not persistent in the environment.

Enzymatic processes linked to the photosynthesis have been the target of many non-selective herbicides: carboximides and dicarboximides (e.g., diflufenican, cinodon-ethyl), diphenylethers and phenylpyrrazoles (e.g., oxyfluorfen, pyraflufen), triazolinones (e.g., amicarbazone, carfentrazone) and uracil herbicides (e.g., bromacil, saflufenacil) inhibit the protoporphyrinogen oxidase (PPO), an enzyme involved in the synthesis of protoporphyrin IX, the precursor of chlorophyll in plants (Theodoridis 2012). Using a different pathway, the systemic herbicide glufosinate inhibits glutamine synthetase, leading to reduced glutamine and elevated ammonia levels in tissues, as a result of which photosynthesis stops; it is broad-spectrum and phytotoxic at very low doses (Carpenter and Bountin 2010). None of these chemicals are systemic, so they must be applied as post-emergence foliar sprays.

The inhibition of acetyl CoA carboxylase (ACCase) stops the synthesis of fatty acids, and this is the typical mode of action of cyclohexanediones (e.g., clethodim, sethoxydim) and the large group of phenoxypropionate herbicides (e.g., fluazifop, haloxyfop). Several other herbicidal groups inhibit more specifically the production of long-chain fatty acids, which are necessary for building the cellular membranes, thus stopping cell division and germination (Duke 1990). Among them are the chloroacetamides (e.g., alachlor, metolachlor), oxyacetamides (e.g., mefenacet), phosphorothioates (anilofos, bensulide), and the organophosphate piperophos. Thiocarbamates (e.g., molinate, thiobencarb) also inhibit the biosynthesis of lipids, but their mechanism is not well known. All these chemicals are systemic pre-emergent and non-selective herbicides.

Several groups of herbicides target the biosynthesis of essential amino acids. The majority of them inhibit acetolactate synthase (ALS), an enzyme required in the production of branched amino acids (valine, leucine and isoleucine), which is only carried out by plants and microorganisms (Shaner et al. 2012). Most of the newly developed herbicides fall into one of the following groups: (a) sulfonylureas (e.g., bensulfuron, monosulfuron, orthosulfamuron), which is the largest group of herbicides currently in the market, with over 40 active ingredients; (b) the diverse pyrimidine group, including the triazolopyrimidines (e.g., metosulam, penoxsulam), benzo-pyrimidine (e.g., pyrithiobac-sodium, pyribambenz-propyl) and pyrazoliums (metazosulfuron, pyroxasulfone); (c) sulfonanilides (e.g., pyrimisulfan, triafamone); or (d) triazolinones (e.g., propoxycarbazone, thiencarbazone-methyl). However, very few herbicides target the biosynthesis of aromatic amino acids (tyrosine, tryptophan, and phenylalanine), more specific to plants than the above pathway. The best known is glyphosate, which inhibits 5-enolpyruvylshikimate-3-phosphate synthase (EPSPS), an enzyme involved in the production of these amino acids. Glyphosate is a non-selective, post-emergence foliar herbicide.

Since the 1980s several agrochemical companies have developed chemicals that inhibit the enzyme 4-hydroxyphenylpyruvate dioxygenase (4-HPPD), which stops the breakdown of the amino acid tyrosine and inhibits carotenoid biosynthesis, thus resulting in bleaching and subsequent death of the treated plants (Hawkes 2012). Among them are the hydroxypyrazoles (e.g., pyrasulfotole, topramezone) and benzopyrazoles (e.g., tolpyralate), some new cyclohexanediones (e.g., fenquinotrione, tefuryltrione) and ketones (e.g., sulcotrione, tembotrione), all of which are applied as post-emergent foliar sprays.

Finally, some new compounds (e.g., indaziflam) inhibit the biosynthesis of cellulose (CBI), the omnipresent and characteristic molecule that envelops all plant cells (Dietrich and Laber 2012). Individual compounds in some chemical groups (amides, anilides, carbamates, organophosphates, oxazoles, pyridaziones) may share one or several of these modes of action above, whereas the mechanism of many others is still unknown. It should be noted that some inorganic compounds (e.g., ammonium sulfamate, sodium chlorate) are also used to kill weeds as well as providing a source of nutrients to the soil (e.g., iron sulphate).

Future Perspectives

Insect pests, fungi and weeds have always responded to the chemical warfare lashed on them by developing resistance. Because the vast majority of current pesticides are organic molecules, it is relatively easy for living organisms to find metabolic pathways that destroy them or render them useless. Indeed, repeated use of the same chemicals is considered to be the main cause behind the development of resistance (Miyata and Saito 1984, Heap 2014), but this does not deter chemical companies in their search for new pesticides. On the contrary, resistance has become the incentive for innovation on their fight against the enemies of the crops (Jeschke 2016).

For instance, widespread pest resistance to the old organochlorine insecticides was the main reason that led to ban DDT and cyclodienes, not the environmental damage they caused. Their replacement with cholinesterase inhibitors was envisaged well before the sublethal effects on birds of prey were noticed. The search for molecules with different MoAs in subsequent years was a necessity to confront resistance mechanisms among insect pests. Despite this, pest resistance developed within a few years of the introduction of the novel neonicotinoids and diamide insecticides (Bass et al. 2015, Uchiyama and Ozawa 2014). The same applies to fungi: the spread of *Fusarium* spp. in many agricultural areas is due to the constant use of broad-spectrum fungicides in crops, which fosters the growth of resistant strains of this and other fungi species, thus requiring the introduction of more selective fungicides. Even more dramatic has been the development of resistance against glyphosate by many weeds due to overuse of this herbicide in genetically modified crop varieties of cotton, soybean and maize (Shaner 2000, Beckie and Hall 2014).

To address the resistance problem, the pesticide chemical industry is looking for new MoAs that may lead to the production of novel chemicals (Jeschke 2016).

In recent years, we have seen a huge growth in the marketing of neonicotinoids, phenyl-pyrazoles and diamide insecticides, strobilurins fungicides and 4-HPPD herbicides (Jeschke 2016). However, it is important to remind us that this chemical warfare cannot win the war against pests or weeds, even if the use of particular pesticides may sometimes help win a few battles. Indeed, the increase in productivity that the world witnessed during the Green Revolution seems to have come to a halt. There is evidence that the routine application of insecticides and fungicides have little or no significant increase in today's crop yields (Lechenet et al. 2017). Most of the productivity increases nowadays appear to correlate with the use of herbicides alone, as these chemicals eliminate the plants that compete with the main crop for nutrients, light and water (Gianessi 2013). However, excessive reliance on more than five modes of action among herbicides means that the efficacy of the new compounds developed by the chemical companies is hampered by resistance mechanisms that are common to one or several of these modes of action (Duke 2012). The solution to this problem, therefore, is not to add new chemicals to the already saturated pesticide market but to find different ways of combating this war against weeds and pests (Owen et al. 2015, Chauhan et al. 2017), without affecting the ecosystem services provided by soil biota and pollinators, which are essential for agricultural productivity (van Hoesel et al. 2017).

Apart from developing products with novel MoA, two modern trends in the pesticide industry are apparent: (i) the shift towards systemic insecticides, and (ii) the development of genetically modified crop varieties (GM-crops).

Systemic insecticides allow more flexibility in their application, avoiding the need for sprays in certain crops and thus a safer use for applicators. However, their very nature implies that residues of the highly toxic chemicals (e.g., neonicotinoids) are present in all plant tissues, including nectar and pollen, and this poses a serious risk to insect pollinators and parasitoids, which are as susceptible to these insecticides as the pests or even more (Pisa et al. 2015). Although neonicotinoids are selective towards arthropods, diamide insecticides are broad spectrum toxins. Besides, whether residues of these compounds in honey (Mitchell et al. 2017), fruits and vegetables have side-effects on humans still requires further studies to be elucidated (Cimino et al. 2017).

Two types of GM-crops have been developed, one for pest control (i.e., Bt-crops) and the other for herbicide resistance (glyphosate-GM crops). Bt-crops produce the toxin of *Bacillus thuringiensis*, which are very effective against caterpillars and grubs (Hutchinson 1999) but appear not to damage natural enemies of the pests (Thomazoni et al. 2010), thus avoiding the use of insecticide sprays against such pests (Wadhwa and Gill 2007); however, this has created secondary pests that require the use of other insecticides. By contrast, glyphosate-GM crops are resistant against this herbicide, so the farmers can apply glyphosate products *ad libitum* without harming the crop plants. Unfortunately, this has led to an overuse of the herbicide that fostered rapid development of resistance among various weeds plus contamination of the environment (Powles 2008, Beckie and Hall 2014). Many now regret what was seen a few years earlier as a technological advance in weed control, so the old residual herbicides are back in fashion (Becki 2011).

References

Ackefors, H. 1971. Effects of particular pollutants—3. Mercury pollution in Sweden with special reference to conditions in the water habitat. Proc. R. Soc. London B 177(1048): 365–87.

Albers, P.H., P.N. Klein, D.E. Green, M.J. Melancon, B.P. Bradley and G. Noguchi. 2006. Chlorfenapyr and mallard ducks: overview, study design, macroscopic effects, and analytical chemistry. Environ. Toxicol. Chem. 25(2): 438–45.

Ambrose, A.M. and H.B. Haag. 1936. Toxicological study of Derris. Ind. and Eng. Chem. 28(7): 815–21.

Arts, G.H., L.L. Buijse-Bogdan, J.D.M. Belgers, C.H.v. Rhenen-Kersten, R.P.v. Wijngaarden, I. Roessink et al. 2006. Ecological impact in ditch mesocosms of simulated spray drift from a crop protection program for potatoes. Integr. Environ. Assess. Manag. 2(2): 105–25.

Bartlett, D.W., J.M. Clough, J.R. Godwin, A.A. Hall, M. Hamer and B. Parr-Dobrzanski. 2002. The strobilurin fungicides. Pest Manag. Sci. 58(7): 649–62.

Bass, C., I. Denholm, M.S. Williamson and R. Nauen. 2015. The global status of insect resistance to neonicotinoid insecticides. Pestic Biochem. Physiol. 121: 78–87.

Beckie, H.J. 2011. Herbicide-resistant weed management: focus on glyphosate. Pest Manag. Sci. 67(9): 1037–48.

Beckie, H.J. and L.M. Hall. 2014. Genetically-modified herbicide-resistant (GMHR) crops a two-edged sword? An Americas perspective on development and effect on weed management. Crop Protection 66: 40–5.

Beketov, M.A., B.J. Kefford, R.B. Schäfer and M. Liess. 2013. Pesticides reduce regional biodiversity of stream invertebrates. PNAS 110(27): 11039–43.

Bernhardt, E.S., E.J. Rosi and M.O. Gessner. 2017. Synthetic chemicals as agents of global change. Front. Ecol. Environ. 15: 84–90.

Boisvert, M. and J. Boisvert. 2000. Effects of *Bacillus thuringiensis* var. *israelensis* on target and nontarget organisms: a review of laboratory and field experiments. Biocontrol Sci. Technol. 10(5): 517–61.

Brock, T.C.M., I. Roessink, J.D.M. Belgers, F. Bransen and S.J. Maund. 2009. Impact of a benzoyl urea insecticide on aquatic macroinvertebrates in ditch mesocosms with and without non-sprayed sections. Environ. Toxicol. Chem. 28(10): 2191–205.

Brown, A.W.A. 1978. Ecology of Pesticides. New York: John Wiley & Sons, Inc. 525 p.

Carpenter, D. and C. Boutin. 2010. Sublethal effects of the herbicide glufosinate ammonium on crops and wild plants: short-term effects compared to vegetative recovery and plant reproduction. Ecotoxicology 19: 1322–36.

Chauhan, B.S., A. Matloob, G. Mahajan, F. Aslam, S.K. Florentine and P. Jha. 2017. Emerging challenges and opportunities for education and research in weed science. Front. Plant Sci. 8(1537).

Chen, W., I. Vermaak and A. Viljoen. 2013. Camphor—A fumigant during the black death and a coveted fragrant wood in ancient Egypt and Babylon—A review. Molecules 18: 5434–54.

Chu, K.H., C.K. Wong and K.C. Chiu. 1997. Effects of the insect growth regulator (S)-methoprene on survival and reproduction of the freshwater cladoceran *Moina macrocopa*. Environ. Pollut. 96(2): 173–8.

Cimino, A.M., A.L. Boyles, K.A. Thayer and M.J. Perry. 2017. Effects of neonicotinoid pesticide exposure on human health: a systematic review. Environ. Health Perspect. 125: 155–62.

Codding, P.W. 1983. Structural studies of sodium channel neurotoxins. 2. Crystal structure and absolute configuration of veratridine perchlorate. J. Am. Chem. Soc. 105(10): 3172–6.

Cole, L.M., R.A. Nicholson and J.E. Casida. 1993. Action of phenylpyrazole insecticides at the GABA-gated chloride channel. Pestic. Biochem. Physiol. 46(1): 47–54.

Crouse, G., J. Dripps, T. Sparks, G. Watson and C. Waldron. 2012. Spinosad and spinetoram, a new semi-synthetic spinosyn. pp. 1108–26. *In*: Krämer, W., U. Schirmer, P. Jeschke and M. Witschel (eds.). Modern Crop Protection Compounds. Weinheim, Germany: Wiley-VCH.

Crouse, G.D. and T.C. Sparks. 1998. Naturally derived materials as products and leads for insect control: the spinosyns. pp. 133–46. *In*: Kuhr, R.J. and N. Motoyama (eds.). Pesticides and the Future. Amsterdam: IOS Publisher.

Davies, P.E. and L.S.J. Cook. 1993. Catastrophic macroinvertebrate drift and sublethal effects on brown trout, *Salmo trutta*, caused by cypermethrin spraying on a Tasmanian stream. Aquat. Toxicol. 27(3-4): 201–24.

Dietrich, H. and B. Laber. 2012. Inhibitors of cellulose biosynthesis. pp. 339–69. *In*: Krämer, W., U. Schirmer, P. Jeschke and M. Witschel (eds.). Modern Crop Protection Compounds. Weinheim, Germany: Wiley-VCH.

Douglas, M.R. and J.F. Tooker. 2016. Meta-analysis reveals that seed-applied neonicotinoids and pyrethroids have similar negative effects on abundance of arthropod natural enemies. PeerJ 4: e2776.

Duke, S.O. 1990. Overview of herbicide mechanisms of action. Environ. Health Perspect. 87: 263–71.

Duke, S.O. 2012. Why have no new herbicide modes of action appeared in recent years? Pest Manag. Sci. 68: 505–12.

Edwards, R., P. Millburn and D. Hutson. 1986. Comparative toxicity of cis-cypermethrin in rainbow trout, frog, mouse, and quail. Toxicol. Appl. Pharmacol. 84(3): 512–22.

Eldefrawi, M.E. and A.T. Eldefrawi. 1990. Nervous-system-based insecticides. *In*: Hodgson, E. and R.J.M. Kuhr (eds.). Safer Insecticides - Development and Use. New York: Dekker Inc.

Fenner, S., G. Koertner and K. Vernes. 2009. Aerial baiting with 1080 to control wild dogs does not affect the populations of two common small mammal species. Wildlife Res. 36(6): 528–32.

Forget, G. 1991. Pesticides and the Third World. J. Toxicol. Environ. Health A 32(1): 11–31.

Freed, V.H., C.T. Chiou and D.W. Schmedding. 1979. Degradation of selected organophosphate pesticides in water and soil. J. Agric. Food Chem. 27(4): 706–8.

Gant, D.B., M.E. Eldefrawi and A.T. Eldefrawi. 1987. Cyclodiene insecticides inhibit GABA—a receptor-regulated chloride transport. Toxicol. Appl. Pharmacol. 88(3): 313–21.

Gianessi, L.P. 2013. The increasing importance of herbicides in worldwide crop production. Pest Manag. Sci. 69(10): 1099–105.

Gisi, U., C. Lamberth, A. Mehl and T. Seitz. 2012. Carboxylic acid amide (CAA) fungicides. pp. 807–30. *In*: Krämer, W., U. Schirmer, P. Jeschke and M. Witschel (eds.). Modern Crop Protection Compounds. Weinheim, Germany: Wiley-VCH.

Harborne, J.B. and H. Baxter. 1993. Phytochemical Dictionary. London, UK: Taylor & Francis.

Hawkes, T. 2012. Hydroxyphenylpyruvate dioxygenase (HPPD): the herbicide target. pp. 225–35. *In*: Krämer, W., U. Schirmer, P. Jeschke and M. Witschel (eds.). Modern Crop Protection Compounds. Weinheim, Germany: Wiley-VCH.

Hayasaka, D., T. Korenaga, K. Suzuki, F. Sánchez-Bayo and K. Goka. 2012. Differences in susceptibility of five cladoceran species to two systemic insecticides, imidacloprid and fipronil. Ecotoxicology 21(2): 421–7.

Hayes, T.B., V. Khoury, A. Narayan, M. Nazir, A. Park, T. Brown et al. 2010. Atrazine induces complete feminization and chemical castration in male African clawed frogs (*Xenopus laevis*). PNAS 107(10): 4612–7.

Heap, I. 2014. Global perspective of herbicide-resistant weeds. Pest Manag. Sci. 70(9): 1306–15.

Hladik, M.L., D.W. Kolpin and K.M. Kuivila. 2014. Widespread occurrence of neonicotinoid insecticides in streams in a high corn and soybean producing region, USA. Environ. Pollut. 193: 189–96.

Hutchison, W.D. 1999. Review and analysis of damage functions and monitoring systems for pink bollworm (Lepidoptera: Gelechiidae) in southwestern United States cotton. Southwestern Entomol. 24(4): 339–62.

Immaraju, J.A. 1998. The commercial use of azadirachtin and its integration into viable pest control programmes. Pestic. Sci. 54(3): 285–9.

Ishaaya, I. and A.R. Horowitz. 1995. Pyriproxyfen, a novel insect growth regulator for controlling whiteflies: mechanisms and resistance management. Pestic. Sci. 43(3): 227–32.

Ishaaya, I., S. Kontsedalov and A.R. Horowitz. 2002. Emamectin, a novel insecticide for controlling field crop pests. Pest Manag. Sci. 58(11): 1091–5.

Iwasa, T., N. Motoyama, J.T. Ambrose and R.M. Roe. 2004. Mechanism for the differential toxicity of neonicotinoid insecticides in the honey bee, *Apis mellifera*. Crop Protection 23(5): 371–8.

Jeschke, P. and R. Nauen. 2008. Neonicotinoids—from zero to hero in insecticide chemistry. Pest Manag. Sci. 64(11): 1084–98.

Jeschke, P. 2016. Progress of modern agricultural chemistry and future prospects. Pest Manag. Sci. 72(3): 433–55.

Jeschke, P. 2016. Propesticides and their use as agrochemicals. Pest Manag. Sci. 72(2): 210–25.

Kellar, C.R., K.L. Hassell, S.M. Long, J.H. Myers, L. Golding, G. Rose et al. 2014. Ecological evidence links adverse biological effects to pesticide and metal contamination in an urban Australian watershed. J. Appl. Ecol. 51(2): 426–39.

Kreutzweiser, D., D. Thompson, S. Grimalt, D. Chartrand, K. Good and T. Scarr. 2011. Environmental safety to decomposer invertebrates of azadirachtin (neem) as a systemic insecticide in trees to control emerald ash borer. Ecotoxicol. Environ. Saf. 74(6): 1734–41.

Kuck, K., K. Stenzel and J. Vors. 2012. Sterol biosynthesis inhibitors. pp. 761–805. *In*: Krämer, W., U. Schirmer, P. Jeschke and M. Witschel (eds.). Modern Crop Protection Compounds. Weinheim, Germany: Wiley-VCH.

Lahm, G., D. Cordova, J. Barry, J. Andaloro, I. Annan, P. Marcon et al. 2012. Anthranilic diamide insecticides: chlorantraniliprole and cyantraniliprole. pp. 1409–25. *In*: Krämer, W., U. Schirmer, P. Jeschke and M. Witschel (eds.). Modern Crop Protection Compounds. Weinheim, Germany: Wiley-VCH.

Langhof, M., A. Gathmann and H.-M. Poehling. 2005. Insecticide drift deposition on noncrop plant surfaces and its impact on two beneficial nontarget arthropods, *Aphidius colemani* Viereck (Hymenoptera, Braconidae) and *Coccinella septempunctata* (Coleoptera, Coccinellidae). Environ. Toxicol. Chem. 24(8): 2045–54.

Lechenet, M., F. Dessaint, G. Py, D. Makowski and N. Munier-Jolain. 2017. Reducing pesticide use while preserving crop productivity and profitability on arable farms. Nature Plants 3: 17008.

Lee, S.-J., P. Caboni, M. Tomizawa and J.E. Casida. 2003. Cartap hydrolysis relative to its action at the insect nicotinic channel. J. Agric. Food Chem. 52(1): 95–8.

Lee, S.J., M. Tomizawa and J.E. Casida. 2003. Nereistoxin and cartap neurotoxicity attributable to direct block of the insect nicotinic receptor/channel. J. Agric. Food Chem. 51(9): 2646–52.

Matsumura, F. 1985. Toxicology of Pesticides. New York, USA: Plenum Press, 598 p.

Mitchell, E.A.D., B. Mulhauser, M. Mulot, A. Mutazabi, G. Glauser and A. Aebi. 2017. A worldwide survey of neonicotinoids in honey. Science 358: 109–11.

Miyata, T. and T. Saito. 1984. Development of insecticide resistance and measures to overcome resistance in rice pests. Pro. Ecol. 7: 183–99.

Nicotine; Product Cancellation Order, Federal Register: 26695–26696 (2009).

Owen, M.D.K., H.J. Beckie, J.Y. Leeson, J.K. Norsworthy and L.E. Steckel. 2015. Integrated Pest Management and weed management in the United States and Canada. Pest Manag. Sci. 71(3): 357–76.

Peakall, D.B. 1993. DDE-induced eggshell thinning: an environmental detective story. Environ. Rev. 1: 13–20.

Phong, T.K., T.D.T. Nhung, T. Motobayashi, D.Q. Thuyet and H. Watanabe. 2009. Fate and transport of nursery-box-applied tricyclazole and imidacloprid in paddy fields. Water, Air Soil Poll. 202(1-4): 3–12.

Pilling, E.D. and P.C. Jepson. 1993. Synergism between EBI fungicides and a pyrethroid insecticide in the honeybee (*Apis mellifera*). Pestic. Sci. 39(4): 293–7.

Pisa, L.W., V. Amaral-Rogers, L.P. Belzunces, J.M. Bonmatin, C.A. Downs, D. Goulson et al. 2015. Effects of neonicotinoids and fipronil on non-target invertebrates. Environ. Sci. Pollut. Res. 22(1): 68–102.

Powles, S.B. 2008. Evolved glyphosate-resistant weeds around the world: lessons to be learnt. Pest Manag. Sci. 64(4): 360–5.

Ragoucy-Sengler, C., A. Tracqui, A. Chavonnet, J.B. Daijardin, M. Simonetti, P. Kintz et al. 2000. Aldicarb poisoning. Hum. Exp. Toxicol. 19(12): 657–62.

Rao, K.S.P., C.H. Trottman, W. Morrow and D. Desaiah. 1986. Toxaphene inhibition of calmodulin-dependent calcium ATPase activity in rat brain synaptosomes. Fund. Appl. Toxicol. 6(4): 648–53.

Ratcliffe, D.A. 1970. Changes attributable to pesticides in egg breakage frequency and shell thickness in some British birds. J. Appl. Ecol. 7: 67–115.

Rattner, B.A., K.E. Horak, S.E. Warner, D.D. Day, C.U. Meteyer, S.F. Volker et al. 2011. Acute toxicity, histopathology, and coagulopathy in American kestrels (*Falco sparverius*) following administration of the rodenticide diphacinone. Environ. Toxicol. Chem. 30(5): 1213–22.

Sánchez-Bayo, F. 2011. Impacts of agricultural pesticides on terrestrial ecosystems. pp. 63–87. *In*: Sánchez-Bayo, F., P.J. van den Brink and R. Mann (eds.). Ecological Impacts of Toxic Chemicals. Online eBook: Bentham Science Publishers.

Sánchez-Bayo, F., K. Goka and D. Hayasaka. 2016. Contamination of the aquatic environment with neonicotinoids and its implication for ecosystems. Front. Environ. Sci. 4: 71.

Sánchez-Bayo, F. 2018. Systemic insecticides and their environmental repercussions. pp. 111–7. *In*: Dellasala, D.A. and M.I. Goldstein (eds.). Encyclopedia of the Anthropocene. Oxford: Elsevier.

Sauter, H. 2012. Strobilurins and other complex III inhibitors. pp. 584–627. *In*: Krämer, W., U. Schirmer, P. Jeschke and M. Witschel (eds.). Modern Crop Protection Compounds. Weinheim, Germany: Wiley-VCH.

Shaner, D., M. Stidham, B. Singh and S. Tan. 2012. Imidazolinone herbicides. pp. 88–99. *In*: Krämer, W., U. Schirmer, P. Jeschke and M. Witschel (eds.). Modern Crop Protection Compounds. Weinheim, Germany: Wiley-VCH.

Shaner, D.L. 2000. The impact of glyphosate-tolerant crops on the use of other herbicides and on resistance management. Pest Manag. Sci. 56(4): 320–6.

Sibly, R.M., I. Newton and C.H. Walker. 2000. Effects of dieldrin on population growth rates of sparrowhawks 1963–1986. J. Appl. Ecol. 37(3): 540–6.

Simon-Delso, N., V. Amaral-Rogers, L.P. Belzunces, J.M. Bonmatin, M. Chagnon, C. Downs et al. 2015. Systemic insecticides (neonicotinoids and fipronil): trends, uses, mode of action and metabolites. Environ. Sci. Pollut. Res. 22(1): 5–34.

Stehle, S. and R. Schulz. 2015. Agricultural insecticides threaten surface waters at the global scale. PNAS 112(18): 5750–5.

Stone, W., J. Okoniewski and J. Stedelin. 1999. Poisoning of wildlife with anticoagulant rodenticides in New York. J. Wildl. Dis. 35(4): 187–93.

Theodoridis, G. 2012. Protoporphyrinogen IX oxidase inhibitors. pp. 163–96. *In*: Krämer, W., U. Schirmer, P. Jeschke and M. Witschel (eds.). Modern Crop Protection Compounds. Weinheim, Germany: Wiley-VCH.

Thomazoni, D., P.E. Degrande, P.J. Silvie and O. Faccenda. 2010. Impact of Bollgard® genetically modified cotton on the biodiversity of arthropods under practical field conditions in Brazil. African J. Biotechnol. 9(37): 6167–76.

Tomizawa, M. and J. Casida. 2005. Neonicotinoid insecticide toxicology: mechanisms of selective action. Annu. Rev. Pharmacol. Toxicol. 45: 247–68.

Toquin, V., C. Sirven, L. Assmann and H. Sawada. 2012. Host defense inducers. pp. 739–48. *In*: Krämer, W., U. Schirmer, P. Jeschke and M. Witschel (eds.). Modern Crop Protection Compounds. Weinheim, Germany: Wiley-VCH.

Uchiyama, T. and A. Ozawa. 2014. Rapid development of resistance to diamide insecticides in the smaller tea tortrix, *Adoxophyes honmai* (Lepidoptera: Tortricidae), in the tea fields of Shizuoka Prefecture, Japan. Jpn. J. Appl. Entomol. Zool. 49(4): 529–34.

Usherwood, P.N.R. and H. Vais. 1995. Towards the development of ryanoid insecticides with low mammalian toxicity. Toxicol. Lett. 82: 247–54.

Van Hoesel, W., A. Tiefenbacher, N. König, V.M. Dorn, J.F. Hagenguth, Ua Prah et al. 2017. Single and combined effects of pesticide seed dressings and herbicides on earthworms, soil microorganisms, and litter decomposition. Front. Plant Sci. 8: 215.

Vörösmarty, C.J., P.B. McIntyre, M.O. Gessner, D. Dudgeon, A. Prusevich, P. Green et al. 2010. Global threats to human water security and river biodiversity. Nature 467: 555–61.

Wadhwa, S. and R.S. Gill. 2007. Effect of Bt-cotton on biodiversity of natural enemies. J. Biol. Control 21(1): 9–15.

Walker, C.H. 2001. Organic Pollutants. Glasgow, UK: Taylor & Francis, 282 p.

Walker, H. 2012. Pyrazole carboxamide fungicides inhibiting succinate dehydrogenase. pp. 175–93. *In*: Krämer, W., U. Schirmer, P. Jeschke and M. Witschel (eds.). Modern Crop Protection Compounds. Weinheim, Germany: Wiley-VCH.

Waters, C.N., J. Zalasiewicz, C. Summerhayes, A.D. Barnosky, C. Poirier, A. Gałuszka et al. 2016. The Anthropocene is functionally and stratigraphically distinct from the Holocene. Science 351(6269).

Watt, B.E., A.T. Proudfoot, S.M. Bradberry and J.A. Vale. 2005. Anticoagulant rodenticides. Toxicol. Revi. 24(4): 259–69.

Wenger, J., T. Niderman and C. Mathews. 2012. Acetyl-CoA carboxylase inhibitors. pp. 447–77. *In*: Krämer, W., U. Schirmer, P. Jeschke and M. Witschel (eds.). Modern Crop Protection Compounds. Weinheim, Germany: Wiley-VCH.

Weston, D.P., Y. Ding, M. Zhang and M.J. Lydy. 2013. Identifying the cause of sediment toxicity in agricultural sediments: the role of pyrethroids and nine seldom-measured hydrophobic pesticides. Chemosphere 90(3): 958–64.

Wettstein, F.E., R. Kasteel, M.F. Garcia Delgado, I. Hanke, S. Huntscha, M.E. Balmer et al. 2016. Leaching of the neonicotinoids thiamethoxam and imidacloprid from sugar beet seed dressings to subsurface tile drains. J. Agric. Food Chem. 64(33): 6407–15.

Yamamoto, I. 1998. Nicotine—old and new topics. pp. 61–9. *In*: Kuhr, R.J. and N. Motoyama (eds.). Pesticides and the Future. Amsterdam: IOS Publisher.

Yaméogo, L., E.K. Abban, J.M. Elouard, K. Traore and D. Calamari. 1993. Effects of permethrin as *Simulium* larvicide on non-target aquatic fauna in an African river. Ecotoxicology 2(3): 157–74.

Young, D. 2012. Fungicides acting on mitosis and cell division. pp. 739–48. *In*: Krämer, W., U. Schirmer, P. Jeschke and M. Witschel (eds.). Modern Crop Protection Compounds. Weinheim, Germany: Wiley-VCH.

Chapter 2

Fertilization Strategies for Agroenvironmental Protection and Sustainable Agriculture Production

Ahammad M. Kamal and *Joann K. Whalen**

INTRODUCTION

Agricultural land supporting the production of annual crops, managed grasslands and perennial crops in agroforestry and bioenergy systems covers about 40–50% of the Earth's land surface (Smith et al. 2007). Demand for food, feed, fiber, fuel and other goods from agricultural land will continue to increase as the world's population grows to an estimated 9.8 billion by 2050 (UN DESA 2017). Since there are physical limits to the expansion of agricultural land, future demand for food, feed, fiber and fuel production will be achieved by adopting more intensive agricultural practices, including greater fertilizer use. This is already anticipated in regional estimates of fertilizer demand (Table 1).

Fertilizers are used in agriculture to supplement the soil nutrient supply, which is insufficient to achieve maximum crop yields in most regions of the world. Judicious use of fertilizer also compensates for soil nutrient removal that occurs when biomass, including grain and non-edible plant residues such as straw and woody biomass, is harvested and taken away from fields. Fertilizers contain nutrients that support crop

Natural Resources Sciences, Faculty of Agricultural and Environmental Sciences, McGill University, Canada.
* Corresponding author: joann.whalen@mcgill.ca

Table 1. Regional fertilizer demand for nitrogen (N), phosphorus (P_2O_5) and potassium (K_2O) from 2005 to 2050 (adapted from Drescher et al. 2011).

Region	2005	2015	2030	2050	Growth (%) from 2015–2050
	Million Mt. nutrients				
Sub-Saharan Africa	1.4	1.9	2.7	7.7	307
Near East and North Africa	8.5	11.8	15.8	32.9	179
South Asia	-	38	-	71.6	88
East Asia	-	68	-	94.2	39
Latin America & Caribbean	12.7	13.9	15.5	36.7	164
Industrial Countries	44.2	62	73.7	104.7	69

growth, but farmers must add the right amount of nutrients to meet crop requirements and avoid negative environmental consequences (e.g., leaching or runoff of excessive nutrients from agricultural land to waterways, greenhouse gas emissions, acidification of terrestrial and aquatic ecosystems). This involves the selection of the appropriate fertilizer source, which may be a chemical fertilizer, compost, animal manure or plant residue that is applied in the right rate, at the right time and in the right place for the crop.

Fertilization of agricultural land is consistent with the United Nations sustainable development goals that aim to eradicate hunger and extreme poverty by 2030 (UN General Assembly 2015) because the fertilizer nutrients are required to increase global food production. However, fertilization strategies must be economically feasible and avoid environmental pollution. For instance, recycling of animal- and plant-based residues as fertilizer is preferable to burning these nutrient-rich residues, which has significant, negative impacts on the environment, local economies and human health (Gupta et al. 2016). The objective of this chapter is to present fertilization strategies that promote sustainable food production, and minimize the negative environmental impacts of fertilizer use.

Agroenvironmental Consequences of Fertilizer Use

The agroenvironmental consequences of fertilizer use are seen on agricultural land and in non-agricultural ecosystems, including waterways and the atmosphere. Fertilizers increase soil fertility because they supply plant-available nutrients, but they have variable effects on soil organic matter and associated soil properties like aggregation and water-holding capacity (Blanco-Canqui et al. 2015, Nunes et al. 2017). Fertilization generally increases biomass production, which increases the amount of unharvested organic components (e.g., roots, stubble and other plant residues) in the field after grain and other edible products are removed. As well, organic fertilizers such as compost, green manure and animal manure contain 50% or more organic matter, which contributes to the soil organic matter pool. Fertilization practices that maintain or increase the soil organic matter content are beneficial for soil structure and water retention, and may alleviate soil compaction (Mujdeci et al. 2017). However, fertilizer sources that contain ammonium (NH_4^+), such as urea, ammonium sulfate and NH_4-rich animal manure, are expected to contribute to soil acidification. This occurs because plant roots maintain electrical neutrality during ion adsorption, so the NH_4^+ ions assimilated into the root must be balanced by releasing an equal amount of hydrogen (H^+) ions into the soil solution. Also, the microbially-mediated processes of ammonia oxidation and nitrification convert NH_4^+ in soil solution to nitrate (NO_3^-), which releases H^+ ions into soil solution. Plant uptake or biological transformations of NO_3^-, as well as NO_3^- leaching below the root zone removes this anion from the soil solution, resulting in a net accumulation of H^+ ions that contribute to soil acidity (Scott et al. 2000).

The impact of fertilizer use on non-agricultural ecosystems occurs when excess nutrients are transferred to non-agricultural land, aquatic systems and the atmosphere. Nitrogen (N) and phosphorus (P) pose the greatest agroenvironmental risks for several reasons. Due to the essential role of these nutrients in plant nutrition, crop growth is most likely to be limited by N deficiency, followed by P deficiency, which means that N and P fertilizers are applied in larger quantities and to more agricultural land than other essential macronutrients (K, Ca, Mg and S; FAO 2015). Fertilizers contain water-soluble nutrients that can be assimilated by plant roots, such as NH_4^+, NO_3^- and ortho-phosphate ($H_2PO_4^-/HPO_4^{2-}$), and nutrient management practices aim to maximize the uptake of these and other essential nutrients by the crop (Fig. 1). However, water-soluble NH_4^+ may be chemically deprotonated to release ammonia, NH_3 (g), while NH_4^+ and NO_3^- can be transformed by soil microorganisms into several gaseous products, including nitric and nitrous oxides, NO (g) and N_2O (g), and dinitrogen, N_2 (g). Dissolved NO_3^- and $H_2PO_4^-/HPO_4^{2-}$ may leach through the soil profile or be lost in surface runoff. Long-term fertilizer applications often contribute to a build-up of N and P in soil organo-mineral fractions, and erosion of these nutrient-rich sediments by wind and water can lead to their deposition in

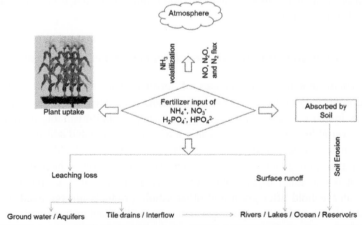

Fig. 1. Fertilizer is a source of water-soluble nitrogen (NH_4^+ and NO_3^-) and phosphorus ($H_2PO_4^-/HPO_4^{2-}$) that is biologically available for crop uptake. These ions may be lost from agricultural soils through gaseous emissions, soil erosion, leaching and runoff of dissolved nutrients.

surface waters. The most serious risks occur when fertilizer nutrients are transferred to aquifers, aquatic ecosystems and the atmosphere.

Risk of Fertilizer Nutrients to Aquifers and Aquatic Ecosystems

Commercial N fertilizers are generally water-soluble NH_4^+-based compounds or salts of NH_4^+ and NO_3^-. Compost, animal manure and plant residues contain a mixture of water-soluble NH_4^+ and organic N compounds, mainly water-insoluble proteins that are converted enzymatically to NH_4^+ by soil microorganisms. Fertilizer sources containing NH_4^+ will undergo biological oxidation to NO_3^- within a few days to weeks of application in warm, well-drained soils (Kowalchuk and Stephen 2001). Due to its negative charge, the NO_3^- is sorbed weakly to negatively-charged soil colloids and remains in the soil solution, where it is vulnerable to leach through the soil profile. An estimated 25 to 67 kg N ha^{-1} y^{-1} of NO_3^- is leached from intensive grain and forage systems in the United States (Power et al. 2001, Basso and Ritchie 2005).

Aquifers are at risk of pollution from NO_3^- that moves through the vadose zone and enters the non-saturated zone. Burkart and Kolpin (1993) reported that 29% of water samples from 303 wells in near-surface unconsolidated and bedrock aquifers of midcontinental United States had NO_3^- concentrations $\geq$ 3.0 mg L^{-1}, and an additional 6% of water samples contained more than 10 mg NO_3-N L^{-1}. The risk of NO_3^- leaching is greater when higher N fertilizer rates are applied, due to the fact that crop N demand may be less than the N fertilizer supplied, and soluble NO_3^- moves with water by mass flow processes (Dinnes et al. 2002). Globally, only 47% of the N input to cropland from synthetic fertilizers, manure and symbiotic nitrogen fixation is converted into harvested products, and the remainder is subject to losses of up to 50 kg N ha^{-1} yr^{-1}, depending on the climate, soils and agricultural practices (Lassaletta et al. 2014). Field drainage systems can have a significant impact on N leaching because they behave like shallow, direct pipelines to surface waters that increase the speed with which water moves from agricultural land to waterways. The NO_3^- concentration in drainage water can be appreciable, with annual loads of 4 to 66 kg NO_3-N ha^{-1} reported in drainage from annually-cropped agroecosystems in the US Midwest (Hatfield et al. 1998). This affects the quality of drinking water in both rural and urban areas. The maximum concentration limit for NO_3^- in drinking water is 10 mg NO_3-N L^{-1}, and this level was exceeded in 0.42% of the public water systems in the United States that served over 2 million people in 2009 (Pennino et al. 2017). Although the quality of potable water derived from surface waterways is improving, violations continue to be reported in groundwater sources due to agricultural intensification and persistent NO_3^- loading in aquifers that supply drinking water (Schindler et al. 2006, Pennino et al. 2017).

Nutrient loading in surface waterways is associated with a higher incidence of algal blooms, eutrophication of lakes and reservoirs, and species extinction in riverine ecosystems. The contribution of dissolved NO_3^- and particulate N in eroded sediments to eutrophication is undisputed, but is overshadowed by the impact of P, which is generally the limiting nutrient for algal growth in freshwater systems. In Canada, water quality guidelines of 10–35 µg total P L^{-1} are set for rivers and lakes,

Table 2. Sources of phosphorus contamination to European lakes (data from European Fertilizer Manufacturers Association 2000).

Sources	% Contribution
Point sources (urban contribution)	50–75
Agriculture	20–40
Natural loading	5–15

since this range is an indication of a change in the trophic status from mesotrophic to meso-eutrophic states.

Agricultural activities are a major source of P entering aquatic ecosystems (Table 2) due to greater use of P fertilizers and more manure generated from intensive livestock rearing facilities since the 1960s (Smith and Schindler 2009). Leaching of dissolved reactive P, along with NO_3^-, can be predicted from animal numbers on a regional scale (Dymond et al. 2013). Appreciable quantities of particulate P associated with eroded sediments are also transported through drainage systems, and the particulate P concentration reached up to 1346 µg P L^{-1} in tile drainage from maize and perennial hayfield agroecosystems in Quebec, Canada (Poirier et al. 2012). Particulate P is the dominant form of P lost during overland runoff events, and contributes more to annual P loads at the watershed scale than tile drainage (Gentry et al. 2007, Poirier et al. 2012). Best management practices to prevent N and P losses from agricultural land to aquifers and surface waterways include balanced fertilization to avoid over-application of N and P fertilizers, soil erosion controls, and end-of-field treatment of drainage and runoff water in natural and constructed wetlands, bioreactors and adsorption filters (Dinnes et al. 2002, Husk et al. 2018).

Risk of Fertilizer Nutrients to the Atmosphere

Several gaseous N compounds are produced by chemical and biological reactions in the N cycle. Agriculture is the major source of most NH_3 (g) released into the atmosphere, with the remainder coming from industrial sources, such as fertilizer manufacturers, municipal wastewater treatment plants, petroleum extraction and refining industries. In countries with confined livestock production, the animal husbandry sector is responsible for much of the NH_3 (g) entering the atmosphere. For example, animal husbandry accounted for nearly 56% of the annual NH_3 (g) emissions, with an additional 35% of the annual NH_3 (g) emissions coming from fertilized agricultural land (Environment Canada/Health Canada 2001). Most of the atmospheric NH_3 (g) is converted to NH_4^+ in particulate and aerosol forms through liquid-phase and gas-phase reactions. While particulate NH_4^+ is typically deposited through dry deposition within 5 km of its origin, aerosol NH_4^+ forms may be transported from tens to thousands of kilometers away (Environment Canada/Health Canada 2001).

Animal manure is an important source of NH_3 (g) because urea, $(NH_2)_2CO$, in animal urine is enzymatically transformed into NH_3 (g) by urease produced by bacteria in animal feces, and a secondary production of this reaction is carbamate

(NH$_2$COOH) that undergoes chemical hydrolysis to release NH$_3$ (g) and CO$_2$ (Huijsmans et al. 2003). These reactions occur in livestock raising facilities, in manure storages and after manure is applied to agricultural land. For instance, volatilization was responsible for the loss of 24–33% of NH$_4^+$ from liquid dairy cattle manure applied to the soil surface and unincorporated, after 6 to 7 days (Beauchamp et al. 1982). Mixing manure into soil causes NH$_3$ (g) to react with excess H$^+$ ions present in soil solution and be protonated to NH$_4^+$, which is biologically available for plants and microorganisms and also reacts with soil minerals through adsorption and fixation reactions. Therefore, best practices for fertilization with manure dictate that manure be incorporated as soon as possible to retain NH$_4^+$ in the soil-plant system (Sommer and Hutchings 1995).

Urea fertilizer manufactured through chemical synthesis undergoes the same reaction when applied to agricultural land, due to the presence of urease produced by soil microorganisms. In a cold humid climate, NH$_3$ (g) losses were 50% of applied N when urea was banded at the surface, and incorporation of urea decreased emissions by an average of 7% cm^{-1}, based on incorporation depths of 0, 2.5, 5, 7.5, and 10 cm in a silt-loam soil (Rochette et al. 2013). For this reason, it is not recommended to leave urea fertilizer on the soil surface. Any fertilizer containing NH$_4^+$ may be susceptible to volatilization as NH$_3$ (g) if applied to soil with a high calcium carbonate level, since chemical reactions leading to the formation of NH$_2$COOH are more likely in alkaline soils. Volatilization of NH$_3$ (g) may be reduced by incorporating ammonium-based fertilizers, to increase its protonation with H$^+$, as demonstrated by Rao and Batra (1983).

Other gaseous forms of N, namely NO (g), N$_2$O (g) and N$_2$ (g), are released through biologically-mediated reactions in the N cycle (Fig. 2). Most attention has focused on N$_2$O emissions due to the climate forcing and stratospheric ozone-depleting properties of this compound (Ravishankara et al. 2009), and agricultural soils are the source of 60–70% of the global annual N$_2$O production (Reay et al. 2012). Most of the N$_2$O emitted from agricultural soils is attributed to denitrification processes (Bouwman 1996), although N$_2$O is also an intermediate product of ammonia oxidation and nitrifier-denitrification reactions (Fig. 2).

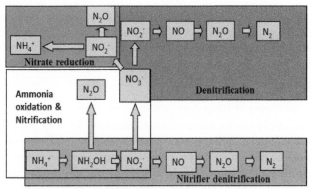

Fig. 2. Nitrogen cycle in agricultural soils, showing several biological reactions that can produce N$_2$O. Soluble NO$_3^-$ is a substrate for both denitrification and nitrate reduction processes.

Soil N_2O emissions are related to the activity of N-cycling microorganisms, and are therefore affected by the concentration of substrates available for the reaction and microbial growth in general (e.g., carbon), as well as environmental factors that control soil biological activity like temperature, moisture, pH, oxygen concentration and porosity (Sahrawat and Keeney 1986). Nitrogen fertilizers that contain NH_4^+ and NO_3^- are expected to contribute to these reactions, which explains the 2-fold increase in N_2O emissions in maize agroecosystems for every 100 kg N ha^{-1} added, although this effect is modulated by the fertilizer source since N_2O emissions were greater with anhydrous ammonia > urea > SuperU fertilizers (Eagle et al. 2017).

Inorganic N fertilizers have a default N_2O emission factor of 1, meaning 1% of the applied N is lost as N_2O (IPCC 2006), but ranged from 0.24 to 2.2% in irrigated maize agroecosystems, depending on the N fertilizer source, application rate and method of fertilizer application (Zhu-Barker et al. 2015). Globally, an N_2O emission factor of 0.57% was reported for organic fertilizers, but this included materials that are unlikely to emit N_2O, such as compost, crop residues and paper mill sludge, as well as animal slurries, wastewater and biosolids that have a mean emission factor of 1.21% (Charles et al. 2017). Co-application of inorganic and organic fertilizers increased the N_2O emission factor substantially, to 0.37% for compost plus inorganic fertilizer and by as much as 2.14% for liquid manure plus inorganic fertilizer (Charles et al. 2017). Physico-chemical characteristics of the fertilizer, as well as its transformations in soil according to the site-specific climatic conditions and cropping system, affect the N_2O loss from fertilizers.

Fertilizer Strategies for Sustainable Agroecosystems

To avoid the agroenvironmental impacts of soluble and reactive N and P, it is necessary to carefully match the amount of nutrients applied in fertilizer with crop nutrient demands. Therefore, fertilization strategies that achieve greater nutrient use efficiency will improve crop productivity, which leads to cost-savings and prevents undesirable environmental impacts. Fertilization strategies are summarized in the 4-R framework, which explains how to apply the right source of fertilizer, at the right rate and the right time, and in the right place to synchronize the nutrient input with the crop nutrient requirements (International Plant Nutrition Institute 2012). Here, we focus on strategies for N and P fertilizer use in sustainable agroecosystems.

Fertilizer Sources

Fertilizer is a substance of natural or synthetic origin that is applied to soil or on plant foliage to supply one or more nutrients essential for crop growth. Fertilizers are typically considered to fall into two broad categories, i.e., inorganic and organic fertilizers. Inorganic fertilizer, also known as chemical fertilizer or mineral fertilizer, may be chemically synthesized or mined from mineral deposits (European Fertilizer Manufacturers Association 2000). Most commercial N fertilizers are produced from the Bosch-Haber process, which combines N_2 from the atmosphere with hydrogen

Table 3. Nutrient concentrations (kg/100 kg) in selected inorganic fertilizers (adapted from Silva 2000).

Nutrient	Fertilizer name	N	P_2O_5	S	Ca
Nitrogen	Urea	45–46			24
	Ammonium sulfate	21		24	
	Sulfur-coated urea	36–38			
Phosphorus	Triple superphosphate		44–53	1–2	13
	Monoammonium phosphate	10–11	50–52	0–2	
	Diammonium phosphate	18–21	46–54	0–2	

from natural gas to yield anhydrous NH_3 as a starting compound for the synthesis of urea, ammonium nitrate ammonium sulfate and other N fertilizers. Phosphate rock is mined and solubilized with concentrated sulfuric acid to produce superphosphate fertilizer, the starting material for commercial P fertilizers. These fertilizers contain nutrients that are water-soluble or readily solubilized when in contact with water. Inorganic fertilizer is typically sold in granular or powder forms and as liquid formulations, although anhydrous NH_3 can be purchased as a supercooled liquid in pressurized tanks.

Inorganic fertilizers that supply N and P are "high-analysis" materials with few impurities, containing water-soluble nutrients that are safe for application on plants and to agricultural land (Table 3). The selection of an inorganic fertilizer is based on the target application rate needed to supply the correct amount of plant nutrients, considering what ionic forms are most readily absorbed by the plant, how the fertilizer reacts in soil and the risk of nutrient loss to the environment.

The cost of purchasing and applying the fertilizer is another important consideration. The price of nutrients contained in an equivalent mass of fertilizer is more expensive for controlled-release fertilizers than other inorganic fertilizers. However, Liu et al. (2011) reported that reducing the number of N fertilizer applications by substituting controlled release fertilizer for conventional urea fertilizer could save $5 to $7/acre in application costs on potato fields of Florida, USA. Obreza et al. (1999) reported that citrus trees receiving one application per year of controlled release fertilizer produced 12–14% greater fruit yield and had 15% gross dollar return than trees receiving conventional fertilizer, which was applied four times per year. However, the cost of fertilizing citrus with controlled release fertilizer was 4 times greater than the cost of conventional fertilizer, which made it economically impractical to apply controlled release fertilizers to commercial citrus orchards in Florida, USA.

Organic fertilizers are the excreta or remains of living organisms and include a wide variety of materials: compost, vermicompost, green manure crops, cow manure, poultry manure, alfalfa, bone meal, fish emulsion and so on. Organic fertilizers provide N, P and K (Table 4) as well as other macronutrients and micronutrients, and are considered to be 'complete' fertilizers capable of replenishing the essential nutrients needed by crops. Organic fertilizers contain a mixture of soluble ions plus insoluble organic compounds, and are available in solid, slurry and liquid forms.

Table 4. Nutrient concentrations (kg/tonne) in selected organic fertilizers (data from Koenig and Johnson 2011).

Source	N	P	K
Compost	15–35	5–10	10–20
Cattle manure	5–15	2–7	5–20
Poultry manure	30–40	10–20	10–20
Fish emulsion	50	10	10
Bone meal	20–60	150–270	0
Blood meal	120	15	6
Alfalfa	25	5	20
Cottonseed meal	40–60	25–30	16

The availability of nutrients from solid organic fertilizers is expected to be lower than other fertilizer sources because the organic compounds must be solubilized and hydrolyzed through biological processes before soluble ions are released into soil solution.

Selecting a fertilizer source starts with an assessment of which nutrients are needed for crop growth. This information generally comes from soil testing and plant tissue analysis. A suitable fertilizer source should provide a balanced supply of essential nutrients, considering the relative quantities and forms of plant-available nutrients present in the fertilizer. Knowledge of nutrient synergies and fertilizer reactions in soil can inform the selection. Three P fertilizer sources are triple superphosphate, monoammonium P (MAP) and diammonium P (DAP). The MAP and DAP fertilizers contain NH_4^+, which is known to enhance P uptake when co-applied, either in a single granule or in a concentrated band (Olson and Drier 1956). However, MAP and DAP have different reactions in the soil solution, since the dissolution of the MAP granule lowers the surrounding pH to < 4 and the solution around a dissolving DAP granule is about pH 8 (Johnston and Bruulsema 2014). In alkaline soils, it is preferable to apply MAP because soluble DAP will release NH_4^+, which may deprotonate to NH_3 (aq) and cause damage to germinating seedlings.

Integrated Soil Fertility Management (ISFM) is a method to make use of the most cost-effective, nutrient sources. Locally available soil amendments (for instance, crop residues, compost, animal manure, green manure) and inorganic fertilizers are applied to increase productivity while maintaining or enhancing the agricultural resource base. Targeted selection and application of inorganic fertilizers together with organic nutrient resources is expected to improve nutrient use efficiency and crop productivity. Although ISFM recognizes the absolute necessity of inorganic fertilizers to supply enough water-soluble nutrients for crop growth, it advocates the best combination of available nutrient management technologies that are economically profitable and socially acceptable to farmers (Place et al. 2003, Vanlauwe et al. 2010). On smallholder farms in central Kenya, maize grain yields were lower with inorganic fertilizer alone compared with organic fertilizers alone or a combination of organic and inorganic fertilizer, due to the low fertility, acidic pH and susceptibility to drought of these soils (Mucheru-Muna et al. 2014).

Cereal production in Hungary required inorganic fertilizer, and optimal yields were achieved with 150 kg N ha^{-1} for wheat, 120 kg N ha^{-1} for barley and 210 kg N ha^{-1} for maize (Kismanyoky and Toth 2012). When inorganic fertilizer was co-applied with farmyard manure, the joint effect of mineral and organic fertilizers was positive for soil fertility and produced up to 1000 kg ha^{-1} yield increase in each phase of the crop rotation. A single application of farmyard manure boosted the soil fertility for three years, and the soil organic matter level was maintained by applying organic fertilizer and leaving straw residue in the field (Kismanyoky and Toth 2012).

Fertilizer Rates

Soil testing is an indicator of how much fertilizer is needed to meet the crop nutrient requirements, since it measures the plant-available nutrient supply. Based on crop yield in soil fertility trials, the relationship between the soil test value and the expected crop yield is used to calibrate the fertilizer application rate (International Plant Nutrition Institute 2012). Low soil test values indicate that the soil is deficient in a particular nutrient and probably will respond to fertilizer inputs, whereas high soil test values indicate an adequate or ample supply of the nutrient, which reduces the likelihood that the crop will respond to fertilization. Complementary information, gathered on the crop nutrient removal at harvest and in omission plots where required nutrients are withheld from the crop, provides further support for selecting fertilizer application rates (Roberts 2008). Crop nutrient removal is assessed by multiplying the biomass by the nutrient content of the grain, other edible components, and the non-edible parts of the crop. This determines how much of each nutrient is removed from the field, as well as how much of each nutrient was required to support the crop growth. Omission plots are used to compare the yield in nutrient-limited conditions (e.g., no N fertilizer added, high rate of P and K fertilizer added) to yield in non-limited growth conditions (e.g., high rate of N, P and K fertilizer added). This allows the experimenter to determine how much inherent soil fertility vs. added fertilizers are contributing to yield outcomes.

As each agricultural crop has different nutrient requirements, the right rate of nutrients must be determined separately for each crop grown in a particular region. Over- or under-application of fertilizer will lower the nutrient use efficiency or reduce the crop yield and crop quality. Plant tissue analysis is a complementary technique to evaluate the nutrient balance in the crop, since plant nutrients rarely work in isolation. Data from on-farm field experiments in southeast Asia illustrates the positive interaction between N and other nutrients, primarily P and K, that increased N use efficiency and boost the yield of rice, wheat, maize and pearl millet (Table 5). A review of 241 site-years of wheat, maize and rice in China, India and North America showed that balanced fertilization with N, P, and K increased first-year recoveries by 54%, on average, compared to recoveries of only 21% when N was applied alone (Fixen et al. 2005). This concept extends to balanced fertilization of N with other macronutrients and micronutrients. Salvagiotti et al. (2009) found that correcting sulfur deficiency increased N recovery in wheat from 42% to 70%, while Messick (2002) reported that the application of S together with NPK led to yield increases of up to 60% for cereal and oil crops grown in India.

Table 5. Balanced fertilization increases yield and improves N use efficiency of cereal crops (adapted from Roberts (2008), assuming a N harvest index of 56%).

Crop	Yield (t ha^{-1})			N use efficiency (kg grain kg N^{-1})		
	Control	N alone	+PK	N alone	+PK	Increase (%)
Rice	2.74	3.28	3.82	13.5	27.0	13.5
Wheat	1.45	1.88	2.25	10.8	20.0	9.2
Maize	1.67	2.45	3.23	19.5	39.0	19.5
Pearl millet	1.05	1.24	1.65	4.7	15.0	10.3

Fertilizer Timing

The nutrient requirements of a crop vary throughout its lifespan and during the growing season. In annual crops, nutrient demands are greater during the mid-vegetative to early reproduction growth stages than in early vegetative growth or late reproduction stages. While it may be easier to apply fertilizer early in the growing season, before seeds are planted, the nutrients applied at this time are susceptible to loss, so there are economic and environmental consequences of applying nutrients that are not used by the crop. Optimal timing of fertilizer applications aims to ensure an adequate supply of nutrients during the periods of crop growth with high nutrient demand, thereby reducing nutrient loss from the agroecosystem (International Plant Nutrition Institute 2012).

Factors to be considered in choosing the right time for fertilizer applications are: determining nutrient needs of the crop, available fertilizer sources, crop nutrient uptake pattern, soil characteristics, and weather. Timing may be further influenced by the availability of equipment and labor on the farm, and fertilizer availability. In-season assessment of the plant nutrient status by plant tissue analysis confirms that the crop has sufficient nutrients to meet yield targets, and can also diagnose nutrient deficiencies and excesses. Hand-held or tractor-mounted chlorophyll meters may prove useful in fine-tuning in-season N applications at the field scale, although leaf color charts are still used to evaluate the in-season N demand of rice and maize crops on smaller farms. Remote sensing (Cilia et al. 2014) and machine-based canopy sensors (Amaral et al. 2018) are increasingly popular methods to evaluate crop N requirements and adjust variable-rate N fertilizer applications. These precision agriculture techniques provide detailed information on crop nutrition, which is transmitted along with georeferenced coordinates to the applicator, so the N fertilizer is applied at the right time and place that it is needed by the crop, for a more efficient, economic and environmentally friendly fertilizer program.

Fertilizer Placement

Fertilizer placement must be chosen to ensure nutrients are used efficiently by the crop. As discussed earlier, the placement of N fertilizer determines its susceptibility to loss in the environment. Placement methods could include: broadcasting fertilizer with or without incorporation, banding fertilizer at various depths, and applying liquid fertilizer to foliage at a suitable growth stages. Nutrients must be placed in a location that is accessible to the plant, which is generally in the root zone or on the foliage, although care must be taken to avoid applying fertilizers with a high salt index (e.g., NH_4^+- and K-based fertilizers) too close to young seedlings, as this causes osmotic stress to the crop (International Plant Nutrition Institute 2012). Before the crop is planted, fertilizers may be incorporated into the soil with tillage implements, but in-season applications are restricted to the soil surface or bands near the crop row, to avoid damaging the roots. Fertilizer placement is important to prevent NH_3 (g) volatilization and N_2O emissions, as noted by Rochette et al. (2013) and Zhu-Baker et al. (2015). Equipment used for variable-rate N fertilizer applications must be calibrated so the fertilizer is placed in the vicinity of growing roots, and this resulted in comparable citrus yield with 40% less fertilizer than the conventional technique in widely-spaced orchards in Florida, USA (Zaman et al. 2005).

Conclusions

Sustainable agriculture aims to conserve resources and protect environmental quality while maintaining or improving farm productivity. Fertilizers are needed to produce food and other materials for the growing global population, but they must be used efficiently. The 4R nutrient stewardship approach, combined with emerging knowledge from the fields of precision agriculture, balanced fertilization, integrated soil fertility management and controlled release fertilizer technologies, provide strategies for agroenvironmental protection and sustainable agriculture production (Fig. 3). These fertilization strategies are suitable for adoption on farms around the

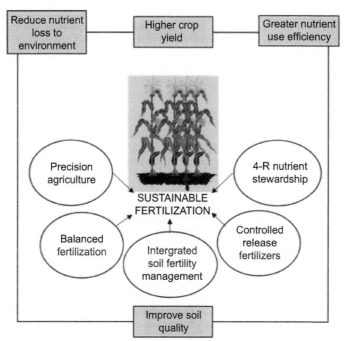

Fig. 3. Fertilization strategies for sustainable agroecosystems that are more efficient, productive and protect agroenvironmental quality.

world, and merit support from policy makers, producers, researchers, and the general public to secure our future food supply.

References

Amaral, L.R., R.G. Trevisan and J.P. Molin. 2018. Canopy sensor placement for variable-rate nitrogen application in sugarcane fields. Precision Agric. 19: 147–160.

Basso, B. and J.T. Ritchie. 2005. Impact of compost, manure and inorganic fertilizer on nitrate leaching and yield for a 6-year maize–alfalfa rotation in Michigan. Agric. Ecosys. Environ. 108: 329–341.

Beauchamp, E.G., G.E. Kidd and G. Thurtell. 1982. Ammonia volatilization from liquid dairy-cattle manure in the field. Can. J. Soil Sci. 62: 11–19.

Blanco-Canqui, H., G.W. Hergert and R.A. Nielsen. 2015. Cattle manure application reduces soil compactibility and increases water retention after 71 years. Soil Sci. Soc. Am. J. 79: 212–223.

Bouwman, A.F. 1996. Direct emission of nitrous oxide from agricultural soils. Nutr. Cycl. Agroecosyst. 46: 53–70.

Burkart, M.R. and D.W. Kolpin. 1993. Hydrologic and land-use factors associated with herbicides and nitrate in near-surface aquifers. J. Environ. Qual. 22: 646–656.

Charles, A., P. Rochette, J.K. Whalen, D.A. Angers, M.H. Chantigny and N. Bertrand. 2017. Global nitrous oxide emission factors from agricultural soils after addition of organic amendments: a meta-analysis. Agric. Ecosys. Environ. 236: 88–98.

Cilia, C., C. Panigada, M. Rossini et al. 2014. Nitrogen status assessment for variable rate fertilization in maize through hyperspectral imagery. Remote Sens. 6: 6549–6565.

Dinnes, D.L., D.L Karlen, D.B. Jaynes et al. 2002. Nitrogen management strategies to reduce nitrate leaching in tile-drained midwestern soils. Agron. J. 94: 153–171.

Drescher, A., R. Glaser, C. Richert and K.-R. Nippes. 2011. Demand for key nutrients (NPK) in 2050. Freiburg, Germany: University of Freiburg. https://esdac.jrc.ec.europa.eu/projects/NPK/Documents/Freiburg_Demand_for_key_nutrients_in_2050_Drescher.pdf.

Dymond, J.R., A.G.E. Ausseil, R.L. Parfitt, A. Herzig and R.W. McDowell. 2013. Nitrate and phosphorus leaching in New Zealand: a national perspective. New Zealand J. Agric. Res. 56: 49–59.

Eagle, A.J., L.P. Olander, K.L. Locklier, J.B. Heffernan and E.S. Bernhardt. 2017. Fertilizer management and environmental factors drive N_2O and NO_3 losses in corn: A meta-analysis. Soil Sci. Soc. Am. J. 81: 1191–1202.

Environment Canada/Health Canada. 2001. Ammonia in the aquatic environment. Priority Substances List Assessment Report. Canadian Environmental Protection Act, 1999. Minister of Public Works and Government Services Canada, Ottawa, ON.

European Fertilizer Manufacturers Association. 2000. Understanding phosphorus and its use in agriculture. ed. A.E. Johnston and I. Steen. Brussels: European Fertilizer Manufacturer's Association http://www.fertilizerseurope.com/fileadmin/user_upload/publications/agriculture_publications/EFMA_Phosphorus_booklet__2_.pdf.

Fixen, P.E., J.Y. Jin, K.N. Tiwari and M.D. Stauffer. 2005. Capitalizing on multi-element interactions through balanced nutrition—A pathway to improve nitrogen use efficiency in China, India and North America. Sci. China Ser. C Life Sci. 48: 780–790.

Food and Agricultural Organization of the United Nations (FAO). 2015. World fertilizer trends and outlook to 2018. Rome: FAO. http://www.fao.org/3/a-i4324e.pdf.

Gentry, L.E., M.B. David, T.V. Royer, C.A. Mitchell and K.M. Starks. 2007. Phosphorus transport pathways to streams in tile-drained agricultural watersheds. J. Environ. Qual. 36: 408–415.

Gupta, S., R. Agarwal and S.K. Mittal. 2016. Respiratory health concerns in children at some strategic locations from high PM levels during crop residue burning episodes. Atmos. Environ. 137: 127–134.

Hatfield, J.L., J.H. Prueger and D.B. Jaynes. 1998. Environmental impacts of agricultural drainage in the Midwest. In Drainage in the 21st century: Food production and the environment, 28–35.

Proc. Annu. Drainage Symp., 7th, Orlando, FL. 8–10 Mar. 1998. Am. Soc. of Agric. Eng., St. Joseph, MI.

Huijsmans, J.F.M., J.M.G. Hol and G.C. Vermeulen. 2003. Effect of application method, manure characteristics, weather and field conditions on ammonia volatilization from manure applied to arable land. Atmos. Environ. 37: 3669–3680.

Husk, B.R., J.S. Sanchez, B.C. Anderson, J.K. Whalen and B.C. Wootton. 2018. Removal of phosphorus from agricultural subsurface drainage water with woodchip and mixed-media bioreactors. J. Soil Wat. Conserv. 73: 263–273.

Intergovernmental Panel on Climate Change (IPCC). 2006. Guidelines for national greenhouse gas inventories. Vol. 4. Agriculture, forestry and other land use. Prepared by the National Greenhouse Gas Inventories Program, ed. H.S. Eggleston, L. Buendia, K. Miwa, T. Ngara and K. Tanabe, 5.1–5.50. Hayama, Kanagawa, IGES.

International Plant Nutrition Institute (IPNI). 2012. 4R plant nutrition: A manual for improving the management of plant nutrition. Norcross, GA: International Plant Nutrition Institute.

Johnston, A.M. and T.W. Bruulsema. 2014. 4R nutrient stewardship for improved nutrient use efficiency. Procedia Eng. 83: 365–370.

Kismanyoky, T. and Z. Toth. 2012. Mineral and organic fertilization to improve soil fertility and increase biomass production and N utilization by cereals. Soil Fertility Improvement and Integrated Nutrient Management: a Global Perspective, ed. J.K. Whalen, 183–200. InTech Open Access Publisher. doi: 10.5772/28985.

Koenig, R. and M. Johnson. 2011. Selecting and using organic fertilizers. Utah State University Cooperative Extension. https://extension.usu.edu/waterquality/files-ou/Agriculture-and-Water-Quality/Fertilizer/HG-510.pdf.

Kowalchuk, G.A. and J.R. Stephen. 2001. Ammonia-oxidizing bacteria: A model for molecular microbial ecology. Ann. Rev. Microbiol. 55: 485–529.

Lassaletta, L., G. Billen, B. Grizzetti, J. Anglade and J. Garnier. 2014. 50 years trends in nitrogen use efficiency of world cropping systems: the relationship between yield and nitrogen input to cropland. Environ. Res. Lett. 9: 105011. doi:10.1088/1748-9326/9/10/105011.

Liu, G.D., L. Zotarelli, Y.C. Li, D. Dinkins, Q.G. Wang and M. Orzores-Hampton. 2017. Controlled-release and slow-release fertilizers as nutrient management tools. Publication HS1255. Gainesville: University of Florida Institute of Food and Agricultural Sciences. http://edis.ifas.ufl.edu/hs1255.

Messick, D.L. 2002. Addressing sulphur deficiency in India. Fertil. Focus 19: 29–32.

Mucheru-Muna, M.D. Mugendi, P. Pypers et al. 2014. Enhancing maize productivity and profitability using organic inputs and mineral fertilizer in central Kenya small-hold farms. Exp. Agric. 50: 250–269.

Mujdeci, M., A.A. Isildar, V. Uygur et al. 2017. Cooperative effects of field traffic and organic matter treatments on some compaction-related soil properties. Solid Earth 8: 189–198.

Nunes, M.R., A.P. da Silva, J.E. Denardin et al. 2017. Soil chemical management drives structural degradation of Oxisols under a no-till cropping system. Soil Res. 55: 819–831.

Obreza, T.A., R.E. Rouse and J.B. Sherrod. 1999. Economics of controlled-release fertilizer use on young citrus trees. J. Prod. Agric. 12: 69–73.

Olson, R.A. and A.F. Drier. 1956. Fertilizer placement for small grains in relation to crop stand and nutrient efficiency in Nebraska. Soil Sci. Soc. Am. J. 1: 19–24.

Pennino, M.J, J.E. Compton and S.G. Leibowitz. 2017. Trends in drinking water nitrate violations across the United States. Environ. Sci. Technol. 51: 13450–13460.

Place, F., C.B. Barrett, H.A. Freeman, J.J. Ramisch and B. Vanlauwe. 2003. Prospects for integrated soil fertility management using organic and inorganic inputs: evidence from smallholder African agricultural systems. Food Policy 28: 365–378.

Poirier, S.-C., A.R. Michaud and J.K. Whalen. 2012. Bioavailable phosphorus in fine-sized sediments transported from agricultural fields. Soil Sci. Soc. Am. J. 76: 258–267.

Power, J.F., R. Wiese and D. Flowerday. 2001. Managing farming systems for nitrate control: a research review from management systems evaluation areas. J. Environ. Qual. 30: 1866–1880.

Rao, D.L.N. and L. Batra. 1983. Ammonia volatilization from applied nitrogen in alkali soils. Plant Soil 70: 219–228.

Ravishankara, A.R., J.S. Daniel and R.W. Portmann. 2009. Nitrous oxide (N_2O): The dominant ozone-depleting substance emitted in the 21st century. Science 326: 123–125.

Reay, D.S., E.A. Davidson, K.A. Smith et al. 2012. Global agriculture and nitrous oxide emissions. Nature Climate Change 2: 410–416.

Roberts, T.L. 2008. Improving nutrient use efficiency. Turk. J. Agric. For. 32: 177–182.

Rochette, P., D.A. Angers, M.H. Chantigny et al. 2013. Ammonia volatilization and nitrogen retention: How deep to incorporate urea? J. Environ. Qual. 42: 1635–1642.

Sahrawat, K.L. and D.R. Keeney. 1986. Nitrous oxide emission from soils. *In*: Stewart, B.A. (ed.). Advances in Soil Science 4: 103–148. New York: Springer.

Salvagiotti, F., J.M. Castellarin, D.J. Miralles et al. 2009. Sulfur fertilization improves nitrogen use efficiency in wheat by increasing nitrogen uptake. Field Crops Research 113: 170–177.

Schindler, D.W., P.J. Dillon and H. Schreier. 2006. A review of anthropogenic sources of nitrogen and their effects on Canadian aquatic ecosystems. Biogeochem. 79: 25–44.

Scott, B.J., A.M. Ridley and M.K. Conyers. 2000. Management of soil acidity in long-term pastures of south-eastern Australia: a review. Aust. J. Exp. Agric. 40: 1173–1198.

Silva, J.A. 2000. Inorganic fertilizer materials. pp. 117–120. *In*: Silva, J.A. and R.S. Uchida (eds.). Plant Nutrient Management in Hawaii's Soils. Manoa: University of Hawai'i Press.

Smith, P.D., M.Z. Cai, D. Gwary et al. 2007. Agriculture. pp. 497–540. *In*: Metz, B., O.R. Davidson, P.R. Bosch, R. Dave and L.A. Meyer (eds.). Climate Change 2007: Mitigation. Contribution of Working Group III to the Fourth Assessment Report of the Intergovernmental Panel on Climate Change. Cambridge: Cambridge Univ. Press.

Smith, V.H. and D.W. Schindler. 2009. Eutrophication science: where do we go from here? Trends Ecol. Evol. 24: 201–207.

Sommer, S.G. and N. Hutchings. 1995. Techniques and strategies for the reduction of ammonia emission from agriculture. Water Air Soil Poll. 85: 237–248.

United Nations Department of Economic and Social Affairs (UN DESA). 2017. World population prospects: The 2017 revision. Key findings and advance tables. Working Paper No. ESA/P/WP/248. https://esa.un.org/unpd/wpp/publications/Files/WPP2017_KeyFindings.pdf.

United Nations General Assembly (UN General Assembly). 2015. Transforming our world: the 2030 agenda for sustainable development. Working Paper No. A/RES/70/1. https://sustainabledevelopment.un.org/content/documents/21252030%20Agenda%20for%20Sustainable%20Development%20web.pdf.

Vanlauwe, B., A. Bationo, J. Chianu et al. 2010. Integrated soil fertility management: Operational definition and consequences for implementation and dissemination. Outlook Agric. 39: 17–24.

Zaman, Q.U., A.W. Schumann and W.M. Miller. 2005. Variable-rate nitrogen application in Florida citrus based on ultrasonically sensed tree size. Appl. Eng. Agric. 21: 331–335.

Zhu-Barker, X., W.R. Horwath and M. Burger. 2015. Knife-injected anhydrous ammonia increases yield-scaled N_2O emissions compared to broadcast or band-applied ammonium sulfate in wheat. Agric. Ecosys. Environ. 212: 148–157.

Chapter 3

Microplastics in Agricultural Soils
Are They a Real Environmental Hazard?

Andrés Rodríguez-Seijo[1,2] and *Ruth Pereira*[2,]*

INTRODUCTION

Since the 1950s, plastic polymers became essential raw materials in the food industry and agriculture due to their low-cost and high variety of applications. In agriculture, greenhouses were constructed using glass materials until that decade, but the development of plastic industry brought a significant advance to agricultural technology under greenhouse. This technological revolution has allowed that infertile lands characterized by a drastic desertification process, adverse climatic conditions, or with a high pest pressure, be transformed in highly productive areas able to boost local/regional economic growth (Espí et al. 2006, Kyrikou and Briassoulis 2007, Thompson et al. 2009, Kasirajan and Ngouajio 2012, Mormile et al. 2017, Orzolek 2017). Thus, the called "plasticulture era" marked a turning point in the development of agriculture, particularly the horticulture (Kyrikou and Briassoulis 2007, Kasirajan and Ngouajio 2012, Mormile et al. 2017). In 2015, the annual plastic production was already around 322 million tonnes, and two percent of this production attended the demand from agriculture and horticulture, representing a global market of 5.8 billion US dollars in 2012 (PlasticsEurope 2015, Horton et al. 2017) (Fig. 1).

A recent study by Mormile et al. (2017), concluded that plastic film consumption increased exponentially in the last decade, being Asia (70%) and Europe (16%) the primary consumers. In horticulture, the most used plastic polymers are polyolefin, polyethene, polypropylene, ethylene-vinyl acetate copolymer, polyvinyl chloride

[1] Department of Plant Biology and Soil Science, Universidade de Vigo, Spain.
  Email: andresrodriguezseijo@hotmail.com
[2] GreenUPorto & Department of Biology, Faculty of Sciences of the University of Porto, Portugal.
* Corresponding author: ruth.pereira@fc.up.pt

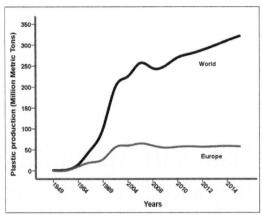

Fig. 1. Evolution of plastic production between 1949 and 2015 (data taken from PlasticsEurope 2015 and Geyer et al. 2017).

and, less frequent, polycarbonate and poly-methyl-methacrylate. These materials are used in greenhouse covers, low and high tunnels, mulching films, plastic reservoirs, irrigation systems, silage films, boxes and packaging materials, harvesting nets and films, and in post-harvesting operations to deliver products to the market (Scarascia-Mugnozza et al. 2011, Kasirajan and Ngouajio 2012, PlasticsEurope 2015, Steinmetz et al. 2016, Vox et al. 2016, Brodhagen et al. 2017).

Plastics have revolutionised our daily lives. Their versatility, low-cost of production, light weight and resistance justify their presence in a vast variety of products. In agriculture, plastics are mainly used to develop and maintain specific microclimate conditions needed for improving plant development (Lamont 2005, Orzolek 2017). For example, mulch plastic films are used to maintain soil temperature and moisture within a narrow range of variation with multiple beneficial effects such as increase of plant growth, reduced runoffs and subsequent retention of nutrients, improved soil physical properties as bulk density and aggregate stability, control of weeds and pests, higher yield and early marketing of seasonal crops, water saving, and reduced applications of plant protection products (Kyrikou and Briassoulis 2007, Scarascia-Mugnozza et al. 2011, Kasirajan and Ngouajio 2012, Steinmetz et al. 2016, Vox et al. 2016, Mormile et al. 2017, Ngouajio et al. 2007, Orzolek 2017). Despite all these benefits, the use of plastics in agriculture may pose severe environmental risks. Some studies have documented that an excessive use of plastics may lead to nutrient depletion by enhanced microbial activity, imbalanced C/N ratio, increased repellency of soil to water, increase of greenhouse gas emissions, alterations in the biogeochemical cycles of nutrients, contamination by plastic debris and adverse impacts on soil biota, among other potential effects (Huerta Lwanga et al. 2016, Steinmetz et al. 2016, Kader et al. 2017, Maaß et al. 2017, Rodríguez-Seijo et al. 2017).

Plastics usually have several additives incorporated in plastic manufacturing to improve its mechanical and chemical properties for UV resistance, durability and malleability (e.g., phthalates, alkylphenols, brominated flame retardants, antioxidant chemicals, and metals). Therefore, releasing of these toxic chemicals during plastic degradation may be a potential risk for soil biota (Kasirajan and Ngouajio 2012,

Li et al. 2016, Steinmetz et al. 2016, Kader et al. 2017, Orzolek 2017). This assumption seems true with plastic film mulches, which landfill discharge is the primary disposal form, representing a point-source of soil contamination. The potential release of environmental contaminants from degradation of plastic film mulch has led Liu et al. (2014) to suggest the concept of "white pollution" related to plastic mulch farming. When plastic wastes left on agricultural soils start to degrade through different mechanisms and processes, they are progressively converted from large pieces into small fragments or microplastics, which can persist in the surface or be incorporated into soil profiles during farming, especially in agriculture soils managed by more conventional practices, or even by biota (Rillig et al. 2017). These plastic debris can be transported to long distances through run-offs from lands to aquatic ecosystems sinking in the oceans (Kyrikou and Briassoulis 2007, Jambeck et al. 2015, Steinmetz et al. 2016, Brodhagen et al. 2017, Kader et al. 2017, Hurley and Nizzetto 2018). Until now, most of studies have dealt with microplastic pollution in the oceans, where the amount of drifting plastics is estimated in more than 236,000 tons (van Sebille et al. 2015, Nizzetto et al. 2016a). Only very recently, the scientific community start to pay attention to terrestrial environments, and in particular to agricultural soils, as significant reservoirs of plastics and microplastics (Nizzetto et al. 2016a,b, Sforzini et al. 2016). However, analytical detection of microplastics in soil is still a challenge, so accumulation and transport of microplastic and biological effects from plastic degradation remain to be investigated.

A literature survey using the ISI® Web of Knowledge and Scopus® databases on microplastic issues provides a quick vision on the research trend on this topic of current concern. In both databases, a literature search was performed using the following keywords: "microplastic", "microplastics", "synthetic fibres", "plastic debris" and "plastic pellets" in combination with "sediment or beach", "soil or terrestrial" and "water, sea, river or oceans". The searching output generated a list of 969 peer-reviewed publications covering a period between 1972 and the beginning of 2018. Most of these publications (74%) were devoted to aquatic environments (rivers, lakes, seas and oceans), a 23% of them dealth with sediments (sediments from lakes, rivers and seas; beaches, estuaries and municipal sludges) and only a 3% were studies related to terrestrial environments (several types of soils, ecotoxicological assays with soil organisms and agricultural areas), which were published after 2012 (Fig. 2).

Traditionally, soil physicochemical properties have been the primary endpoints to assess the impact of agricultural plastics on soil (Liu et al. 2014, Steinmetz et al. 2016, Jiang et al. 2017). However, the effects of plastics and microplastics on soil biota is gaining a growing concern (Kiyama et al. 2012, Huerta Lwanga et al. 2016, 2017, Nizzetto et al. 2016a, Sforzini et al. 2016, Steinmetz et al. 2016, Horton et al. 2017, Rillig et al. 2017, Rodríguez-Seijo et al. 2017). This chapter discusses the current knowledge on the potential impact of microplastics on agricultural soils under a biological perspective. The first section describes the main applications of polymeric-based materials in agriculture, particularly horticulture, which may pose a significant source of plastic contamination. A second section summarises the main routes of plastic input in agricultural soil. The last section will discuss the main

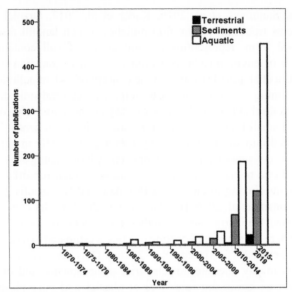

Fig. 2. Global scientific production on microplastic contamination. The bars represent the number of publications (scientific articles and reviews) published in the corresponding 5-years period. Literature survey was performed using the following keywords: "microplastic*" "synthetic fibres", "plastic debris" and "plastic pellets" in combination with "sediment or beach", "soil or terrestrial" and "water, sea, river or oceans" (ISI® Web of Knowledge and Scopus®). Papers published up to October 2017.

uncertainties regarding the occurrence of microplastics in the soil system, and what are the research areas that will require further investigation.

Main Applications of Polymer-Based Materials in Agriculture

Greenhouse Covers and Tunnel Films

Greenhouses are permanent structures for plant culturing, which are covered mainly by plastic films allowing the entrance of solar radiation and, simultaneously, reduces gas emissions and radiation from the soil as well as the risk for plant infestation by pests. Moreover, these structures maintain specific environmental conditions (e.g., temperature and humidity) needed for an optimal plant growth and development (Hussain and Hamid 2003, Espí et al. 2006) (Fig. 3).

Although different types of polymers are used in the manufacturing of greenhouse plastic films, more than 80% of the worldwide production is based on three types of polymers: low-density polyethylene, ethyl acetate and vinyl, and polyvinyl chloride (Espí et al. 2006, Mormile et al. 2017). One of the limitations in the use of plastic films in greenhouses is their useful half-life, which is around 6 and 45 months depending on UV stabilizers that form part of their chemical structure, the local climate conditions and the use of agrochemicals, among other factors (Espí et al. 2006). Therefore, greenhouse plastic films need necessarily to be replaced by new films, contributing thereby to generate continually plastic wastes. These plastic wastes can be incinerated, recycled, or, in the worst of cases, discharged in landfills

Fig. 3. A typical walk-in greenhouse for cool and humid climates. Photography courtesy by Rosa María Piñeiro-Lijó.

or even remained in the agricultural soils generating about 2,400 kg year⁻¹ ha⁻¹ of plastic residues (Blázquez 2003).

Low tunnel films constitute another polymer-based material widely used in agriculture, which may contaminate the soil with plastic debris. Basically, these structures have the same functional characteristics than greenhouses, except for their complexity and height. Low tunnel films are used mainly in the cultivation of asparagus, watermelon, and strawberries. For this application, the same plastic materials of those from greenhouse film manufacturing are used, although the production of plastic wastes is lower than greenhouse covers (around 190 kg year⁻¹ ha⁻¹ of plastic wastes) (Blázquez 2003).

Plastic Mulching

Mulching techniques are very ancient (2500–3000 years B.C.) and consist in leaving a layer of organic material (e.g., plant litter) on the soil surface. Traditionally, mulching covers are made by straw, dried leaves, natural fibres, crop residues or organic wastes (Lal 2015, Mormile et al. 2017). However, the introduction of polymer-based films in the mulching technology brought significant improvements, because of its easier, cheaper and broader applicability compared with the most traditional methods (Espí et al. 2006, Mormile et al. 2017) (Fig. 4).

The use of plastic films for mulching is the second largest application of plastic polymers after greenhouse covers and consists of putting a plastic film directly over the soil (Espí et al. 2006, Mormile et al. 2017). The scope of this strategy is to maintain soil moisture, improve structure and fertility, reduce soil erosion, limit weed growth, stimulates early root growth and nutrient uptake, harvest crop faster and with a higher economic yield (Hussain and Hamid 2003, Kasirajan and Ngouajio 2012, Liu et al. 2014, Kader et al. 2017).

Low-density polyethylene (12–80 μm thickness) is the most used polymer in mulching. Its usage half-life is as short as 2 or 4 months, and occasionally, it is photo- or biodegradable with the purpose of reducing the costs associated with withdrawal after each crop season. However, the use of biodegradable plastics is a

Fig. 4. A typical geotextile-mulching cover used for gardens. These mulching covers are composed by polypropylene and polyethene. In addition to agricultural uses, they can also be used on urban gardens to reduce the use of water, and it partially blocks light and reduces the growth speed re-wild weeds.

recent and expensive strategy. As indicated by Sintim and Flury (2017), "the main advantage of biodegradable plastic mulch is that it can be tilled into the soil after use and that it would be degraded by the action of soil organisms, thus saving labor and disposal costs". However, fully biodegradable plastic material has not been yet obtained. Besides, more rigorous toxicological assays are needed to ensure that their use is environmentally safe for soil organisms and causes no adverse effects on soil properties (Hussain and Hamid 2003, Sforzini et al. 2016, Steinmetz et al. 2016, Brodhagen et al. 2017, Horton et al. 2017, Huerta Lwanga et al. 2017, Sintim and Flury 2017).

Nowadays, around 80% of the plastic mulching crop production is located in China, with about 19,791,000 hm^2 in 2011, and a forecasted 25% annual growth (Espí et al. 2006, Liu et al. 2014). Huerta Lwanga et al. (2017) showed that plastic mulch used in the Chinese agricultural soils represent a 60–100% of soil coverage, and it is often not removed after harvesting. In most cases, plastic wastes remain in the field, mainly in the topsoil (0–20 cm), where they may attain levels of plastic contamination up to 317 kg ha^{-1} of residues (Liu et al. 2014, Zhao et al. 2015, Zou et al. 2017).

Solarisation Films

A variant of the mulching techniques is the solarisation films. According to FAO (Abu-Irmaileh 2003), is a hydrothermal process that takes place in moist soil covered by plastic films and exposed to sunlight during the warm months for 4–5 weeks. The purpose of this technique is to sterilise the soil by heating and increasing the soil temperature at different depths, to reduce or even eliminate plant pathogens that may have been accumulated during successive crop seasons (Abu-Irmaileh 2003, Mormile et al. 2017). In addition, solarium films can contribute to the degradation of pesticide residues in soil (Gopal et al. 2000, Fenoll et al. 2017).

However, this technique may cause several impacts on the soil structure, as well as reduces some beneficial soil organism communities. As for plastic mulch, high amounts of plastic wastes can remain on the soil surface between crop seasons, thus contributing to contaminate topsoil (Abu-Irmaileh 2003, Bonanomi et al. 2008). Moreover, an increase of soil temperature by this technique can also facilitate the release of plastic additives to the soil as well as change the pesticide behavior in soil (Flores et al. 2008, Fenoll et al. 2010).

Plastic Reservoirs, Irrigation Systems, Silage, and other uses

Plastic films have an essential role in storage and irrigation systems, due to their low cost, flexibility to generate interesting technical solutions, easy installation and efficiency. These films have considerable importance in the silage of grasses or crops for the feeding of cattle during winter, with the aim to maintain nutritional values and inhibit the fermentation processes. In addition, there is a great use of plastic in packing crates for vegetables/fruit transportation (Hussain and Hamid 2003). Except for the use of PVC in irrigation systems, the other applications described above usually rely on low or high-density polyethylene plastics, which depending on their use and climatic conditions, may contain organic or inorganic additives (e.g., metals) to improve the strength and resistance to photodegradation, among other properties.

Sewage Sludges: A Potential Source of Microplastics

According to Nizzetto et al. (2016a), up to 63,000–430,000 and 44,000–300,000 tonnes of microplastics are discharged annually to farmlands in Europe and North America, respectively, by the application as soil amendments of sewage sludge from municipal wastewater treatment plants (Lambert et al. 2014, Ramos et al. 2016, Steinmetz et al. 2016). Likewise, Horton et al. (2017) estimated a plastic burden discharge to soils between 4 and 23 times higher than that discharged to oceans, which is a much higher quantity than that estimated by Nizzetto et al. (2016a).

In the recent years, plastic contamination research has acquired a new dimension with the study of microplastics (Rillig et al. 2012). Microplastics are a heterogeneous group of particles that differ in their size, density, shape, and chemical composition, which originates from different sources. Although some guidelines remark that the term microplastic denotes those particles with a size less than 5 mm, recently, some studies also include plastic particles < 1 mm size (Rillig 2012, GESAMP 2015, Horton et al. 2017, Rodríguez-Seijo and Pereira 2017).

In terrestrial systems, microplastics can enter through sewage sludges, aerial transport, mismanagement of plastic wastes on landfills, as "microbeads" from personal care applications which are not captured by wastewater treatment plants (microbeads size between 1.7 and 30.6 μm), and/or by the degradation of plastic from agricultural uses (Fendall and Sewell 2009, Hohenblum et al. 2015, UNEP 2015, Steinmetz et al. 2016, Horton et al. 2017, Huerta Lwanga et al. 2017, Mahon et al. 2017, Hurley and Nizzetto 2018).

The application of sewage sludge on agricultural soils is a significant source of microplastics, mainly in the form of microfibres and microbeads (Zubris and Richards 2005, Hopewell et al. 2009, Duis and Coors 2016, Mahon et al. 2017), although this organic amendment is recommended by the European (European Commission 1986) and North American legislation (EPA 1993). For example, synthetic fibres were identified using polarised light microscopy in agricultural soils from the United States (Habib et al. 1998). Moreover, Zubris and Richards (2005) also detected fibres in agricultural soils 15 years after sewage sludge application, with some evidence of vertical translocation in the soil profile. More recently, Mahon et al. (2017) analysed microplastics in sludge samples from seven wastewater treatment plants in Ireland. They found that, despite a reduction of microplastic content in sewage sludges treated by anaerobic digestion, these materials were still present in all analysed samples independently of the sludge treatment (anaerobic digestion, thermal drying and lime stabilisation).

The studies by Nizzetto et al. (2016a,b) also showed that the use of sewage sludge as soil amendments represents a significant input of microplastics to agricultural soils. They estimated that between 125 and 850 tonnes microplastics/million of inhabitants had been added annually to European agricultural soils either by direct application of sewage sludges or by processed biosolids (Nizzetto et al. 2016a). Accordingly, these studies suggest that the regulation on the application of sewage sludges on agriculture soils needs to incorporate microplastics as an additional contaminant to establish regulatory limits in sewage sludge and recommended application rates to environmental protection (Nizzetto et al. 2016a,b).

Uncertainties Regarding Microplastics in Soils and their Environmental Hazards

Despite the benefits of polymer-based materials in agriculture, their widespread use has generated significant amounts of plastic residues that remain in soil systems and are becoming an environmental issue of scientific and public concern (Liu et al. 2014, Jiang et al. 2017). These recalcitrant residues may pose a considerable risk to soil properties, terrestrial biota and ultimately to food webs (Hopewell et al. 2009, Steinmetz et al. 2016, Horton et al. 2017, Jiang et al. 2017). Moreover, disposal of plastic residues on landfill sites may be significant point-sources of microplastic contamination towards other environmental compartments such as rivers and oceans (Hurley and Nizzetto 2018), or even the atmosphere (Gasperi et al. 2018).

Impact of Plastic Debris and Microplastics on Soil Properties

Accumulation of plastic debris on the soil surface and topsoil can lead to adverse effects on soil physicochemical properties. These impacts can be summarized as follow: (i) changes in moisture and nutrient mobility because plastic wastes may reduce water infiltration and movement, ultimately cause secondary soil salinization (Liu et al. 2014, Zhang et al. 2014); (ii) reduction of soil pH by excessive N mineralisation (Jiang et al. 2017, Wang et al. 2017); (iii) changes in bulk soil density, aggregate stability, soil porosity and aeration (Jiang et al. 2017, Zou et al. 2017); (iv) increase of soil organic matter decomposition because of plastic-induce enhancement of soil temperature (Steinmetz et al. 2016, Jiang et al. 2017); (v) effect on seed emergence, root length and root surface (Liu et al. 2014, Zou et al. 2017); (vi) increase of soil temperature (Jiang et al. 2017); and finally (vii) soil contamination by releasing of plastic additives such as phthalate esters, furans, dioxins or metals (Kasirajan and Ngouajio 2012, Chen et al. 2013, Liu et al. 2014, Steinmetz et al. 2016, Jiang et al. 2017). Despite these adverse effects, some authors still argue that the benefits derived from the use of polymer-based materials in agriculture outweigh their side-effects (Jiang et al. 2017). Moreover, these undesirable environmental impacts are dependent on soil properties, climate conditions and/or polymer types.

Impact of Microplastics on Soil Organisms

To date, a few studies have dealt with the real effects of microplastics on soil biota. Microplastics can be ingested accidentally or intentionally by terrestrial organisms such as earthworms. In general, ingested plastic particles may transit the gastrointestinal tract being excreted in faeces. But occasionally, and depending on the particle size, microplastics may cross the gastrointestinal epithelium and accumulate in multiple tissues. As indicated by Horton et al. (2017), this hypothesis is based on: (i) a high similarity and taxonomical proximity between terrestrial and aquatic organisms (e.g., nematodes, annelids, mollusks, or arthropods), and (ii) some terrestrial organisms also have a significant role as detritivorous, so there is a reliable possibility that organisms such as soil-dwelling earthworms (*Lumbricus terrestris*) may transport or disperse microplastics initially accumulated on soil surface to deeper layers of soil through burrowing and casting activities (Huerta Lwanga et al. 2016, 2017, Rillig et al. 2017).

Unlike aquatic organisms, a few studies have assessed the effects of microplastics on terrestrial organisms such as the nematode *Caenorhabditis elegans* (Kiyama et al. 2012), the earthworm species *L. terrestris* (Huerta Lwanga et al. 2016, 2017, Rillig et al. 2017) and *Eisenia andrei* (Sforzini et al. 2016, Rodríguez-Seijo et al. 2017), and the collembola species *Folsomia candida* and *Proisotoma minuta* (Maaß et al. 2017).

Kiyama et al. (2012) observed that *C. elegans* discriminated physically and chemically food from Fluoresbrite® Polystyrene Carboxylate microspheres (< 10 µm diameter). However, ingestion of these microspheres was reported, which could mean a risk for *C. elegans* survival and the soil functions in which these organisms are involved. Although some nematode species are plant parasitic or pathogens, most of them are beneficial with several key functions in soil processes. These organisms contribute to nitrogen mineralisation and distribution of biomass within plants, with subsequent influences on soil fertility and agricultural production (Neher 2001, Khan and Kim 2007, Yeates et al. 2009). They have a crucial role in controlling nutrient availability through modification of some nutrients into more bioavailable forms (e.g., NH_4^+), and a role in increasing the release of carbon-rich root exudates into the soil (Savin et al. 2001, Gebremikael et al. 2016). Nematodes can also stimulate microbial growth and reduce microbial competition, resulting in an increment of soil organic matter decomposition (Gebremikael et al. 2016).

Collembolans also play an important role in terrestrial ecosystems, because of their activity in the degradation of organic matter, recycling of C and N, mobilisation of available micronutrients such as Ca, or in the control of soil microbial activity (Rusek 1998, Culliney 2013). Also, they can improve the soil microstructure by creating micro-tunnels (Culliney 2013) or by the formation of soil agglomerates and water-stable aggregates (Siddiky et al. 2012). Maaß et al. (2017) observed that two collembolan species, *F. candida* and *P. minuta*, did not ingest microplastics (urea-formaldehyde and polyethylene terephthalate fibres of < 100 and 100–200 µm sizes).

However, they were able to act as carriers of these particles, and *F. candida* was able to disperse larger particles more quickly and over greater distances than *P. minuta*. Furthermore, due to their role in soil microstructure, it is possible that collembolans can contribute to the transport and accumulation of microplastics in the soil profile, facilitating their bioavailability to other soil organisms.

Earthworms have a prominent role in the soil with a significant influence on its edaphic properties and soil microbial communities. These invertebrates can improve the soil structure through increasing macroporosity, the formation of stable aggregates, ultimately improving soil bulk density (Blouin et al. 2013, Bertrand et al. 2015). Moreover, they can change the availability of some nutrients such as C and P, with particular attention to the acceleration of soil organic matter decomposition and nitrogen mineralisation (Blouin et al. 2013, van Groenigen et al. 2014). In addition, these annelids have a role in the remediation of soils contaminated by organic and inorganic contaminants. This environmental service is due to the stimulatory effect that earthworms cause on microbial degraders of organic contaminants, and changes in metal distribution and speciation. Furthermore, earthworm activity may also increase metal mobility and availability, improving the efficiency of phytoremediation (Sizmur and Hodson 2009, Blouin et al. 2013, Lemtiri et al. 2016).

Only five studies have been performed on the effects of microplastics on earthworms so far. The earthworm species used in these investigations have been the anecic earthworm *L. terrestris* (Huerta Lwanga et al. 2016, 2017, Rillig et al. 2017) and the epigeic earthworm *E. andrei* (Sforzini et al. 2016, Rodríguez-Seijo et al. 2017). Huerta Lwanga et al. (2016, 2017) showed that exposure of *L. terrestris* to polyethene particles (< 400 μm) increased the mortality of earthworms by 8% and up to 25% at particle concentrations of 450 and 600 mg kg^{-1} (in overlying leaf litter), respectively. Furthermore, these researchers found a reduction of earthworm weight and adverse effects on burrowing capability. They also observed that *L. terrestris* could transport and incorporate small plastic particles from the surface litter into the soil profile. This finding has been recently confirmed by Rillig et al. (2017), which demonstrated that the same earthworm species can transport plastic spheres to a depth of 10 cm. The study by Rillig et al. (2017) also demonstrated that *L. terrestris* could increase the risk of microplastic leaching through the soil profile. Because hydrophobic organic pollutants are increasingly being detected sorbed onto microplastics, the risk of groundwater contamination should not be ignored.

Toxic effects of microplastic exposure have also been studied in other earthworm species. For example, Rodríguez-Seijo et al. (2017) found severe histological damages, gut inflammation, and other apparent immune system responses in *E. andrei* exposed to several concentrations of polyethene microplastics (0, 62, 125, 250, 500, and 1000 mg kg^{-1}) with a size less than 1 mm. Although mortality was not observed, they also observed an increase of total proteins, lipids, and polysaccharides as the microplastic content in soil increased. *Eisenia andrei* was also exposed to biodegradable plastics (based on cornstarch and biodegradable polyesters) used in packaging (Sforzini et al. 2016), but no acute and chronic toxic effects were observed. In addition, Sforzini et al. (2016) analysed the ecotoxicity of the biodegradation by-products of these plastics after a 6-months incubation period with artificial soils spiked with three types of bioplastics. They concluded that these plastics, and their potentially toxic degradation by-products, were not toxic as revealed by multiple ecotoxicity tests that involved elutriates of plastic-incubated soils and the bulk soil. Nevertheless, these results did not discourage the need for assessing the impact on soil properties of biodegradable plastic polymers, especially under different soil and climatic conditions (Sintim and Flury 2017).

Finally, microplastics disposition in landfill or agricultural soils is a serious hazard to other terrestrial organisms. For example, microplastics ingestion has been widely described in more than one hundred seabird species, with traces of plastics found in their gut contents (Gregory 2009, Gall and Thompson 2015, Wilcox et al. 2015, Zhao et al. 2016). Birds inhabiting agroecosystems may also be at risk of microplastic ingestion and toxicity. Although there are a few reports on this subject, it has already been demonstrated that seabirds ingest accidentally high amounts of plastic debris and microplastics because of their inability of discriminating them from food items (Teuten et al. 2009, Tanaka et al. 2013, Savoca et al. 2015). However, birds can also be exposed through food chain transferences, a route which has been demonstrated in marine food webs (Farrel and Nelson 2013, Romeo et al. 2015, Kärrman et al. 2016). Zhao et al. (2016) indicated that microscopic anthropogenic litter in carnivorous birds could be derived from the assimilation of plastic debris previously consumed by their preys. Townsend and Barker (2014) showed that 85% of the 106 nests of the American crow (*Corvus brachyrhynchos* Brehm, 1822) sampled from both agricultural and urban areas of California, had anthropogenic materials (synthetic twine, string and rope, plastic strips, tape and wire), including plastic wastes. These authors hypothesised that these residues constituted a severe risk to bird health. Zhao et al. (2016) also found a high amount of plastic/synthetic microfibres in the digestive tract of 12 bird species caught in the Shanghai urban environment and covering multiple trophic niches (eight carnivorous species, three omnivorous species and one herbivorous species). Taken together, these studies suggest that it is highly probable that birds inhabiting rural environments can ingest plastic debris from agricultural uses, in a comparable way than that widely described for ingestion of lead pellet from shooting and hunting areas (Thomas 1997, de Francisco et al. 2003, Haig et al. 2014, Pain et al. 2015, Arnemo et al. 2016). However, further studies are still required to validate this hypothesis as well as to determine the real impact of microplastic exposure to other organisms and populations.

Conclusions

Current studies suggest that plastic wastes and microplastics are affecting soil properties, soil biota and organisms that forage on soil resources, including humans. Several agriculture practices are contributing to plastic waste input in agriculture soils. Among the soil organisms studied for microplastic contamination, the toxic effects seemed to be species-specific. Interestingly, soil organisms may contribute to the transport of plastic debris and microplastics in the soil, with subsequent changes in soil physicochemical and biological properties.

There are some knowledge gaps on the occurrence and environmental impact of microplastics in the terrestrial ecosystems that need urgent attention to induce changes in the policies addressed to regulate the use and disposal of plastic materials for sustainable agriculture. The issues that form the basis for future research are: (1) Short- and long-term effects of microplastics on soil organic matter dynamics, soil microbiome and soil enzymatic activity. (2) Impact of global warming and soil characteristics on the biodegradation rate of plastic polymers used in agriculture, with particular emphasis in adverse effects on soil biota. (3) Impact of plastic debris and microplastics on the environmental fate of agrochemicals, and other potential environmental pollutants such as metal, pharmaceuticals, polycyclic aromatic hydrocarbons currently found in agricultural soils. This issue is of particular concern in the case of using sewage sludge as a soil amendment, where pollutants and microplastics may co-exist.

Acknowledgements

A. Rodríguez-Seijo would like to thank the Universidade de Vigo for his predoctoral fellowship (P.P. 00VI 131H 64102) and BEV1 research group (Agrobioloxía ambiental: Calidade, solos e plantas, Universidade de Vigo) for his postdoctoral contract (P.P. V534 131H 6450211).

References

Abu-Irmaileh, B. 2003. Soil solarization. *In*: Labrada, R. (ed.). Weed Management for Developing Countries. FAO Plant Production and Protection Paper 120 Add. 1, FAO, Rome, Italy.

Arnemo, J.M., O. Andersen, S. Stokke, V.G. Thomas, O. Krone, D.J. Pain et al. 2016. Health and Environmental risks from lead-based ammunition: Science versus socio-politics. Ecohealth. 13(4): 618–622.

Bertrand, M., S. Barot, M. Blouin, J. Whalen, T. de Oliveira and J. Roger-Estrade. 2015. Earthworm services for cropping systems. A review. Agron. Sustain. Dev. 35: 553–567.

Blázquez, M.A. 2003. Capítulo X. Los residuos plásticos agrícolas. pp. 306–326. *In*: Soria Tonda, J.M. (ed.). Los Residuos Urbanos y Asimilables, Junta de Andalucía, Spain (In Spanish).

Blouin, M., M.E. Hodson, E.A. Delgado, G. Baker, L. Brussaard, K.R. Butt et al. 2013. A review of earthworm impact on soil function and ecosystem services. Eur. J. Soil Sci. 64: 161–182.

Bonanomi, G., M. Chiurazzi, S. Caporaso, G. Del Sorbo, G. Moschetti and S. Felice. 2008. Soil solarization with biodegradable materials and its impact on soil microbial communities. Soil Biol. Biochem. 40(8): 1989–1998.

Brodhagen, M., J.R. Goldberger, D.G. Hayes, D.A. Inglis, T.L. Marsh and C. Miles. 2017. Policy considerations for limiting unintended residual plastic in agricultural soils. Environ. Sci. Policy 69: 81–84.

Chen, Y., C. Wu, H. Zhang, Q. Lin, Y. Hong and Y. Luo. 2013. Empirical estimation of pollution load and contamination levels of phthalate esters in agricultural soils from plastic film mulching in China. Environ. Earth Sci. 70: 239–247.

Culliney, T.W. 2013. Role of Arthropods in maintaining soil fertility. Agriculture 3(4): 629–659.

de Francisco, N., J.D. Ruiz Troya and E.I. Agüera. 2003. Lead and lead toxicity in domestic and free living birds. Avian Pathol. 32(1): 3–13.

Duis, K. and A. Coors. 2016. Microplastics in the aquatic and terrestrial environment: sources (with a specific focus on personal care products), fate and effects. Environ. Sci. Eur. 28: 2.

EPA. 1993. 40 CFR Part 503. Standards for the use of disposal of sewage sludges. Environmental Protection Agency, Washington D.C., USA.

Espí, E., A. Salmerón, A. Fontecha, Y. García and A.I. Real. 2006. Plastic Films for Agricultural Applications. J. Plast. Film Sheeting 22: 85–102.

European Commission. 1986. Council Directive 86/278/EEC of 12 June 1986 on the protection of the environment, and in particular of the soil, when sewage sludge is used in agriculture. Official Journal L 181, 04/07/1986. pp. 0006-0012.

Farrel, P. and K. Nelson. 2013. Trophic level transfer of microplastic: *Mytilus edulis* (L.) to *Carcinus maenas* (L.). Environ. Pollut. 177: 1–3.

Fendall, L.S. and M.A. Sewell. 2009. Contributing to marine pollution by washing your face: Microplastics in facial cleansers. Mar. Poll. Bull. 58(8): 1225–1228.

Fenoll, J., E. Ruiz, P. Hellín, S. Navarro and P. Flores. 2010. Solarization and biosolarization enhance fungicide dissipation in the soil. Chemosphere 79(2): 216–220.

Fenoll, J., I. Garrido, N. Vela, C. Ros and S. Navarro. 2017. Enhanced degradation of spiro-insecticides and their leacher enol derivatives in soil by solarization and biosolarization techniques. Environ. Sci. Pollut. Res. 24(10): 9278–9285.

Flores, P., A. Lacasa, P. Fernández, P. Hellín and J. Fenoll. 2008. Impact of biofumigation with solarization on degradation of pesticides and heavy metal accumulation. J. Environ. Sci. Health B 43(6): 513–518.

Gall, S.C. and R.C. Thompson. 2015. The impact of debris on marine life. Mar. Pollut. Bull. 92: 170–179.

Gasperi, J., S.L. Wright, R. Dris, F. Collard, C. Mandin, M. Guerrouache et al. 2018. Microplastics in air: Are we breathing it in? Curr. Opin. Environ. Sci. Health 1: 1–5.

Gebremikael, M.T., H. Steel, D. Buchan, W. Bert and S. De Neve. 2016. Nematodes enhance plant growth and nutrient uptake under C and N-rich conditions. Sci. Rep. 6: 32862.

GESAMP. 2015. Sources, fate and effects of microplastics in the marine environment: a global assessment. IMO/FAO/UNESCO-IOC/UNIDO/WMO/IAEA/UN/UNEP/UNDP Joint Group of Experts on the Scientific Aspects of Marine Environmental Protection. Rep. Stud. GESAMP No. 90. IMO, London, UK.

Geyer, R., J.R. Jambeck and K. Lavender Law. 2017. Production, use, and fate of all plastics ever made. Sci. Adv. 3(7): e1700782.

Gopal, M., I. Mukherjee, D. Prasad and N.T. Yaduraju. 2000. Soil solarization: Technique for decontamination of an organophosphorus pesticide from soil and nematode control. Bull. Environ. Contam. Toxicol. 64: 40.

Gregory, M.R. 2009. Environmental implications of plastic debris in marine settings—entanglement, ingestion, smothering, hangers-on, hitch-hiking and alien invasions. Philos. Trans. R. Soc. B. 364: 2013–2025

Habib, D., D.C. Locke and L.J. Cannone. 1998. Synthetic fibers as indicators of municipal sewage sludge, sludge products, and sewage treatment plant effluents. Water Air Soil Pollut. 103: 1–8.

Haig, S.M., J. D'Elia, C. Eagles-Smith, J.M. Fair, J. Gervais, G. Herring et al. 2014. The persistent problem of lead poisoning in birds from ammunition and fishing tackle. Condor 116(3): 408–428.

Hohenblum, P., B. Liebman and M. Liedermann. 2015. Plastic and Microplastic in the environment. Austrian Federal Ministry of Agriculture, Forestry, Environment and Water Management, Viena, Austria.

Hopewell, J., R. Dvorak and E. Kosior. 2009. Plastics recycling: challenges and opportunities. Philos. Trans. R. Soc. B 364(1526): 2115–2126.

Horton, A.A., A. Walton, D.J. Spurgeon, E. Lahive and C. Svendsen. 2017. Microplastics in freshwater and terrestrial environments: Evaluating the current understanding to identify the knowledge gaps and future research priorities. Sci. Total Environ. 586: 127–141.

Huerta Lwanga, E., H. Gertsen, H. Gooren, P. Peters, T. Salánki, M. van der Ploeg et al. 2016. Microplastics in the terrestrial ecosystem: Implications for *Lumbricus terrestris* (Oligochaeta, Lumbricidae). Environ. Sci. Technol. 50(5): 2685–2691.

Huerta Lwanga, E., H. Gertsen, H. Gooren, P. Peters, T. Salánki, M. van der Ploeg et al. 2017. Incorporation of microplastics from litter into burrows of *Lumbricus terrestris*. Environ. Pollut. 220: 523–531.

Hurley, R.R. and L. Nizzetto. 2018. Fate and occurrence of micro(nano)plastics in soils: Knowledge gaps and possible risks. Curr. Opin. Environ. Sci. Health 1: 6–11.

Hussain, I. and H. Hamid. 2003. Plastics in Agriculture. pp. 185–209. *In*: Andrady, A.L. (ed.). Plastics and the Environment. John Wiley & Sons, Inc., Hoboken, NJ, USA.

Jambeck, J.R., R. Geyer, C. Wilcox, T.R. Siegler, M. Perryman, A. Andrady et al. 2015. Plastic waste inputs from land into the ocean. Science 347(6223): 768–771.

Jiang, X., W. Liu, E. Wang, T. Zhou and P. Xin. 2017. Residual plastic mulch fragments effects on soil physical properties and water flow behavior in the Minqin Oasis, northwestern China. Soil Tillage Res. 166: 100–107.

Kader, M.A., M. Senge, M.A. Mojid and K. Ito. 2017. Recent advances in mulching materials and methods for modifying soil environment. Soil Tillage Res. 168: 155–166.

Kärrman, A., C. Schönlau and M. Engwall. 2016. Exposure and Effects of Microplastics on Wildlife. A review of existing data. Report case for Swedish Environmental Protection Agency. Örebro University, Sweden.

Kasirajan, S. and M. Ngouajio. 2012. Polyethylene and biodegradable mulches for agricultural applications: a review. Agron. Sustain. Dev. 32: 501–529.

Khan, Z. and Y.H. Kim. 2007. A review on the role of predatory soil nematodes in the biological control of plant parasitic nematodes. Appl. Soil Ecol. 35: 370–379.

Kiyama, Y., K. Miyahara and Y. Ohshima. 2012. Active uptake of artificial particles in the nematode *Caenorhabditis elegans*. J. Exp. Biol. 215: 1178–1183.

Kyrikou, I. and D. Briassoulis. 2007. Biodegradation of Agricultural Plastic Films: A Critical Review. J. Polym. Environ. 15: 125–150.

Lal, R. 2015. Restoring Soil Quality to Mitigate Soil Degradation. Sustainability 7: 5875–5895.

Lambert, S., C. Sinclair and A. Boxall. 2014. Occurrence, degradation, and effect of polymer-based materials in the environment. pp. 1–53. *In*: Whitacre, D. (ed.). Rev. Environ. Contam. Toxicol. Vol. 227. Springer, Cham, Switzerland.

Lamont, W.J., Jr. 2005. Plastics: Modifying the microclimate for the production of vegetable crops. HortTechnology 15(3): 477–481.

Lemtiri, A., A. Liénard, T. Alabi, Y. Brostaux, D. Cluzeau, F. Francis et al. 2016. Earthworms *Eisenia fetida* affect the uptake of heavy metals by plants *Vicia faba* and *Zea mays* in metal-contaminated soils. Appl. Soil Ecol. 104: 67–78.

Li, K., D. Ma, J. Wu, C. Chai and Y. Shi. 2016. Distribution of phthalate esters in agricultural soil with plastic film mulching in Shandong Peninsula, East China. Chemosphere 164: 314–321.

Liu, E.K., W.Q. He and C.R. Yan. 2014. 'White revolution' to 'white pollution'—agricultural plastic film mulch in China. Environ. Res. Lett. 9: 091001.

Maaß, S., D. Daphi, A. Lehmann and M.C. Rillig. 2017. Transport of microplastics by two collembolan species. Environ. Pollut. 225: 465–459.

Mahon, A.M., B. O'Connell, M.G. Healy, I. O'Connor, R. Nash et al. 2017. Microplastics in sewage sludge: Effects of treatment. Environ. Sci. Technol. 51(2): 810–818.

Mormile, P., N. Stahl and M. Malinconico. 2017. The World of plasticulture. pp. 1–21. *In*: Malinconico, M. (ed.). Soil Degradable Bioplastics for a Sustainable Modern Agriculture. Springer, Springer Berlin Heidelberg.

Neher, D.A. 2001. Role of nematodes in soil health and their use as indicators. J. Nematol. 33(4): 161–168.

Ngouajio, M., G. Goldy, B. Zandstra and D. Warncke. 2007. Plasticulture For Michigan Vegetable Production. Michigan State University Extension. East Lansing, Michigan, U.S.A.

Nizzetto, L., M. Futter and S. Langaas. 2016a. Are agricultural soils dumps for microplastics of urban origin? Environ. Sci. Technol. 50(20): 10777–10779.

Nizzetto, L., G. Bussi, M.N. Futter, D. Butterfield and P.G. Whitehead. 2016b. A theoretical assessment of microplastic transport in river catchments and their retention by soils and river sediments. Environ. Sci.: Processes Impacts 18: 1050–1059.

Orzolek, M.D. 2017. A Guide to the Manufacture, Performance, and Potential of Plastics in Agriculture. William Andrew, Elsevier.

Pain, D.J., R. Cromie and R.E. Green. 2015. Poisoning of birds and other wildlife from ammunition-derived lead in the UK. pp. 58–84. *In*: Delahay, R.J. and C.J. Spray (eds.). Proceedings of the

Oxford Lead Symposium: Lead Ammunition: Understanding and Minimizing the Risks to Human and Environmental Health. Edward Grey Institute: Oxford University, UK.

PlasticsEurope. 2015. Plastics—The Facts, 2015, An analysis of European plastics production, demand and waste data from 2014, Brussels, Belgium.

Ramos, L., G. Berenstein, E.A. Hughes, A. Zalts and J.M. Montserrat. 2016. Polyethylene film incorporation into the horticultural soil of small periurban production units in Argentina. Sci. Total Environ. 523: 74–81.

Rillig, M.C. 2012. Microplastic in Terrestrial ecosystems and the soil? Environ. Sci. Technol. 46: 6453–6454.

Rillig, M., L. Ziersch and S. Hempel. 2017. Microplastic transport in soil by earthworms. Sci. Rep. 7: 1362.

Rodríguez-Seijo, A. and R. Pereira. 2017. Morphological and physical characterization of microplastics. pp. 49–66. *In*: Rocha-Santos, T.A.P. and A.C. Duarte (eds.). Comprehensive Analytical Chemistry, Vol. 75. Elsevier, B.V.

Rodríguez-Seijo, A., J. Lourenço, T.A.P. Rocha-Santos, J. da Costa, A.C. Duarte, H. Vala et al. 2017. Histopathological and molecular effects of microplastics in *Eisenia andrei* Bouché. Environ. Pollut. 220: 495–503.

Romeo, T., B. Pietro, C. Pedà, P. Consoli, F. Andaloro and M.C. Fossi. 2015. First evidence of presence of plastic debris in stomach of large pelagic fish in the mediterranean sea. Mar. Pollut. Bull. 95(1): 358–361.

Rusek, J. 1998. Biodiversity of Collembola and their functional role in the ecosystem. Biodivers. Conserv. 7: 1207–1219.

Savin, M.C., J.H. Görres, D.A. Neher and J.A. Amador. 2001. Uncoupling of carbon and nitrogen mineralization: role of microbivorous nematodes. Soil Biol. Biochem. 33(11): 1463–1472.

Savoca, M.S., M.E. Wohlfeil, S.E. Ebeler and G.A. Nevitt. 2016. Marine plastic debris emits a keystone infochemical for olfactory foraging seabirds. Sci. Adv. 2(11): e1600395.

Scarascia-Mugnozza, G., C. Sica and G. Russo. 2011. Plastic materials in european agriculture: actual use and perspectives. J. Agric. Eng.—Riv. di Ing. Agr. 3: 15–28.

Sforzini, S., L. Oliveri, S. Chinaglia and A. Viarengo. 2016. Application of biotests for the determination of soil ecotoxicity after exposure to biodegradable plastics. Front. Environ. Sci. 4: 68.

Siddiky, M.R.K., J. Schaller, T. Caruso and M.C. Rillig. 2012. Arbuscular mycorrhizal fungi and Collembola non-additively increase soil aggregation. Soil Biol. Biochem. 47: 93–99.

Sintim, H.Y and M. Flury. 2017. Is biodegradable plastic mulch the solution to agriculture's plastic problem? Environ. Sci. Technol. 51(3): 1068–1069.

Sizmur, T. and M.E. Hodson. 2009. Do earthworms impact metal mobility and availability in soil?—A review. Environ. Pollut. 157: 1981–1989.

Steinmetz, Z., C. Wollmann, M. Schaefer, C. Buchmann, J. David, J. Tröger et al. 2016. Plastic mulching in agriculture. Trading short-term agronomic benefits for long-term soil degradation? Sci. Total Environ. 550: 690–705.

Tanaka, K., H. Takada, R. Yamashita, K. Mizukawa, M.A. Fukuwaka and Y. Watanuki. 2013. Accumulation of plastic-derived chemicals in tissues of seabirds ingesting marine plastics. Mar. Pollut. Bull. 69(1-2): 219–22.

Teuten, E.L., J.M. Saquing, D.R. Knappe, M.A. Barlaz, S. Jonsson, A. Björn et al. 2009. Transport and release of chemicals from plastics to the environment and to wildlife. Philos. Trans. R. Soc. B 364(1526): 2027–2045.

Thomas, V.G. 1997. The environmental and ethical implications of lead shot contamination of rural lands in North America. J. Agric. Environ. Ethics 10(1): 41–54.

Thompson, R.C., C.J. Moore, F.S. vom Saal and S.H. Swan. 2009. Plastics, the environment and human health: current consensus and future trends. Philos. Trans. R. Soc. B. 364(1526): 2153–2166.

Townsend, A.K. and C.M. Barker. 2014. Plastic and the nest entanglement of urban and agricultural crows. PLoS ONE 9(1): e88006.

UNEP. 2015. Plastic in cosmetics. Are we polluting the environment through our personal care? United Nations Environment Programme. United Nations Environment Programme, Naironi, Kenya.

van Groenigen, J.W., I.M. Lubbers, H.M.J. Vos, G.G. Brown, G.B. De Deyn and K.J. van Groenigen. 2014. Earthworms increase plant production: a meta-analysis. Sci. Rep. 4: 6365.

van Sebille, E., C. Wilcox, L. Lebreton, N. Maximenko, B.D. Hardesty, J.A. Van Franeker et al. 2015. A global inventory of small floating plastic debris. Environ. Res. Lett. 10(12): 124006.

Vox, G., R.S. Loisi, I. Blanco, G.S. Mugnoza and E. Schettini. 2016. Mapping of agriculture plastic waste. Agric. Agric. Sci. Procedia 8: 583–591.

Wang, L., G. Li, J. Lv, T. Fu, Q. Ma, W. Song et al. 2017. Continuous plastic-film mulching increases soil aggregation but decreases soil pH in semiarid areas of China. Soil Tillage Res. 167: 46–53.

Wilcox, C., E. van Sebille and B.D. Hardesty. 2015. Threat of plastic pollution to seabirds is global, pervasive, and increasing. Proc. Natl. Acad. Sci. USA 112(38): 11899–11904.

Yeates G.W., H. Ferris, T. Moens and W.H. Van Der Putten. 2009. The role of nematodes in ecosystems. pp. 1–44. *In*: Wilson, J.W. and T. Kakouli-Duarte (eds.). Nematodes as Environmental Indicators. CAB International, London, UK.

Zhang, Z., H. Hu, F. Tian, H. Hu, X. Yao and R. Zhong. 2014. Soil salt distribution under mulched drip irrigation in an arid area of northwestern China. J. Arid Environ. 104: 23–33.

Zhao, F.J., Y. Ma, Y.G. Zhu, Z. Tangm and S.P.M. Grath. 2015. Soil contamination in China: Current status and mitigation strategies. Environ. Sci. Technol. 49(2): 750–759.

Zhao, S., L. Zhu and D. Li. 2016. Microscopic anthropogenic litter in terrestrial birds from Shanghai, China: Not only plastics but also natural fibers. Sci. Total Environ. 550: 1110–1115.

Zou, X., W. Niu, J. Liu, Y. Li, B. Liang, L. Guo et al. 2017. Effects of residual mulch film on the growth and fruit quality of tomato (*Lycopersicon esculentum* Mill.). Water Air Soil Pollut. 228: 71.

Zubris, K.A. and B.K. Richards. 2005. Synthetic fibers as an indicator of land application of sludge. Environ. Pollut. 138(2): 201–211.

Chapter 4

Pollutants in Urban Agriculture
Sources, Health Risk Assessment and Sustainable Management

Camille Dumat,[1,2,]* *Antoine Pierart,*[3]
Muhammad Shahid[4] and *Sana Khalid*[4]

INTRODUCTION

Sustainable food production has become a crucial issue in the current agricultural system to endure the growing global population and urbanization. According to the Food and Agriculture Organization (FAO 2015), 60% of humanity lives in urban areas and the forecast for 2050 is 80% (with > 3 billion inhabitants on the planet). Consequently, urban agriculture has recently received a considerable attention worldwide for local fresh food production and sustainable cities promotion (Dumat et al. 2016). This agricultural system is regarded as one of the key strategies to attend the high food demand in the coming years (Hugo 2017). However, urban agriculture has a serious threat: natural and anthropogenic sources of pollution that release into the (peri)urban environment a hazardous cocktail of organic and inorganic contaminants

[1] Centre d'Etude et de Recherche Travail Organisation Pouvoir (CERTOP), UMR5044, Université J. Jaurès–Toulouse II, 5 allée Antonio Machado, 31058 Toulouse, France.
[2] Université de Toulouse, INP-ENSAT, Avenue de l'Agrobiopole, 31326 Auzeville-Tolosane.
[3] Ecotoxicology Lab, Fac. Environmental Science and Biochemistry, University of Castilla-La Mancha, Toledo, Spain.
[4] Department of Environmental Sciences, COMSATS Institute of Information Technology, Vehari-61100, Pakistan.
* Corresponding author: camille.dumat@ensat.fr

(Xiong et al. 2017). To illustrate these issues under a scientific perspective, Fig. 1 illustrates the estimated change in urban population over 100 years (expressed in %), and the global scientific concern dealing with urban agriculture, polluted urban soils, and remediation actions in urban soils.

Despite the considerable progress by scientists, the production of healthy and pollutant-free food is nowadays a pursuit of producers, researchers, scientists and policy makers. Issues related to the autonomy and quality of food in cities and, more generally, to their efficiency and resilience are becoming more and more frequent (Barthel and Isendahl 2013). How to produce sufficient quantities of healthy food to attend the world population with an acceptable impact on the environment? This is a simple question that has gained considerable attention by multiple sectors of society including science, engineering, economy and politics. Science and engineering can suggest paths and guidelines for developing a more efficient and eco-friendly agriculture. However, these guidelines are subject to regulatory aspects of regional economy, traditional agriculture policy and other political constraints. Indeed, different stakeholders are involved in the development of urban agriculture, from citizens and elected officials to academia and other professionals in this productive sector (Mougeot 2010). Urban agriculture projects are, therefore, at the heart of science and society dynamics, and it is fundamental to better understand the mechanisms underlying food production in urban areas and their consequences on human and environmental health.

In order to meet the current demographic trend, large areas of natural habitats and agriculture are converted into buildings or channels of communication (Morefield et al. 2016). This process of intense urbanization has major environmental consequences at the global scale (Cui and Shi 2012). It is largely responsible for improper resource use, greenhouse gas emissions, habitat destruction, increased solid waste production, and landscape fragmentation, therefore contributing to the decline

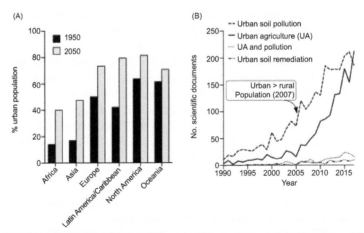

Fig. 1. (A) Estimated change in urban population over 100 years (expressed in %) among continents (data taken from Lal 2018). (B) Global scientific production (papers and reviews) dealing with UA, polluted urban soils, and remediation actions in urban soils. Survey was performed in February 18, 2018 using the bibliographic search engine Scopus™. The time which human urban population exceeded the rural population is indicated (Orsini et al. 2013).

of biodiversity and the increase of environmental contamination (Grimm et al. 2008). For example, it is estimated that two-thirds of the global annual greenhouse gaseous emissions are generated in urban areas because of the strong carbon footprint of the materials used to build homes and infrastructure (Hoornweg 2012).

Cities are also areas of numerous and continuous anthropogenic activities, which generate a high pressure on the quality of the urban environment (Cecchi et al. 2008, Khalid et al. 2016, Saeed et al. 2017). Moreover, because cities are densely populated, the likelihood of human exposure to pollutants is high. Human exposure to urban pollutants come from the air, ingestion of polluted water from local wells and reservoirs, and the consumption of vegetables grown on polluted urban soils or even having intercepted metal-rich atmospheric particles (Uzu et al. 2010, Schreck et al. 2012, Pascaud et al. 2014, Xiong et al. 2014a, 2014b, Shahid et al. 2016, Mombo et al. 2016a, Khalid et al. 2017a, Dumat et al. 2018, Tabassum et al. 2018).

This chapter will discuss the different factors involved in urban agriculture, the major pollution sources of urban agriculture as well as their potential human health hazards. The first two sections provide an overview on the major challenges for a sustainable urban agriculture and safety crops. The following section will identify the major chemical threats coming from pollutants currently detected in the urban environment. The final section of the chapter deals with the health risk assessment of pollutants in urban agriculture, and identifies areas of emerging research in this kind of emerging agriculture.

Urban Agriculture: Production and Challenges

The term urban agriculture refers to cultivation of plants and the rearing of livestock within and around cities and towns (Mougeot 2010). Many cities around the world develop some form of agriculture within and around urban areas. However, only few studies provide reliable data concerning the proportion of food consumed and produced in urban areas. For example, Royte (2015) estimated more than 15% and foresees a sharp increase in this proportion within the next twenty years. It has been estimated that by 2020, the population relying on food produced from urban agriculture will reach up to 40–45% in Africa and Asia, and 85% in Latin America (Hoornweg and Munro-Faure 2008). Nevertheless, some cities and countries perform an intensive urban agriculture (UAN 2017): about 72% of all urban households in Russia and 68% in Tanzania perform urban farming. In China, the 14 largest cities produce more than 85% of vegetables that the population consumes. In this way, urban agriculture contributes to about 2% of the city Gross Domestic Product— defined as total dollar value of all the output and services produced over a certain course of time—in Shanghai (China) and 4% in Lima (Peru). Keeping in view the future growth and considerable contribution of urban farming towards human food, it is of utmost importance to monitor the environmental pollutants in urban agriculture. Table 1 indicates the proportion of different urban agriculture food products.

The surface of the planet covered by cities is currently increasing, and it is expected to pass from 3% to 6% of the total land area by 2030 (d'Amour et al. 2016). In France, urban areas and cities now cover 20% of the territory. However, pavilion

Table 1. Contribution of urban agriculture in food production.

City/Country	Product	Share (%)	Agriculture type	References
Hanoi/Vietnam	Vegetables	80% of city production	Urban and periurban	Nguyen Tien Dinh 2000
	Pork, poultry and fresh water fish	50% of city production		
	Eggs	40% of city production		
Shanghai/China	Vegetables	60% of city production	Urban and periurban	Cai Yi-Zhang and Zhang Zhangen in Bakker et al. 2000
	Milk	100% of city production		
	Eggs	90% of city production		
	Pork and poultry meat	50% of city production		
Jakarta/Indonesia	Rice	1.2% of country production	Urban agriculture	Purnomohadi 2000
	Vegetables	0.5% of country production		
	Fruits	19.6% of country production		
	Caloric consumption	18% of caloric consumption	Home gardens	
	Proteins	14% of protein consumption		
Dakar/Senegal	Vegetables	65% of national vegetable consumption	Urban agriculture	Mbaye and Moustier 1999
	Poultry	60% of national vegetable consumption		
	Milk	60% of national vegetable consumption		
Mexico City	Milk	100% of household income	Urban cowshed	Pablo Torres Lima, L.M.R. Sanchez, B.I.G. Uriza in Bakker et al. 2000
	Swine production	10-40% of household earnings		
	Maize	10-30% of household income	Urban and periurban	
	Vegetable and legume	80% of household income		
Accra/Ghana	Fresh vegetable	90% of national vegetable consumption	Urban and periurban	Cencosad 1994
Shanghai/China	Vegetables	85% of national vegetable consumption	Urban and periurban	FAO 2017
Kathmandu/Nepal	Vegetables	30% of national vegetable consumption	Urban and periurban	FAO 2017
Karachi/Pakistan	Vegetables	50% of national vegetable consumption	Urban and periurban	FAO 2017

gardens are often considered as green spaces as they are planted with hedges and trees (Goddard et al. 2010). In addition, many local and regional authorities have now understood the social and ecological benefits (green and blue ways) of urban agriculture. They have also identified the opportunity to organize therein another economic sector, thereby generating new jobs. If the installation of urban agriculture questions the spatial organization to be implemented to find an ideal place for its development, it also questions the role of this new form of nature in the city of the coming years, in terms of landscapes and the wellbeing of populations through the reduction of ecological inequalities (Chaumel and La Branche 2008).

One of the strengths of urban agriculture is its integration with numerous sectors of the city functionality such as social, economic and ecologic sectors (Mougeot 2010). However, the most interesting aspect in urban agriculture is the participation of city inhabitants, the use of urban resources such as city wastewater for crop irrigation and organic waste as soil amendment or food for farm animals, and the impact on contaminants in urban crops. Urban agriculture includes food products from different types of crops (vegetables, fruits, grains, and mushrooms) and animals (pigs, poultry, sheep, goats, cattle, fish, etc.) as well as non-food products (ornamental, medicinal, and aromatic plants) or a combination of them (Van Veenhuizen and Danso 2007).

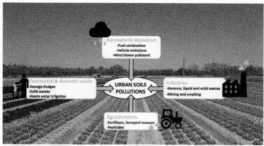

Fig. 2. Sources of urban soil pollution.

Pollution is perhaps the most limiting factor in urban agriculture. Many human activities and a vast variety of environmental chemicals contaminate the urban soil and threaten the quality of crops (Fig. 2).

Pesticides and fertilizers coming from agricultural activities, volatile organic and inorganic pollutants from vehicle exhaust and industry, and hydrocarbons are the most frequent detected pollutants in urban agriculture (Mombo et al. 2016a). The major anthropogenic sources of metal(loid)s in the environment are mining and smelting activities with the production and dispersion of huge amounts of wastes and particles enriched with pollutants (Shahid et al. 2012), inducing health-related consequences through food chain accumulation (Schreck et al. 2012). However, irrigation and fertilization of urban soils with wastewater and sewage sludge, respectively, are a significant source of organic contaminants such as detergents, hydrocarbons, pharmaceuticals and cosmetics (Margenat et al. 2017). Furthermore, a new threat appears with the use of this type of fert-irrigation, i.e., microplastics, whose impact on soil fertility and quality is still unknown (Bläsing and Amelung 2018). Taken together, these studies suggest that one of the most important challenges in consolidating urban agriculture as a form of sustainable and safety food production is the control of soil pollution. The actual lack of regulatory guidelines for developing safe crops in urban agriculture threats the future of this promising agriculture.

Urban Agriculture Projects

In essence, urban agriculture projects are multidisciplinary and multifunctional. These characteristics have implications on the forms of urban agriculture that can lastingly develop: multi-actor, efficient, multifunctional participatory projects with low inputs and bringing about ecological improvements and human values such as autonomy and solidarity. Nowadays, numerous urban agriculture projects are being developed around the world, with innovations, thereby increasing the sustainability of the urban "metabolism" in relation to circular economy and ecosystems services (Duchemin et al. 2008, Kennedy et al. 2011, Madre et al. 2014, Dumat et al. 2016).

For many reasons such as the global financial crisis in the last decade, or the uncertainty about the quality and origin of food, growing social pressure for gardening activities has induced the development of numerous collective gardens across the world (Ghose and Pettygrove 2014, Madre et al. 2014). Producing healthy edible plants is the main objective of urban gardeners, even though the areas allocated to urban agriculture are frequently contaminated (Pierart et al. 2015, Mombo et al. 2016a). According to Menozzi (2014), urban gardens are a real tool to think and develop tomorrow's cities. Besides real environmental issues and social cohesion, the association "Green Garden" in Britain aims to offer a credible economic urban gardening alternative in connection with the food issue. Furthermore, in an increasingly industrialized food system, children are disconnected from opportunities to generate their own food. Consequently, current and future generations of young people may lack both the experience of gardening and a holistic appreciation of food and nutrition because of an insufficient understanding of our food system, its ecology and risks (Devine et al. 1998). Hence, Hale et al. (2011) and Dumat et al. (2018) consider that gardens are a potential urban resource for active and passive learning about ecological processes. As demonstrated by recent research using geographic information and mathematical methods (Ghose and Pettygrove 2014), the development of gardening activities could contribute to preserving the environment and reducing human exposure to pollutants.

Types of Environmental Pollution

Urban, industrial and agricultural advance is linked not only with achieving better living standards, but also with negative environmental consequences. The modern civilization and its industrial and agricultural activities are heavily dependent on the consumption of fossil fuels and the alteration of the geochemical cycles of numerous chemical elements, which therefore profoundly affects our urban environment. Contamination of food products generated via urban agriculture comes mainly from both polluted soils and wet/dry deposition of contaminated particles from the air (Table 2).

Table 2. Different urban soil pollutants and their sources.

Urban soil contaminant	General source	References
Heavy metals (As, Pb, Hg, Zn, Cd, Cr)	Pesticides, sewage sludge, vehicle emissions, manure, industrial effluents	(Wuana and Okieimen 2011)
POPs (Dioxins, DDT, toxaphene, heptachlor, chlordane)	Incineration, pesticides, organic chemicals, pharmaceuticals	(Ritter et al. 2007)
PAHs (benzene, toluene, xylene, ethyl, benzene)	Petroleum spills, burning waste, industrial sites, oil spills, incomplete combustion of wood	(Abdel-Shafy and Mansour 2016)
PCBs	paint, plastics and adhesives, pharmaceuticals, accidental spills, leakage from old appliances, electric equipment	(Ritter et al. 2007)
Metals and cyanides	Metal(loid) finishing operations	(EPA 2011)
PBTs	Sewage sludge, fly ash, fertilizers	(Shayler et al. 2009)

POPs: Persistent organic pollutants, PCBs: Polychlorinated biphenyls, PAHs: Polycyclic aromatic hydrocarbons, PBTs: persistent bio-accumulative toxins.

Urban Soil Pollution

In urban cultivated areas, there is a continuous addition and build-up of pollutants into the soil. These pollutants are deposited in urban soils via different sources such as application of wastewater, sewage sludge, and pesticides, as well as the fertilizing with problematical mineral compounds (Tables 3 to 6) (Shahid et al. 2013a, Pascaud et al. 2014, Xiong et al. 2014a). According to European commission, the fecal matter contains, on an average, 2, 10, 70 and 250 mg/kg of Cd, Pb, Cu and Zn, respectively (Kirchmann et al. 2017). Even, in some areas, high metal concentrations have been reported in surface and ground water (Shakoor 2018, Tabassum et al. 2018). Therefore, the use of surface and ground water for crop irrigation causes build-up of these toxic metal(loid)s in urban soil. The application of sewage sludge, therefore, results in the build-up of these toxic metals in (peri)urban soils. Globally, the use of wastewater for crop irrigation has increased about 10–29% per year in Europe, the United States, and China and by up to 41% in Australia (Aziz and Farissi 2014). Global wastewater discharge reaches 400 billion m^3/year, polluting $\sim$ 5500 billion m^3 of water of per year (Zhang and Shen 2017). Several previous studies reported high metal concentration in soils irrigated with sewage or municipal wastewater. For example, Avci and Deveci (2013) found high concentrations of Ni, Cr, Zn and Pb in the wastewater-irrigated soils compared to controls. Similarly, Tiwari et al. (2011) revealed high concentrations of Fe, Zn, Cd, Cu, Cr and Pb in soils irrigated with effluent than tube well water irrigated soils.

In addition, urban soil also frequently receives pollutants from anthropogenic deposits such as accidental spills, heaps and dumps. These compounds belong to different categories of pollutants such as inorganic (heavy metals), organic (petroleum hydrocarbons, pesticides, PCB and PAH), and microbial (bacteria, viruses, etc.) (Ahmad et al. 2016, Khalid et al. 2016, Natasha et al. 2018). Among these toxic chemicals, metal(loid)s have gained profound consideration globally because of their high persistency, build-up in food chain, potential danger to food safety and human health, and negative effects on ecosystems (Pascaud et al. 2014, Khalid et al. 2017a, Rafiq et al. 2017a,b, Shahid et al. 2017a).

The biogeochemical behavior of metals in soil depends on several factors. For example, their speciation (i.e., the different chemical forms in which a metal can exist in an environmental compartment, via combination with ions and organic molecules) affects the bioavailable fraction of these chemicals, therefore being the most important predictor for its environmental and public health risks (Shahid et al. 2014, 2017a,b, Shamshad et al. 2018). In general, the bioavailability of a pollutant is defined as its capacity to be absorbed by living organisms (Xiong et al. 2016).

Different interconnected soil factors greatly affect the chemical speciation and behavior of metal(loid)s in the soil-plant system, including soil pH and mineralogy, redox conditions, cation-exchange capacity, form and applied levels of metal(loid)s, level of competing cations, organic and inorganic ligands, biological and microbial conditions, and the type of plant species (Rafiq et al. 2017). For example, a soil that receives sludge from a wastewater treatment plant may contain higher concentrations of total lead and copper than a nearby soil that does not receive sludge (Jamali et al. 2007, Khalid et al. 2017). However, if a soil has a more acidic pH than the other one,

Table 3. Urban soil contamination by metals (Cd, Pb, Zn, Ni and Cu).

Metal	Concentration (mg/kg)	Source	Location	References
Cd	2.7	Wastewater	USA	(Sidle et al. 1977)
	5.8	Wastewater	Isfahan, Iran	(Abedi-Koupai et al. 2015)
	9.6	Fertilizer and sewage sludge	USA	(Sterrett et al. 1996)
	10	Sewage sludge	USA	(Baker et al. 1979)
	19.25	Industrial effluents	Gujarat, India	(Tiwari et al. 2011)
Pb	31	Wastewater	Vehari-Pakistan	(Khalid et al. 2017a)
	33.3	Digested sludge	Tai Po, Hong Kong	(Wong 1985)
	41.56	Wastewater	Tianjin China	(Meng et al. 2016)
	46	Industrial effluents	Gujarat, India	(Tiwari et al. 2011)
	49.4	Wastewater	Beijing, China	(Khan et al. 2008)
	140	Wastewater	USA	(Sidle et al. 1977)
	411.7	Wastewater	Shiraz suburban, Iran	(Qishlaqi et al. 2008)
	5210	Fertilizer and sewage sludge	USA	(Sterrett et al. 1996)
Zn	86	Wastewater	North Temperate Zone, China	(Lu et al. 2015)
	134.22	Wastewater	Nicosia, Cyprus	(Christou et al. 2014)
	157	Wastewater	Beijing, China	(Khan et al. 2008)
	167	Wastewater	Kano, Bobo Dioulasso and Sikasso (West Africa)	(Abdu et al. 2011)
	376	Pig manure compost	Tai Po, Hong Kong	(Wong 1985)
	753	Digested sludge	Tai Po, Hong Kong	(Wong 1985)
	3490	Fertilizer and sewage sludge	USA	(Sterrett et al. 1996)
Ni	11.3	Wastewater	Gothenburg, Sweden	(Farahat and Linderholm 2015)
	13.30	Wastewater	North Temperate Zone, China	(Lu et al. 2015)
	14.97	Industrial effluents	Gujarat, India	(Tiwari et al. 2011)
	24.9	Wastewater	Beijing, China	(Khan et al. 2008)
	28.13	Wastewater	Barueri, Brazil	(de Melo et al. 2007)
	35	Wastewater	China	(Hu et al. 2014)
	50	Wastewater	USA	(Sidle et al. 1977)
	200	Wastewater	Gaziantep, Turkey	(Avci and Deveci 2013)
	276.6	Wastewater	Shiraz suburban, Iran	(Qishlaqi et al. 2008)

Table 3 contd. ...

... Table 3 contd.

Metal	Concentration (mg/kg)	Source	Location	References
Cu	7.4	Wastewater	Kumasi, Ghana	(Akoto et al. 2015)
	32.8	Wastewater	Beijing, China	(Khan et al. 2008)
	52	Pig manure compost	Tai Po, Hong Kong	(Wong 1985)
	94.38	Wastewater	Nicosia, Cyprus	(Christou et al. 2014)
	105	Digested sludge	Tai Po, Hong Kong	(Wong 1985)

Table 4. Metal concentrations in soil from cities around the world. Concentrations are expressed as the mean or range (mg/kg).

City	Cd	Cu	Ni	Pb	Zn
Europe					
Sevilla (Spain)	-	9–365	16–62	15–977	21–443
Madrid (Spain)	-	11.3–171	4.89–15.03	15–598	76–309
Vigo (Spain)	-	22.6–208	11.5–60	34.4–259	59–234
Torino (Italy)	-	15–430	77–830	14–1,440	53–880
Palermo (Italy)	-	12.2–89.4	-	36.3–472	57.7–851
Siena (Italy)	0.05–0.48	20.4–82.1	-	16.3–178	57.2–207
Stockolm (Sweden)	-	7.3–153	2.1–21.3	2.4–444	18–408
Berlin (Germany)	0.35	31	8	77	129
Oslo (Norway)	0.34	24	24	34	130
Szczecin (Poland)	0.5–1.8	13.8–32.4	8.2–13.6	48–140	63–398
Sofia (Bulgaria)	0.11–0.33	32.6–65.7	-	30.9–34.3	37.9–81.0
Berger (Norway)	< 0.1–1.5	4–2,850	1–310	3–5,780	8–998
Trondheim (Norway)	< 0.1–11.3	1.7–706	6–231	9–976	7–3,420
Aberdeen (UK)	-	24	-	94.4	58.4
London (UK)	-	73	-	294	183
Zagreb (Croatia)	0.5	18	49	23	70
Izmit (Turkey)	0.23	37	39	35	72
Novi Sad (Serbia)	-	4.4–459	10.2–74.2	8.9–1,000	46.2–194
Ljubljana (Slovenia)	-	14–135	14–45	10–387	56–581
North America					
New York (USA)	0.1–11	5–1,186	2–333	3–8,912	35–2,352
Chicago (USA)	20	150	36	395	397
Connecticut (USA)	< 0.5	40	12	176	163
Baltimore (USA)	0.56	17	2.8	100	92
Asia					
Bangkok (Thailand)	0.05–2.53	5.1–283	4.1–52.1	12.1–269	3–814
Moscow (Russia)	2.0	59	19	37	208
Hong Kong (China)	0.33	10	4	71	78
Beijing (China)	< 0.01–0.97	2–282	2.8–169	5–117	22–400
Islamabad (Pakistan)	0–8.6	1.1–41.6	2.7–220	69–973	101–3,256
Sialkot (Pakistan)	46	19	83	122	78

Table 4 contd. ...

... Table 4 contd.

City	Cd	Cu	Ni	Pb	Zn
Ghaziabad (India)	0.2–4	30–444	206–403	28–341	42–477
Ulaanbaatar (Mongolia)	0.80	36	19	64	159
Seoul (South Korea)	1.0–4.4	11–471	-	93–1,636	55–596
Malayer (Iran)	0.01–2.39	1.82–30.41	-	6.72–40	63.9–218
Africa					
Ibadan (Nigeria)	0.15	32	17	47	94
Latin America					
Mexico city (Mexico)	-	15–398	20–146	5–452	36–1,641
Talcahuano (Chile)	-	4.6–113	-	0.3–201	20.8–347
Dutch Soil Guidelines[a]	0.8	36	35	85	140

[a] Values for soil remediation proposed by Dutch Ministry of Housing, Spatial Planning and Environment (VROM 2000).

VROM. 2000. Circular on target values and intervention values for soil remediation Annex A: target values, soil remediation intervention values and indicative levels for serious contamination. Dutch Ministry of Housing, Spatial Planning and Environment.

Table 5. Urban soil contamination by organic pollutants in different areas.

Organic pollutant	Concentration (µg/kg)	Source	Country	References
PAHs	260	Industrial emissions	USA	(Bradley 2011)
	1,000	Vehicle emission	Spain	(García-Alonso et al. 2003)
	450–5,650	Industry	France	(Motelay-Massei et al. 2004)
	184–10,279	Traffic and industrial emissions	Nepal	(Aichner et al. 2007)
	18,000	Diffuse pollution	UK	(Vane et al. 2014)
PAH-Naphthalene	1,625.10	Agricultural practices	Nigeria	(Rose et al. 2013)
PCBs	0.1	Vehicle emission	Spain	(García-Alonso et al. 2003)
	0.09–150	Industry	France	(Motelay et al. 2004)
	10.58	Transport, fuel combustion, industry	Bulgaria	(Dimitrova et al. 2011)
	3.55 and 23.64	Burning of old appliances	Nigeria	(Rose et al. 2013)
	11	Electrical products	China	(Wu et al. 2011)
	0.356 to 447.10	Industrial activity	Nepal	(Aichner et al. 2007)
	56 ± 160	Urban water runoff	USA	(Martinez et al. 2012)
	123	Industrial/commercial spills, disposal	UK	(Vane et al. 2014)
Dioxin like PCBs	0.04–0.93	Industrial activity	Brazil	(Pussente et al. 2017)
Non-dioxin like PCBs	0.22–7.48	Industrial activity	Brazil	(Pussente et al. 2017)

Table 5 contd. ...

... Table 5 contd.

Organic pollutant	Concentration (µg/kg)	Source	Country	References
Chlordanes	130 ± 920	Urban water runoff	USA	(Martinez et al. 2012)
Dioxins	19	Industrial emissions	USA	(Bradley 2011)
Dioxin-TCDD	0.4–5.5	-	USA	(Paustenbach et al. 2006)
DDT	0.54–37.42	Anaerobic DDT degradation and contaminations from aged DDT	India	(Sujatha et al. 1994)
	2.22–1,440	Agricultural practices	Moscow	(Brodskiy et al. 2016)
	16,900	Agricultural practices	Chile	(Henriquez et al. 2006)
DDE	117.98	Agriculture, pest control	Nigeria	(Rose et al. 2013)
	600	Agricultural practices	Chile	(Henriquez et al. 2006)
Aldrin	1,000	Agricultural practices	Chile	(Henriquez et al. 2006)
Dieldrin	2,000	Agricultural practices	Chile	(Henriquez et al. 2006)
HCHs	0.56–8.52	Historical HCHs contamination	India	(Sujatha et al. 1994)
OCs	0.13 to 117.98	Pest/vector control	Nigeria	(Rose et al. 2013)

PAHs: Polycyclic aromatic hydrocarbons, PCBs: Polychlorinated biphenyls, DDT: Dichlorodiphenyltrichloroethane, DDE: Dichlorodiphenyldichloroethylene, HCHs: Hexachlorocyclohexanes, OCs: Organochlorine compounds, TCDD: Tetrachlorodibenzo-*p*-dioxin.

it is likely that higher lead concentrations would be available for plants grown on the first one because acid conditions promotes the solubility of the metals and their root transfer, especially lead for which phytoavailability is highly dependent on soil pH. This environmental risk may therefore be reduced by liming (i.e., soil pH adjustment by adding basic amendments), which is a well-known ancestral technique used by farmers to avoid aluminum phytotoxicity.

The rational management of chemical substances in Europe to be marketed as consumer goods is driven by the European Chemical Agency (ECHA), which is based on the European regulation REACH, in force since June, 1st of 2007 (https://echa.europa.eu/en/regulations/reach). REACH is the acronym for "Registration, Evaluation, Authorization and Restriction of Chemicals". This regulation has been adopted to enhance both human health and environmental protection from the risk of (eco)toxic chemicals while promoting the competitiveness of the EU chemical industry. It applies to all chemical substances (used in industrial processes or present in the everyday life, such as in cleaning products, paints, clothing or furniture) and imputes the burden of proof to companies that have to identify and propose procedures for the use of the substances they manufacture and market in the EU for which risks are acceptable. REACH promotes alternative methods for the assessment

Table 6. Concentration (mean or range) of selected organic pollutants in urban soils from several cities around the world.

City	ΣPAHs (μg g^{-1})	ΣPCBs (ng g^{-1})
Bangkok (Thailand)[1]	0.017–3.26	0.3–6.17
Beijing (China)[1]	0.093–365.9	1.63–3,883
Brastislava (Slovakia)[1]	0.045–12.15	-
Kurukshetra (India)[1]	0.019–2.54	3.3–35
Lisbon (Portugal)[1]	0.0063–22.67	0.18–34
Dhanbad (India)[2]	1,019–10,856	-
London (UK)[3]	4.0–67	1–750
Madrid (Spain)[3]	-	9.0–66
Bayreuth (Germany)[3]	-	0.82–158
Cedar Rapids (USA)[3]	-	3.0–1,200
Moscow (Russia)[3]	-	3.1–42
Glasgow (UK)[4,1]	1.48–51.82	4.5–78
Ljubljana (Slovenia)[4,1]	0.218–4.48	2.8–48
Torino (Italy)[4]	0.148–23.5	

References: (1) Li et al. (2018), (2) Suman et al. (2016), (3) Vane et al. (2014), (4) Morillo et al. (2007).

of the hazards of substances in order to reduce the number of animal tests. For example, *in vitro* bioassays are now commonly used in research projects concerning metal-polluted sites (Pascaud et al. 2014, Xiong et al. 2016).

However, in France, despite the strengthening of environmental-health regulations such as ICPE or REACH (which resulted in a significant reduction in the emissions of chemical substances into the environment), pollution of the environment is still widely observed (Foucault et al. 2013, Leveque et al. 2013), especially because of atmospheric deposition of particulate pollutants over long distances in urban areas as a result of historical pollution by lead, cadmium or mercury (Schreck et al. 2012, Uzu et al. 2014, Xiong et al. 2014, Mombo et al. 2016).

Contrary to relatively homogeneous aquatic or aerial environments that appear as "common goods" and are in consequence protected by European framework directives (with threshold values allowing a simplified management of pollutions), in the case of soils, a draft framework directive was discussed in the 2000s, but it was ultimately unsuccessful. For this environment, pollution management is, therefore, always performed on a case-by-case basis, which is an important part of the expertise and the protection of soil media, which is also included in various non-specific regulations such as "ICPE" in France (which considers that this resource must be preserved and restored). To help in the decision process to conclude whether or not a spot is safe to be converted for food production, modeling has been showing very interesting results over the last years (Lateb et al. 2016). For example, the lowest aerial deposition areas can be identified thanks to urban architecture. Furthermore, Säumel et al. (2012) showed that the presence of plants constituting a barrier between roadside and crop can reduce heavy metals in edible plant parts. In France, the recent ALUR (Access to Housing and Renovated Urbanism) law aims to facilitate and increase the effort to build housing, while curbing the artificializing of land and combating urban sprawl (Fig. 3). The rehabilitation of brownfields, often in the heart of cities, is therefore a major issue.

However, atmospheric and soil pollution is often observed in these urban areas mainly due to road proximity, and agricultural and industrial activities which occurred during previous centuries (Mombo et al. 2016). Actually, many chemicals can flow or accumulate in atmosphere, water, garden soils, and finally vegetables (Schreck et al. 2012, Xiong et al. 2016, 2017). However, there are no French regulatory threshold values for either total or bioaccessible concentrations of pollutants in garden soils (Foucault et al. 2013, Mombo et al. 2016). Indeed, only marketed plants are regulated in Europe only for some targeted inorganic pollutants such as lead, cadmium and mercury (EC, no 466/2001). Arsenic is a persistent, highly (eco)toxic and very often observed metal(loid) in the environment (Shahid et al. 2013) and according to Shakoor et al. (2016), chronic oral As exposure can result in gastrointestinal distress, anemia, peripheral neuropathy, skin lesions, hyperpigmentation, and liver or kidney damage. For such an unregulated inorganic pollutant, a specific quantitative assessment of health risks (QSRA) must then be carried out in order to scientifically access the human exposure in the case of consumption of potentially polluted vegetables. For example, QSRA was necessary in the case of arsenic pollution in collective gardens in order to carry scientific arguments to the authorities in charge of these gardens and to inform the gardeners. The objective of the QSRA is to assess the pollutant quantity potentially ingested by gardeners in the case of consumption of contaminated plants and compare it with reference value (Dumat et al. 2015). It is therefore necessary to both fill the quantities of produced vegetables in the gardens and their use (consumption, donations, and so on) through a survey of gardeners' habits and a measurement of the pollutant concentration in vegetables (Xiong et al. 2014).

Then, the development of new manufactured products usually involves the use of new chemicals (organic and inorganic) whose residues are also dismissed in the environment. Recently, special attention has been dedicated to antimony, an emerging contaminant whose concentration in urban environment and in food has been increasing since the last decade (Pierart et al. 2015, Mombo et al. 2016). An excessive intake of this toxic element by humans may cause vomiting, diarrhea, skin rashes and respiratory symptoms but no legal threshold in vegetables currently

Fig. 3. Wind direction and building arrangement effects on particle dispersion (reproduced from Lateb et al. 2016).

exists for it and its transfer at the soil-plant interface still needs further investigation (Pierart et al. 2018).

Urban Air Pollution

Air quality is also currently considered in order to discuss the vegetable quality when fine particles enriched with pollutants are involved (Schreck et al. 2012, Xiong et al. 2014). It has been reported that the urban air pollution has enhanced rapidly with urban populations and vehicle density, therefore negatively affecting the agricultural areas in (peri)urban regions. One of the major threats to crop cultivation in these areas mainly concerns nitrogen dioxide, sulphur dioxide, and ozone. These toxic gases have negative effects on crop production as well as the livelihoods. The thesis by Xiong T. (INPT 2015, http://theses.fr/189876085) focused on that phenomenon, which is particularly important in the case of certain pollutants such as lead that display low solubility and mobility in soil. Actually, these low mobile pollutants in the soil are generally fixed by the soil compounds and/or adsorbed by plant roots, but in the case of atmospheric foliar deposit, these pollutants can be directly absorbed by plant leaves by entering the stomata (Uzu et al. 2010). Even with careful washing of plant leaves, these pollutants mainly stay adsorbed into intercellular spaces or absorbed in the plants and can be therefore absorbed by humans when the plants are consumed.

Consequently, it is crucial to grow the right type of vegetable (leaf, root or fruit), regarding the pollution source it will face (Fig. 4) (Khalid et al. 2017). For example, leafy vegetables present the highest potential risk of metal(loid) accumulation because they are facing both aerial pollution through deposition of metal-rich particles as shown above, and soil pollution through translocation from roots to leaves. Moreover, they cannot be peeled, and they are generally more difficult to wash properly. Root vegetables are facing all the contaminants present in the soil and the soil solution. These elements are usually either incorporated or adsorbed by the roots and the cultivated vegetables. However, the knowledge related to the role of these microorganisms in the transfer of emerging metal(loid) such as antimony is still scattered (Pierart et al. 2018b).

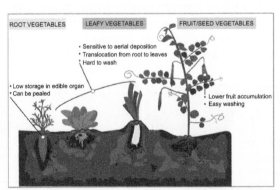

Fig. 4. Choosing the best plant type facing a hazardous contamination.

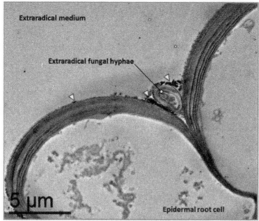

Fig. 5. Picture of leek root cells in transmission electronic microscopy coupled with X-ray analysis. White arrows indicate accumulation of metals bound to root and fungal surface (Pierart 2016).

However, controversial results suggest that the difference in metal(loid) accumulation seems more plant species dependent than plant type (leafy vegetable, root vegetable, fruits, etc.) dependent (Säumel et al. 2012). The hardest element to consider is also that phytoaccumulation variations of metal(loid) usually occur at the cultivar scale (Michalska and Asp 2001). Consequently, the precaution principle suggests growing, preferentially, fruit vegetables in urban area, and to wash the harvest carefully, to limit the ingestion of thin particles of soil and thereby decrease the contaminant concentration by up to 75–100%. However, a fraction of metal(loid) usually remains bound to the root surfaces (Fig. 5), so peeling remains the best option when possible. Finally, efforts are still needed on the effects of food processing (packing, storage, cooking, etc.) as these actions can influence metal(loid) bioavailability.

Management of Pollutants in (Peri)Urban Areas to Promote Health Safety

Operational Collaboration between Researchers and Urban Gardeners

In order to favor sustainable urban agriculture, promoting operational collaboration between researchers and gardeners is a crucial issue as millions of citizens cultivate and consume vegetables in the world. For example, this functional tandem was the main goal of the national French scientific research project "JASSUR" (Associative Urban Gardens in France and sustainable cities: practices, functions and risks). This interdisciplinary project proposed to clarify urban agriculture functions, uses, way of operation, and benefits or potential hazards within cities. The project aimed to identify the necessary action levers to maintain or even restore, develop or evolve these associative gardens in urban areas facing the challenges of sustainability. To do so, it relied on a consortium of 12 research partners and associations in seven French cities. Food and other ecosystem services rendered by UA to the cities, are still poorly understood. Therefore, they were the central node of reflection of the JASSUR program through:

- A bio-physicochemical characterization of soil and products from these gardens: the question of the potential risk of pollution caused by urban environment (soil, atmosphere) is central here because the food supply service could be thwarted (cultural practices, productions and products' locations, measurements of quantities consumed and nutrient intake, gardener's representation regarding the interests and potential hazards);
- A socio-technical characterization of gardener practices, through their crop choices, techniques, and the participation of their garden produces to food and good nutrition of their family;
- A socio-political characterization of the governance of these spaces in urban areas, particularly in terms of management of locations, modes of operation, and the potential environmental and health risks. The food supply services were analyzed and possible pollution management methods by communities were also studied;
- This program was a thorough example of "citizen science" as described by Hand (2010): gardeners were directly implicated in the research program and were actively involved in the risk construction and management.

Human Health Risks in the Context of Urban Agriculture and other Recreational Activities

Food safety in urban agriculture has become the most important public concern because of large concentrations of metal(loid)s are taken up by edible plant parts (Mombo et al. 2016, Rafiq et al. 2017) widely consumed. Several studies reported that the vegetables and crops grown on metal(loid)s-contaminated soils can accumulate these chemicals up to concentrations several folds higher than the threshold level with serious human health hazard (Foucault et al. 2013). Therefore, it has become necessary to assess the biogeochemical behavior of metal(loid)s in the soil-plant system in relation to their potential mobility towards the food chain.

Studies focusing on heavy metal(loid)s risk assessment and environmental pollution are abundant in the literature. However, majority of these studies have considered only total metal(loid) concentrations. Nowadays, this classical approach is being regarded as less appropriate since their biochemical reactivity and (eco) toxicity depend on the chemical speciation of the metal rather than just its total concentration (Pourrut et al. 2011, Pascaud et al. 2014, Rafiq et al. 2017). It is evident that biochemical behavior (bioavailability, soil-plant transfer, translocation to aerial tissues and toxicity) of metal(loid)s depends on their chemical form (Abbas et al. 2016), which can also influence their solubility, mobility and bioavailability (Schreck et al. 2014).

In a case study performed in urban and peri-urban areas of Vehari-Pakistan, Khalid et al. (2017a) assessed the effect of irrigation using untreated city waters on vegetables. Soil and wastewater samples were collected from three different wastewater collection locations of Vehari city. The collected soils were filled in pots and were irrigated with wastewater at regular intervals. Lead concentrations were 0.31 mg/l in wastewater samples. Concentrations of metal(loid)s were higher in all vegetables irrigated with contaminated wastewater than those of vegetables irrigated with groundwater (control treatment). For example, concentrations of Pb in roots (range 21–28 mg/kg^{-1}), and shoot (10–29 mg/kg^{-1}) were higher than the permissible Pb limit of the FAO (5 mg/kg^{-1} dry weight). The health risk index values were higher than 1 for radish and cauliflower, but lower than 1 for spinach. These results demonstrated that vegetable type can greatly affect the risk associated with the use of polluted wastewater for vegetable irrigation.

The behavior of chemical elements is usually assessed by a sequential or selective extraction procedure recommended by the Community Bureau of Reference (BCR), which provides semi-quantitative information on their compartmentalization in soils and their mobility and bioavailability in relation to the stability of the interactions developed with soil components or with the solid metal(loid) compounds formed (Quenea et al. 2009, Austruy et al. 2014). The coupling of this BCR chemical approach with a mineralogical methodology (e.g., scanning electron microscopy (SEM), X-ray diffraction (XRD), transmission electron microscopy (TEM)) allows determining the various bearing phases of metal(loid)s in soils and then leads to predicting their stability. As stated earlier, the use of modeling tools could allow the aggregation of these large variety of datasets and pedo-climatic conditions to help in the decision-making process.

According to Pelfrêne et al. (2012), soils enriched with metal(loid)s can pose a potential threat to human health if directly ingested or transferred through food. However, the conventional BCR extraction procedure is not suitable to assess the bioavailable fraction of pollutants in the case of accidental soil ingestion (e.g., children through outdoor hand-to-mouth activities). Indeed, the human bioavailability of a pollutant, whose contaminant fraction is absorbed through the gastrointestinal tract and reaches systemic circulation (Türkdoğan et al. 2003, Foucault et al. 2013), is initially controlled by its release from the solid soil phase into the stomach and intestine, which defines the bioaccessible fraction. The Bioaccessibility Research Group of Europe (BARGE) has developed a standardized procedure (Cave et al. 2006) known as the Unified Bioaccessibility Method (UBM) for current use in contaminated soil studies (Pelfrêne et al. 2012).

When contaminated vegetables are consumed, the same phenomenon occurs. However, recent insights revealed that the bioaccessible fraction of some metal(loid)s (in particular antimony) is affected by symbiotic fungi within plants under certain soil conditions (Pierart et al. 2018b). Efforts are currently needed in this particular field of investigation as these microorganisms, along with bacteria, are already used as biofertilizers in agriculture while legal thresholds are still either non-existent or only based on total metal(loid) content.

Conclusions

Urban agriculture plays a vital role in food production and quality. But, rapid and uncontrolled urbanization and industrialization have seriously polluted (peri)urban areas. Both natural and anthropogenic sources are contributing to the contamination of the urban environment. The quality of food produced in urban agriculture depends on the type of pollutant as well as agricultural activities. Moreover, soil type and metal(loid)s speciation play a key role in pollutant transport through the food web. In the light of metal and organic pollutant concentrations in the urban environment, it is highly recommended for the development and implementation of health, and environmental, risk assessment frameworks in urban agriculture projects. For example, practices such as growing local vegetables accumulating less metal(loid)s in their edible portions or the use of mix of vegetables can greatly reduce health risks. Moreover, the awareness of urban gardeners about the biogeochemistry of pollutants and their possible health risks is of prime importance. In this way, the main objective of the Reseau-Agriville association (http://reseau-agriville.com/) is to increase the interaction between citizens and scientists in order to promote sustainable urban agriculture. Indeed, it is crucial to widely develop the critical and scientific spirit of citizens to favor democratic and sustainable decisions in cities at the global scale. Raising the awareness of very young people to sustainable food, through educational projects in kindergarten, is also a guarantee of success. Actually, producing sufficient quantities of healthy food in urban areas requires multidisciplinary expertise, including social science, economy and politics.

References

Abbas, G., M. Saqib, J. Akhtar, G. Murtaza, M. Shahid and A. Hussain. 2016. Relationship between rhizosphere acidification and phytoremediation in two acacia species. J. Soils Sediments 16: 1392–1399.

Abdel-Shafy, H.I. and M.S.M. Mansour. 2016. A review on polycyclic aromatic hydrocarbons: Source, environmental impact, effect on human health and remediation. Egypt J. Pet. 25: 107–123.

Abdu, N., A. Abdulkadir, J.O. Agbenin and A. Buerkert. 2011. Vertical distribution of heavy metals in wastewater-irrigated vegetable garden soils of three West African cities. Nutr. Cycl. Agroecosys 89: 387–397.

Abedi-Koupai, J., R. Mollaei and S.S. Eslamian. 2015. The effect of pumice on reduction of cadmium uptake by spinach irrigated with wastewater. Ecohydrology Hydrobiology 15: 208–214.

Ahmad, I., M.J. Akhtar, H.N. Asghar, U. Ghafoor and M. Shahid. 2016. Differential effects of plant growth-promoting rhizobacteria on maize growth and cadmium uptake. J. Plant Growth Regul. 35: 303–315.

Aichner, B., B. Glaser and W. Zech. 2007. Polycyclic aromatic hydrocarbons and polychlorinated biphenyls in urban soils from Kathmandu, Nepal. Org. Geochem. 38: 700–715.

Akoto, O., D. Addo, E. Baidoo, E.A. Agyapong, J. Apau and B. Fei-Baffoe. 2015. Heavy metal accumulation in untreated wastewater-irrigated soil and lettuce (Lactuca sativa). Environ. Earth Sci. 74: 6193–6198.

Austruy, A., M. Shahid, T. Xiong, M. Castrec, V. Payre, N.K. Niazi et al. 2014. Mechanisms of metal-phosphates formation in the rhizosphere soils of pea and tomato: environmental and sanitary consequences. J. Soil. Sediment 14: 666–678.

Avci, H. and T. Deveci. 2013. Assessment of trace element concentrations in soil and plants from cropland irrigated with wastewater. Ecotoxicol. Environ. Saf. 98: 283–291.

Aziz, F. and M. Farissi. 2014. Reuse of treated wastewater in agriculture: Solving water deficit problems in arid areas. Annales of West University of Timisoara. Ser. Biol. 17: 95.

Baker, D., M. Amacher and R. Leach. 1979. Sewage sludge as a source of cadmium in soil–plant–animal systems. Environ. Health Perspect. 28: 45.

Barthel, S. and C. Isendahl. 2013. Urban gardens, agriculture, and water management: Sources of resilience for long-term food security in cities. Ecol. Econ. 86: 224–234.

Bläsing, M. and W. Amelung. 2018. Plastics in soil: Analytical methods and possible sources. Sci. Total Environ. 612: 422–435.

Boutaric, F. 2013. La méthode de l'évaluation des risques sanitaires en France: représentations, évolutions et lectures plurielles. [VertigO] Larevue électronique en sciences de l'environnement 13.

Bradley, D. 2011. Washington Soil Dioxin Study. Department of Ecology.

Brodskiy, E., A. Shelepchikov, D. Feshin, G. Agapkina and M. Artukhova. 2016. Content and distribution pattern of dichlorodiphenyltrichloroethane (DDT) in soils of Moscow. Moscow Univ. Soil Sci. 71: 27–34.

Cave, M., J. Wragg, B. Klinck, C. Gron, A. Oomen, T. Van de Wiele et al. 2006. Preliminary assessment of a unified bioaccessibility method for potentially harmful elements in soils. Epidemiol. 17: S39.

Cecchi, M., C. Dumat, A. Alric, B. Felix-Faure, P. Pradère and M. Guiresse. 2008. Multi-metal contamination of a calcic cambisol by fallout from a lead-recycling plant. Geoderma 144: 287–298.

Chaumel, M. and S. La Branche. 2008. Inégalités écologiques: vers quelle définition? Espace populations sociétés. Space Populations Societies: 101–110.

Cheng, Z., A. Paltseva, I. Li, T. Morin, H. Huot, S. Egendorf, Z. Su et al. 2015. Trace metal contamination in New York City garden soils. Soil Sci. 180: 167–174.

Christou, A., E. Eliadou, C. Michael, E. Hapeshi and D. Fatta-Kassinos. 2014. Assessment of long-term wastewater irrigation impacts on the soil geochemical properties and the bioaccumulation of heavy metals to the agricultural products. Environ. Monit. Assess. 186: 4857–4870.

Cui, L. and J. Shi. 2012. Urbanization and its environmental effects in Shanghai, China. Urban Clim. 2: 1–15.

d'Amour, C.B., F. Reitsma, G. Baiocchi, S. Barthel, B. Güneralp, K.-H. Erb et al. 2016. Future urban land expansion and implications for global croplands. Proc. Natl. Acad. Sci. U.S.A. 201606036.

de Melo, W.J., P. de Stefani Aguiar, G.M.P. de Melo and V.P. de Melo. 2007. Nickel in a tropical soil treated with sewage sludge and cropped with maize in a long-term field study. Soil Biol. Biochem. 39: 1341–1347.

Devine, C.M., M. Connors, C.A. Bisogni and J. Sobal. 1998. Life-course influences on fruit and vegetable trajectories: qualitative analysis of food choices. J. Nutr. Educ. Behav. 30: 361–370.

Dimitrova, A., Y. Stoyanova and A. Tachev. 2011. Distribution and risk assessment of polychlorinated biphenyls (PCBs) in urban soils of Sofia City, Bulgaria. Recent Advances in Chemical Engineering, Biochemistry and Computational Chemistry.

Duchemin, E., F. Wegmuller and A.-M. Legault. 2008. Urban agriculture: multi-dimensional tools for social development in poor neighbourhoods. Field Actions Science Reports. The Journal of Field Actions 1: 43–52.

Dumat, C., J.T. Wu, A. Pierart and L. Sochacki. 2015. Interdisciplinary and participatory research for sustainable management of arsenic pollution in French collective gardens: collective process of risk manufacture, in conference. Journées de Recherches en Sciences Sociales, Nancy-France: 10–11.

Dumat, C., T. Xiong and M. Shahid. 2016. Agriculture urbaine durable: opportunité pour la transition écologique. Presses Universitaires Européennes.

Dumat, C., A. Pierart, M. Shahid and J. Wu. 2018. Collective conceptualization and management of risk for arsenic pollution in gardens. Rev. Agric. Food Environ. Stud. 99: 167–187.

EPA. 2011. Reusing Potentially Contaminated Landscapes: Growing Gardens in Urban Soils. Environmental Protection Agency.

Farahat, E. and H.W. Linderholm. 2015. The effect of long-term wastewater irrigation on accumulation and transfer of heavy metals in Cupressus sempervirens leaves and adjacent soils. Sci. Total Environ. 512: 1–7.

Foucault, Y., T. Lévêque, T. Xiong, E. Schreck, A. Austruy, M. Shahid et al. 2013. Green manure plants for remediation of soils polluted by metals and metalloids: Ecotoxicity and human bioavailability assessment. Chemosphere 93: 1430–1435.

García-Alonso, S., R. Pérez-Pastor and M.L. Sevillano-Castaño. 2003. Occurrence of PCBs and PAHs in an urban soil of Madrid (Spain). Toxicol. Environ. Chem. 85: 193–202.

Ghose, R. and M. Pettygrove. 2014. Urban community gardens as spaces of citizenship. Antipode 46: 1092–1112.

Goddard, M.A., A.J. Dougill and T.G. Benton. 2010. Scaling up from gardens: biodiversity conservation in urban environments. Trends Ecol. Evol. 25: 90–98.

Grimm, N.B., S.H. Faeth, N.E. Golubiewski, C.L. Redman, J. Wu, X. Bai et al. 2008. Global change and the ecology of cities. Sci. 319: 756–760.

Hale, J., C. Knapp, L. Bardwell, M. Buchenau, J. Marshall, F. Sancar et al. 2011. Connecting food environments and health through the relational nature of aesthetics: Gaining insight through the community gardening experience. Soc. Sci. Med. 72: 1853–1863.

Henriquez, M., J. Becerra, R. Barra and J. Rojas. 2006. Hydrocarbons and organochlorine pesticides in soils of the urban ecosystem of chillan and chillan viejo, Chile. J. Chil. Chem. Soc. 51: 938–944.

Hoornweg, D. and P. Munro-Faure. 2008. Urban agriculture for sustainable poverty alleviation and food security. Position paper, FAO. Africa.

Hoornweg, D. 2012. Cities and climate change: An urgent agenda. Sustainable Low-Carbon City Development in China: 3.

Hu, H., Q. Jin and P. Kavan. 2014. A study of heavy metal pollution in China: Current status, pollution-control policies and countermeasures. Sustainability 6: 5820–5838.

Hugo, G. 2017. New forms of urbanization: beyond the urban-rural dichotomy. Routledge 1: 1–448.

Jamali, M., T. Kazi, M. Arain, H. Afridi, N. Jalbani and A. Memon. 2007. Heavy metal contents of vegetables grown in soil, irrigated with mixtures of wastewater and sewage sludge in Pakistan, using ultrasonic-assisted pseudo-digestion. J. Agron. Crop Sci. 193: 218–228.

Kennedy, C., S. Pincetl and P. Bunje. 2011. The study of urban metabolism and its applications to urban planning and design. Environ. Pollut. 159(8–9): 1965–1973.

Khalid, S., M. Shahid, N.K. Niazi, B. Murtaza, I. Bibi and C. Dumat. 2016. A comparison of technologies for remediation of heavy metal contaminated soils. J. Geochem. Explor.

Khalid, S., M. Shahid, C. Dumat, N.K. Niazi, I. Bibi, H.F.S. Gul Bakhat et al. 2017. Influence of groundwater and wastewater irrigation on lead accumulation in soil and vegetables: Implications for health risk assessment and phytoremediation. Int. J. Phytoremediat. 19: 1037–1046.

Khalid, S., M. Shahid, N.K. Niazi, M. Rafiq, H.F. Bakhat, M. Imran et al. 2017. Arsenic behaviour in soil-plant system: Biogeochemical reactions and chemical speciation influences. pp. 97–140. In: Enhancing Cleanup of Environmental Pollutants. Springer.

Khan, S., Q. Cao, Y. Zheng, Y. Huang and Y. Zhu. 2008. Health risks of heavy metals in contaminated soils and food crops irrigated with wastewater in Beijing, China. Environ. Pollut. 152: 686–692.

Kirchmann, H., G. Börjesson, T. Kätterer and Y. Cohen. 2017. From agricultural use of sewage sludge to nutrient extraction: A soil science outlook. Ambio 46: 143–154.

Lal, R. 2018. Urban agriculture in the 21st century. pp. 1–13. In: Lal, R. and B.A. Steward (eds.). Urban Soils, CRC Press, Francis & Taylor Group, boca Raton, Florida.

Lateb, M., R. Meroney, M. Yataghene, H. Fellouah, F. Saleh and M. Boufadel. 2016. On the use of numerical modelling for near-field pollutant dispersion in urban environments—A review. Environ. Pollut. 208: 271–283.

Leveque, T., Y. Capowiez, E. Schreck, C. Mazzia, M. Auffan and Y. Foucault. 2013. Assessing ecotoxicity and uptake of metals and metalloids in relation to two different earthworm species (Eiseina hortensis and Lumbricus terrestris). Environ. Pollut. 179: 232–241.

Li, G., G. Sun, Y. Ren, X. Luo and Y. Zhu. 2018. Urban soil and human health: A review. Eur. J. Soil Sci. 69: 196–215.

Lu, Y., H. Yao, D. Shan, Y. Jiang, S. Zhang and J. Yang. 2015. Heavy metal residues in soil and accumulation in maize at long-term wastewater irrigation area in Tongliao, China. J. Chem.

Luo, X., S. Yu, Y. Zhu and X. Li. 2012. Trace metal contamination in urban soils of China. Sci. Total Environ. 421-422: 17–30.

Madre, F., A. Vergnes, N. Machon and P. Clergeau. 2014. Green roofs as habitats for wild plant species in urban landscapes: first insights from a large-scale sampling. Landsc. Urban Plan. 122: 100–107.

Margenat, A., V. Matamoros, S. Díez, N. Cañameras, J. Comas and J.M. Bayona. 2017. Occurrence of chemical contaminants in peri-urban agricultural irrigation waters and assessment of their phytotoxicity and crop productivity. Sci. Total Environ. 599: 1140–1148.

Martinez, A., N.R. Erdman, Z.L. Rodenburg, P.M. Eastling and K.C. Hornbuckle. 2012. Spatial distribution of chlordanes and PCB congeners in soil in Cedar Rapids, Iowa, USA. Environ. Pollut. 161: 222–228.

Meng, W., Z. Wang, B. Hu, Z. Wang, H. Li and R.C. Goodman. 2016. Heavy metals in soil and plants after long-term sewage irrigation at Tianjin China: A case study assessment. Agric. Water Manag. 171: 153–161.

Menozzi, M.-J. 2014. Les jardins dans la ville. Entre nature et culture. PU Rennes.

Meuser, H. 2010. Contaminated Urban Soils. Springer, Dordrecht; Heidelberg; London (p. 318).

Michalska, M. and H. Asp. 2001. Influence of lead and cadmium on growth, heavy metal uptake, and nutrient concentration of three lettuce cultivars grown in hydroponic culture. Commun. Soil Sci. Plant Anal. 32: 571–583.

Mombo, S., C. Dumat, M. Shahid and E. Schreck. 2016. A socio-scientific analysis of the environmental and health benefits as well as potential risks of cassava production and consumption. Environ. Sci. Pollut. Res.: 1–15.

Mombo, S., Y. Foucault, F. Deola, I. Gaillard, S. Goix, M. Shahid et al. 2016. Management of human health risk in the context of kitchen gardens polluted by lead and cadmium near a lead recycling company. J. Soils Sediments 16: 1214–1224.

Morefield, P.E., S.D. LeDuc, C.M. Clark and R. Iovanna. 2016. Grasslands, wetlands, and agriculture: the fate of land expiring from the Conservation Reserve Program in the Midwestern United States. Environ. Res. Lett. 11: 094005.

Motelay-Massei, A., D. Ollivon, B. Garban, M. Teil, M. Blanchard and M. Chevreuil. 2004. Distribution and spatial trends of PAHs and PCBs in soils in the Seine River basin, France. Chemosphere 55: 555–565.

Mougeot, L.J. 2010. Agropolis: The Social, Political and Environmental Dimensions of Urban Agriculture. Routledge.

Natasha, M. Shahid, N.K. Niazi, S. Khalid, B. Murtaza, I. Bibi and M.I. Rashid. 2018. A critical review of selenium biogeochemical behavior in soil-plant system with an inference to human health. Environ. Pollut. 234: 915–934.

Orsini, F., R. Kahane, R. Nono-Womdim and G. Gianquinto. 2013. Urban agriculture in the developing world: A review. Agron. Sustain. Dev. 33: 695–720.

Pascaud, G., T. Leveque, M. Soubrand, S. Boussen, E. Joussein and C. Dumat. 2014. Environmental and health risk assessment of Pb, Zn, As and Sb in soccer field soils and sediments from mine tailings: solid speciation and bioaccessibility. Environ. Sci. Pollut. Res. 21: 4254–4264.

Paustenbach, D.J., K. Fehling, P. Scott, M. Harris and B.D. Kerger. 2006. Identifying soil cleanup criteria for dioxins in urban residential soils: how have 20 years of research and risk assessment experience affected the analysis? J. Toxicol. Environ. Health B Crit. Rev. 9: 87–145.

Pelfrêne, A., C. Waterlot, M. Mazzuca, C. Nisse, D. Cuny, A. Richard et al. 2012. Bioaccessibility of trace elements as affected by soil parameters in smelter-contaminated agricultural soils: A statistical modeling approach. Environ. Pollut. 160: 130–138.

Pierart, A., M. Shahid, N. Séjalon-Delmas and C. Dumat. 2015. Antimony bioavailability: knowledge and research perspectives for sustainable agricultures. J. Hazard. Mater. 289: 219–234.

Pierart, A. 2016. Role of arbuscular mycorrhizal fungi and bioamendments in the transfer and human bioaccessibility of Cd, Pb, and Sb contaminant in vegetables cultivated in urban areas.

Ecology, environment. Université Paul Sabatier–Toulouse III, 2016. HAL. English. <NNT : 2016TOU30148>.

Pierart, A., C. Dumat, A.Q. Maes, C. Roux and N. Sejalon-Delmas. 2018. Opportunities and risks of biofertilization for leek production in urban areas: Influence on both fungal diversity and human bioaccessibility of inorganic pollutants. Sci. Total Environ. 624: 1140–1151.

Pierart, A., C. Dumat, A.Q. Maes and N. Sejalon-Delmas. 2018. Influence of arbuscular mycorrhizal fungi on antimony phyto-uptake and compartmentation in vegetables cultivated in urban gardens. Chemosphere 191: 272–279.

Pourrut, B., M. Shahid, C. Dumat, P. Winterton and E. Pinelli. 2011. Lead uptake, toxicity, and detoxification in plants. pp. 113–136. *In*: Rev. Environ. Contam. Toxicol. Volume 213. Springer.

Pussente, I.C., G.t. Dam, S.v. Leeuwen and R. Augusti. 2017. PCDD/Fs and PCBs in Soils: a study of case in the City of Belo Horizonte-MG. J. Braz. Chem. Soc. 28: 858–867.

Qishlaqi, A., F. Moore and G. Forghani. 2008. Impact of untreated wastewater irrigation on soils and crops in Shiraz suburban area, SW Iran. Environ. Monit. Assess. 141: 257–273.

Quenea, K., I. Lamy, P. Winterton, A. Bermond and C. Dumat. 2009. Interactions between metals and soil organic matter in various particle size fractions of soil contaminated with waste water. Geoderma 149: 217–223.

Rafiq, M., M. Shahid, G. Abbas, S. Shamshad, S. Khalid, N.K. Niazi et al. 2017. Comparative effect of calcium and EDTA on arsenic uptake and physiological attributes of Pisum sativum. Int. J. Phytoremediat. 19: 662–669.

Rafiq, M., M. Shahid, S. Shamshad, S. Khalid, N.K. Niazi, G. Abbas et al. 2017. A comparative study to evaluate efficiency of EDTA and calcium in alleviating arsenic toxicity to germinating and young Vicia faba L. seedlings. J. Soil Sediments: 1–11.

Ritter, L., K. Soloman and J. Forget. 2007. Persistent organic pollutants: An assessment report on DDT-Aldrin-Dieldrin-Endrin-Chlordane Heptachlor-Hexachlorobenzene Mirex-Toxaphene polychlorinated biphenyls dioxins furans. The International Program on Chemical Safety (IPCS).

Rose, A., O. Kehinde and A. Babajide. 2013. The level of persistent, bioaccumulative, and toxic (PBT) organic micropollutants contamination of Lagos soils. J. Environ. Chem. Ecotoxicol. 5: 26–38.

Royte, E., O. Kehinde and A. Babajide. 2015. Urban farming is booming, but what does it really yield?

Saeed, M.F., M. Shaheen, I. Ahmad, A. Zakir, M. Nadeem, A.A. Chishti et al. 2017. Pesticide exposure in the local community of Vehari District in Pakistan: An assessment of knowledge and residues in human blood. Sci. Total Environ. 587: 137–144.

Säumel, I., I. Kotsyuk, M. Hölscher, C. Lenkereit, F. Weber and I. Kowarik. 2012. How healthy is urban horticulture in high traffic areas? Trace metal concentrations in vegetable crops from plantings within inner city neighbourhoods in Berlin, Germany. Environ. Pollut. 165: 124–132.

Schreck, E., Y. Foucault, G. Sarret, S. Sobanska, L. Cécillon, M. Castrec-Rouelle et al. 2012. Metal and metalloid foliar uptake by various plant species exposed to atmospheric industrial fallout: mechanisms involved for lead. Sci. Total Environ. 427: 253–262.

Schreck, E., V. Dappe, G. Sarret, S. Sobanska, D. Nowak, J. Nowak et al. 2014. Foliar or root exposures to smelter particles: consequences for lead compartmentalization and speciation in plant leaves. Sci. Total Environ. 476: 667–676.

Shahid, M., M. Arshad, M. Kaemmerer, E. Pinelli, A. Probst, D. Baque et al. 2012. Long-term field metal extraction by Pelargonium: phytoextraction efficiency in relation to plant maturity. Int. J. Phytoremediat. 14: 493–505.

Shahid, M., E. Ferrand, E. Schreck and C. Dumat. 2013. Behavior and impact of zirconium in the soil–plant system: plant uptake and phytotoxicity. *In* Rev. Environ. Contam. Toxicol. Springer 221: 107–127.

Shahid, M., T. Xiong, M. Castrec-Rouelle, T. Leveque and C. Dumat. 2013. Water extraction kinetics of metals, arsenic and dissolved organic carbon from industrial contaminated popular leaves. J. Environ. Sci. 25: 2451–2459.

Shahid, M., E. Pinelli, B. Pourrut and C. Dumat. 2014. Effect of organic ligands on lead-induced oxidative damage and enhanced antioxidant defense in the leaves of Vicia faba plants. J. Geochem. Explor. 144: 282–289.

Shahid, M., C. Dumat, S. Khalid, E. Schreck, T. Xiong and N.K. Niazi. 2016. Foliar heavy metal uptake, toxicity and detoxification in plants: A comparison of foliar and root metal uptake. J. Hazard. Mater..

Shahid, M., C. Dumat, S. Khalid, N.K. Niazi and P.M. Antunes. 2017. Cadmium bioavailability, uptake, toxicity and detoxification in soil-plant system. Rev. Environ. Contam. Toxicol. 241: 73–137.

Shahid, M., M. Rafiq, N.K. Niazi, C. Dumat, S. Shamshad, S. Khalid et al. 2017. Arsenic accumulation and physiological attributes of spinach in the presence of amendments: an implication to reduce health risk. Environ. Sci. Pollut. Res. 1–10.

Shahid, M., S. Shamshad, M. Rafiq, S. Khalid, I. Bibi, N.K. Niazi et al. 2017. Chromium speciation, bioavailability, uptake, toxicity and detoxification in soil-plant system: A review. Chemosphere 178: 513–533.

Shakoor, M.B., N.K. Niazi, I. Bibi, G. Murtaza, A. Kunhikrishnan, B. Seshadri et al. 2016. Remediation of arsenic-contaminated water using agricultural wastes as biosorbents. Crit. Rev. Environ. Sci. Technol. 46: 467–499.

Shakoor, M.B., I. Bibi, N.K. Niazi, M. Shahid, M.F. Nawaz, A. Farooqi et al. 2018. The evaluation of arsenic contamination potential, speciation and hydrogeochemical behaviour in aquifers of Punjab, Pakistan. Chemosphere 199: 737–746.

Shamshad, S., M. Shahid, M. Rafiq, S. Khalid, C. Dumat, M. Sabir et al. 2018. Effect of organic amendments on cadmium stress to pea: A multivariate comparison of germinating vs young seedlings and younger vs older leaves. Ecotoxicol. Environ. Saf. 151: 91–97.

Shayler, H., M. McBride and E. Harrison. 2009. Sources and Impacts of Contaminants in Soils.

Sidle, R.C., J. Hook and L. Kardos. 1977. Accumulation of heavy metals in soils from extended wastewater irrigation. J. Water Pollut. Control Fed.: 311–318.

Sterrett, S., R. Chaney, C. Gifford and H. Mielke. 1996. Influence of fertilizer and sewage sludge compost on yield and heavy metal accumulation by lettuce grown in urban soils. Environ. Geochem. Health 18: 135–142.

Sujatha, C., S. Nair, N. Kumar and J. Chacko. 1994. Distribution of dichlorodiphenyltrichloroethane (DDT) and its metabolites in an Indian waterway. Environ. Toxicol. 9: 155–160.

Tabassum, R.A., M. Shahid, C. Dumat, N.K. Niazi, S. Khalid, N.S. Shah et al. 2018. Health risk assessment of drinking arsenic-containing groundwater in Hasilpur, Pakistan: effect of sampling area, depth, and source. Environ. Sci. Pollut. Res. doi:10.1007/s11356-018-1276-z.

The benefits of city-based agriculture go far beyond nutrition. Ensia magazine (https://ensia.com/features/urban-agriculture-is-booming-but-what-does-it-really-yield/).

Tiwari, K., N. Singh, M. Patel, M. Tiwari and U. Rai. 2011. Metal contamination of soil and translocation in vegetables growing under industrial wastewater irrigated agricultural field of Vadodara, Gujarat, India. Ecotoxicol. Environ. Saf. 74: 1670–1677.

Türkdoğan, M.K., F. Kilicel, K. Kara, I. Tuncer and I. Uygan. 2003. Heavy metals in soil, vegetables and fruits in the endemic upper gastrointestinal cancer region of Turkey. Environ. Toxicol. Pharmacol. 13: 175–179.

UAN. 2017. Urban enfvooirdonment. http://www.cityfarmer.org/.

Uzu, G., S. Sobanska, G. Sarret, M. Muñoz and C. Dumat. 2010. Foliar lead uptake by lettuce exposed to atmospheric fallouts. Environ. Sci. Technol. 44: 1036–1042.

Uzu, G., E. Schreck, T. Xiong, M. Macouin, T. Lévêque, B. Fayomi et al. 2014. Urban market gardening in Africa: foliar uptake of metal(loid)s and their bioaccessibility in vegetables; implications in terms of health risks. Water Air Soil Pollut. 225: 2185.

Van Veenhuizen, R. and G. Danso. 2007. Profitability and sustainability of urban and periurban agriculture. Food & Agriculture Org. 19: 1–95.

Vane, C.H., A.W. Kim, D.J. Beriro, M.R. Cave, K. Knights, V. Moss-Hayes et al. 2014. Polycyclic aromatic hydrocarbons (PAH) and polychlorinated biphenyls (PCB) in urban soils of Greater London, UK. Appl. Geochem. 51: 303–314.

Wong, M. 1985. Heavy metal contamination of soils and crops from auto traffic, sewage sludge, pig manure and chemical fertilizer. Agric. Ecosyst. Environ. 13: 139–149.

Wu, S., X. Xia, L. Yang and H. Liu. 2011. Distribution, source and risk assessment of polychlorinated biphenyls (PCBs) in urban soils of Beijing, China. Chemosphere 82: 732–738.

Wuana, R.A. and F.E. Okieimen. 2011. Heavy metals in contaminated soils: a review of sources, chemistry, risks and best available strategies for remediation. Isrn Ecology.

Xiong, T.-T., T. Leveque, A. Austruy, S. Goix, E. Schreck, V. Dappe et al. 2014. Foliar uptake and metal (loid) bioaccessibility in vegetables exposed to particulate matter. Environ. Geochem. Health 36: 897–909.

Xiong, T., T. Leveque, M. Shahid, Y. Foucault, S. Mombo and C. Dumat. 2014. Lead and cadmium phytoavailability and human bioaccessibility for vegetables exposed to soil or atmospheric pollution by process ultrafine particles. J. Environ. Qual. 43: 1593–1600.

Xiong, T., C. Dumat, A. Pierart, M. Shahid, Y. Kang, N. Li et al. 2016. Measurement of metal bioaccessibility in vegetables to improve human exposure assessments: field study of soil–plant–atmosphere transfers in urban areas, South China. Environ. Geochem. Health 38: 1283–1301.

Xiong, T., C. Dumat, V. Dappe, H. Vezin, E. Schreck, M. Shahid et al. 2017. Copper oxide nanoparticle foliar uptake, phytotoxicity, and consequences for sustainable urban agriculture. Environ. Sci. Technol. 51: 5242–5251.

Zhang, Y. and Y. Shen. 2017. Wastewater irrigation: past, present, and future. Wiley Interdiscip. Rev. Water.

Part 2
In situ Bioremediation

Chapter 5

Metal Contamination in Urban Soils
Use of Nature-Based Solutions for Developing Safe Urban Cropping

Ryad Bouzouidja,[1,2] *Dorine Bouquet,*[2,7] *Antoine Pierart,*[4]
Muhammad Shahid,[5] *Cécile Le Guern,*[7,8] *Liliane
Jean-Soro,*[6,7] *Camille Dumat*[3,4] and *Thierry Lebeau*[2,7,]*

The Urban Soil Specificities

In the urban areas, the soils are, most of the time, stripped, filled, mixed, compacted and supplemented with artificial materials. Soil profiles are enormously modified, leading to high spatial and vertical heterogeneity (Meuser 2010). At the same time, a strong spatial heterogeneity characterizes the urban crop soil from physical, chemical

[1] EPHor, AGROCAMPUS OUEST, IRSTV, 49042, Angers, France.
[2] Université de Nantes, UMR 6112 LPG-Nantes (Laboratoire de Planétologie et Géodynamique de Nantes), 2 rue de la Houssinière, BP 92208, 44322 Nantes cedex 3, France.
[3] Centre d'Etude et de Recherche Travail Organisation Pouvoir (CERTOP), UMR5044, Université J. Jaurès–Toulouse II, 5 allée Antonio Machado, 31058 Toulouse, France.
[4] Université de Toulouse, INP-ENSAT, Avenue de l'Agrobiopole, 31326 Auzeville-Tolosane.
[5] Department of Environmental Sciences, COMSATS Institute of Information Technology, Vehari-61100, Pakistan.
[6] IFSTTAR, GERS, EE, F-44340, Bouguenais, France.
[7] Institut de Recherche en Sciences et Techniques de la Ville, FR CNRS 2488, Ecole Centrale de Nantes, BP 92101, 1 rue de la Noë, 44321 Nantes, France.
[8] BRGM, Regional Geological Survey Pays de la Loire, 1 rue des Saumonières, BP92343, 44323 Nantes cedex 3, France.
* Corresponding author: thierry.lebeau@univ-nantes.fr

and biological aspects (Morel et al. 2005, Béchet et al. 2009). This heterogeneity can be explained by a wide range of applications aimed for the welfare of citizens (support for buildings, road infrastructure, recreational areas, kitchen gardens and parklands) (Blanchart et al. 2017). In addition, urban soils are known to have peculiar characteristics such as unpredictable layering, poor structure, and high concentrations of persistent contaminants such as trace metals (Kabata-Pendias 2010). Eventually, many studies have measured high bulk densities in urban soils (up to 2 g/cm³) due to a compaction phenomenon (Baumgartl 1998, Jim 1998, Morel et al. 2005).

As opposed to agricultural soils, urban soils could have either lost their structures (i.e., soil aggregation) and/or accumulated pollutants because of the presence of large natural- and/or anthropogenic-sourced particles (El Khalil et al. 2008, Nehls et al. 2013). Urban soil also differs from the agricultural one by the fact that the former is more strongly influenced by: (i) continuous and intense anthropogenic contaminating activities, (ii) contamination as the result of a higher loads of contaminants (Biasioli et al. 2006) and (iii) the age of soil (Morel et al. 2005). In effect, soils play an important role in maintaining the environmental quality as they can act as both source and sink for pollutants that can easily affect human health (De Kimpe and Morel 2000). From a chemical point of view, urban crop soils are characterized by heterogeneous values of pH and alkalinity due to carbonates (Morel et al. 2005). Yet, plants require certain chemical elements to complete their life cycles (Da Silva and Williams 2001, Knecht and Göransson 2004). Commonly, soil contains nutrients that are directly absorbed by plants as inorganic compounds, or organic nutrients that need to be mineralized to generate inorganic forms easily assimilable by plants. Nevertheless, nutrient uptake by plants is highly affected not only by the chemical form of inorganic compounds but also by the soil properties. For example, K^+, Na^+, NO_3^-, and NH_4^+ ions are absorbed rapidly, whereas PO_4^{-3}, SO_4^{-2}, Ca^{+2}, and Mg^{+2} ions are absorbed more slowly (Tisdale et al. 1985). Similarly, Scharenbroch et al. (2005) found that old urban soils had significantly greater values for weak Bray P (24%), strong Bray P (51%), and K (45%) than newer urban soils. However, Joimel et al. (2016) observed that the extractable phosphorous ratio of the anthropized urban soils was slightly close to the natural soils, in particular forestry and agricultural soils. These elements are naturally present in the soil. Hence, the chemical composition of a soil is inherited from the geological (named parental) material from which the soil has grown, more or less modified by pedogenic evolution without human intervention (background level) (Baize et al. 2007). The presence and heterogeneity of trace metals in urban soils show the relative influence of background and inputs from external factors due to land use (e.g., industrial activity or traffic emissions) (Bechet et al. 2016, Dumat et al. 2017). The most common trace metals in the urban area are Cd, Cu, Ni, Pb and Zn (Dudka et al. 1996). Overall, urban soils are currently at a low or medium level of metal pollution, as is the case, for instance, of Chinese soils contaminated with Cd (Wu et al. 2016). Jacobs et al. (2017) reported that urban soil contamination with trace metals is a major obstacle to the development of urban agriculture because of the high risk of metal accumulation by plants up to toxic levels for humans.

An alternative to reduce metal bioavailability and toxicity in urban soils is phytoremediation, a gentle strategy that maintains the agronomical properties of soils and preserves its function. Phytoremediation can be suggested, therefore, as a low-cost and environmentally friendly strategy to remediate urban soils contaminated with trace metals with two major aims: (i) to maintain the crop potential of soils, and (ii) to reduce the huge amounts of slightly contaminated soils that are commonly excavated and evacuated from the cities where they become wastes. Nature provided several essential services so-called ecosystem services. Plants provide many of these services, and we can optimize the delivery of some of them by growing the appropriate plants in the appropriate media. Phytoremediation is a way to preserve or restore some of these services: (i) a regulation service, (ii) a supply service owing to raw materials it generates for energy and/or metal recycling, and (iii) a cultural service with its contribution to the greening of cities and contribution to urban landscapes.

Crop Activities in the Urban Context

Urbanization and the urban sprawl led to changes in land use, which can be characterized by the growth of built-up areas and a loss of farmland (Cai 2000, Liu et al. 2003, Tan et al. 2005, Li et al. 2016). Thus, urban agriculture becomes a strategy for reducing the ecological footprint of cities, providing some food to the inhabitants, and reconnecting people with nature. Indeed, urban collective gardens (allotment gardens, community gardens, shared gardens) are growing in the last decades (Pourias et al. 2016). For example, in UK, over 78,827 people waited for a plot to develop urban farming (Campbell and Campbell 2010). Moreover, time for authorization does not facilitate the development of urban agriculture. For instance, in the city of Nantes (France), the time to obtain a plot is between 3 to 5 years. In developing countries, urban agriculture is the main option to produce food, whereas in developed countries its scope is quite different; it also provides recreational activities and an educational function (Dubbeling et al. 2010).

The French national scientific research project "JASSUR" (ANR-12-VBDU-0011) studied practices, functions and risks associated with urban gardens. A socio-technical characterization of gardening practices was carried out to evaluate the potential of urban garden to produce foodstuff. The results showed that the food supply service was workable, but the productivity of the system varied (https://www6.inra.fr/jassur). In some cases, the productivity of urban gardening was equivalent to those typically recorded for agricultural lands.

Besides urban collective gardens, urban microfarm is catching the attention of worldwide society. Microfarms are defined like small-sized commercial market gardens, which share some important characteristics: cultivated acreage smaller than official recommendations for market gardening (below 1.5 ha); community-oriented marketing through short-supplying chains; wide diversity of plants cultivated with more than 30 crops per farm to promote biodiversity; and low level of mechanization and investment (Morel and Léger 2016). Their productions are sold through short supplying chains by directly selling to consumers or with only one intermediary (Aubry and Chiffoleau 2009). The production of these microfarms is often low. Indeed, productivity is the second goal because they are multifunctional and propose multiple social activities. Moreover, a great part of the production is donated or auto consumed (Daniel 2017).

What Can We Do against Metal-Contaminated Urban Soils?

Metal contamination of urban soils is incompatible with its use for cropping or for recreational activities (Pascaud et al. 2014a, Mombo et al. 2016). In this situation, management options must be considered to restore its adequacy through reduction of human exposure to pollutants (MEEM 2017). Among these management strategies, removing the source of pollutants (e.g., excavation of the soil), reducing the concentration of pollutants in the soil, or their transport capability (e.g., immobilization, trapping, precipitation, complexation, or reactive barriers) and reducing pollutant availability (e.g., insulating membranes) are currently the most used to limit human exposure to pollutants.

In case of urban allotment gardens, excavating the contaminated soil (*ex situ* management) and its replacement by uncontaminated topsoil is the most used strategy by urban planners. This approach can be assumed as a form of the precautionary principle. For urban planners, removal of the contaminated soil indeed solves the problem immediately and definitively, thereby avoiding any responsibility in case of not properly cleaning-up the soil. But two issues emerge: (i) the availability and the quality of the marketed topsoil used for the soil replacement and (ii) the disposal of low-contaminated soils after removing from the urban allotment garden. This waste soil can be also used for very specific uses such as road base course. In the context of experimental urban gardens in Paris, Badreddine et al. (2017) developed a procedure to manage safe gardening activities on variously polluted urban soils.

In situ management of polluted soils is undoubtedly more sustainable, especially options friendly to the soil quality devoted to crop activities that allow maintaining their physical, chemical and biological characteristics. Among the possible options, reducing the transfer of metals from soil to crops to guarantee the regulatory threshold is technically feasible, but it is not widely accepted by society and legislators (notably in Europe) as long as trace metals remain in the soil. In China, where 20% of agricultural land is currently contaminated (Zhao et al. 2015) and is insufficient to feed the population, several experiments to exploit contaminated lands were undertaken (Tang et al. 2012). Nonetheless, the consumption of contaminated crops is not the only risk for the consumer. It must be kept in mind that the risk of direct soil consumption (Denys et al. 2007, Pascaud et al. 2014b) and inhalation of soil particles, in particular by children, may be higher than the health risk resulting from consuming contaminated vegetables (see Chapter 5 for more details).

This chapter provides an overview on the gentle methods for managing crop soils contaminated by trace metals in the context of urban farming and allotment gardens, with particular emphasis on *in situ* solutions. These *in situ* methods of remediation will be illustrated by some case studies performed by our research group. Lastly, some innovative experiments performed in agricultural areas are also discussed in the context of their potential application in urban agriculture.

Management of Urban Agricultural Soils

In situ Methods for Reducing Transfer of Trace Elements from Soils to Plants

Phytostabilization is a relevant method to reduce the transfer of trace metals from soils to plants, and to avoid their dispersion in the environment (e.g., erosion, transport of airborne soil particles). However, the use of non-food producing plants (Linger et al. 2002, Khan 2003, 2005) leads to change the soil use unless phytostabilization involves the culture of vegetables (in association or as an intermediate culture).

Recently, several studies have attempted to seek potential solutions that enable growing healthy vegetables. For example, the intervention in the physicochemical properties of the soil such as the pH has an immediate effect on tracing metal mobility (Kalkhajeh et al. 2017, Tedoldi et al. 2017, Brimo et al. 2018). On the other hand, the addition of organic amendments not only improves soil quality but also reduces trace metal mobility (Mench and Martin 1991, Mench et al. 1999, 2000, Tang et al. 2012, Austruy et al. 2016). Trace metals are generally adsorbed by carbonates, organic matter, Fe-Mn oxides and primary or secondary minerals (Ross 1994). However, fertilizers occasionally contain by-products or contaminants in their formulation that may be an environmental risk to soil quality and plant health. This is the case of commercial phosphates that contain Cd in their formulation, although some effort is being addressed to improve the quality of this kind of fertilizer (French agency for environmental quality and human health, ANSES). Bioaugmentation of soil with microorganisms is able to immobilize metals by both sorptive and precipitation mechanisms (Volesky and Holan 1995). Many studies have demonstrated that bioaugmentation with (symbiotic) microorganisms may reduce metal uptake by plants (Joner et al. 2000, Karagiannidis and Nikolaou 2000, Lovely and Lloyd 2000, Tonin et al. 2001, Jézéquel et al. 2005, Jézéquel and Lebeau 2008). In this context, bioaugmentation using mycorrhizae is a promising strategy in the bioremediation of metal-contaminated soils (Lebeau et al. 2008, Aghababaei et al. 2014). These microorganisms are abundant in species diversity, with a marked variation in morphology and physiology (Selosse and Tacon 1998). In general, ectomycorrhizal fungi are associated mostly with trees, whereas about 94% of angiosperms establish symbiosis with endomycorrhizal fungi (Brundrett 2009). The arbuscular mycorrhizal fungi have been the most studied group, which is implied in the solubilization of inorganic phosphorus and subsequent transfer to the host plant together with water and other nutrients. In this symbiotic relationship, the fungi obtain photosynthetic carbon compounds from the plant (Smith and Smith 2011, Li et al. 2013). The arbuscular mycorrhizal fungi have been also been studied as biological vectors in the bioremediation of metal-contaminated soils. They play an important role in the soil-plant interface because of their capability to be either a barrier or an enhancer of metal transfer, using a large variety of metabolic pathways (binding in fungal wall, excretion of organic acids and glomalin, among others) (Amir et al. 2014).

These metabolic pathways suggest two different strategies for reducing mobility and toxicity of trace metals in the soils. First, the so-called phytoextraction of trace metals, which is facilitated by the presence of arbuscular mycorrhizal fungi and requires a certain resistance of the target plants to the metals (Amir et al. 2008, Lebeau et al. 2008, Danh et al. 2014). Second, the so-called phytoimmobilization of trace metals that results in the accumulation of these chemicals in the rhizosphere by the joint action of root secretions and the physical barrier by arbuscular mycorrhizal fungi (Lebeau et al. 2008, Rangel et al. 2014). The latter strategy obviously raises a protective mechanism to the plant, a topic that has been intensively tackled in the context of urban agriculture (Pierart 2016).

Phytoextraction

The scope of phytoextraction is the removal of trace metals from the soil by repeatedly harvesting plant biomass from a polluted site. Harvesting is prolonged in time until metal concentrations in the soil are below the regulatory threshold. In their review, McGrath and Zhao (2003) highlighted that metal-accumulating plants are still being sought, and that the phytoextraction capability of high hyperaccumulator plants still needs to be validated in field conditions (Greenland project, e.g., Cundy et al. 2015).

In situ phytoextraction is a preferred strategy for a set of advantages: (i) it is a gentle remediation method that maintains the agronomical properties of the soil (Gerhardt et al. 2009); (ii) it is not expensive compared to other bioremediation procedures such as bioaugmentation with microorganisms (Garbisu and Alkorta 2001); (iii) it is a relatively simple approach (Angle and Linacre 2005); and (iv) it is a socially welcome method (Lambert et al. 2000).

Phytoextraction has, however, several limitations. First, it takes a long time, which is certainly the main restrictive factor explaining why phytoremediation has not gained popularity as a bioremediation methodology. Nevertheless, the duration of phytoextraction can be shortened by increasing the mobility of metals (e.g., pH modification, using complexing agents such as synthetic agents or naturally produced agents from microorganisms inoculated in soil (Lebeau 2011)). Second, phytoextraction is only workable in the topsoil; otherwise, trees can be used to reach deeper layers of contaminated soils (Gerhardt et al. 2009). Third, disposal of plant residues after harvesting arise an environmental problem. Only the recycling of nickel-accumulating plants is at an advanced stage of development for its disposal (Chaney et al. 2018). Finally, phytoextraction must be validated at full scale, which is still in a premature stage of development (Greenland project, Cundy et al. 2015).

Assessing the Performance of Nature-Based Solutions

Plants and trees in urban areas (green spaces) are being increasingly recognized for their capacity not only to support biodiversity conservation (Goddard et al. 2010), but also to generate additional environmental, economic, and social benefits (Haase et al. 2014, Kabisch et al. 2015). Also, they promote the functioning of ecosystems as essential backbones to climate change mitigation and adaptation (European Commission 2015). Currently, the urban vegetation forms a part of the nature-based solutions (NBS) scheme. Most researchers agree that the NBS concept is a rational strategy for promoting the ecological restoration and enhancement of biodiversity, as well as the maintenance of the urban structure (e.g., Kabisch et al. 2016, Maes et al. 2017). In addition, NBS can be characterized by the use of nature in tackling challenges previously cited and conserving biodiversity in a sustainable manner (Balian et al. 2014).

The concept of urban performance indicators comprises those relevant indicators of changes in the soil and water quality related to environmental stressors such as chemical contamination (Whitford et al. 2001, Dyckhoff and Allen 2001). Environmental performance indicators are predominantly integrated to regulating ecosystem services and refer to biodiversity such as vegetation cover (Kabisch et al. 2016). Many authors defined the term of bioindicators that are living organisms such as plants, planktons, animals, and microbes, which are utilized to screen the health of the natural ecosystem in the environment (Gerhardt 2002, Holt and Miller 2010, Parmar et al. 2016). The abiotic indicator comprises temperature, saltiness, stratification, and pollutants, pH, water content, organic matter content, bulk density of the soil, type of soil (e.g., sand peat, clay), and degree of pollution with metals and organic pollutants, but also the type of management (agriculture, application of manure and/or fertilizer, nature, recreation area, etc.) and vegetation (crop rotation). Regarding biotic indicator, it is defined as the abundance and diversity of nematodes, earthworms, enchytraeids, and micro-arthropods, nitrifying activity, microbial functions, genetic diversity, total activity and numbers of bacterial cells (Breure 2004). Heink and Kowarik (2010) indicated that the term "indicator" is frequently used for the interface between science and policy. There is still a great demand for clear definitions of technical terms in science and policy.

There are many examples in the scientific literature on the use of indicators to assess the performance of phytomanagement using vegetation. Pérez de Mora et al. (2011) proposed the use of soil microbial activity and community composition as suitable indicators of phytoremediation actions in metal-contaminated soils. In addition, Parraga-Aguado et al. (2013) established significant relationships between the outcomes of some ecological indices (e.g., heterogeneity of the plant communities, number of different species) and those coming from (i) some physicochemical properties of soil (electrical conductivity, equivalent calcium carbonate, total nitrogen, organic carbon), and (ii) the water extractable ions and dissolved organic carbon. Regarding trace metal phytoextraction, McGrath and Zhao (2003) demonstrated that both high biomass yields and metal hyperaccumulation are required to make the process efficient.

Case Study: Reducing the Trace Metal Transfer from Soil to Crops

Arbuscular Mycorrhizal Fungi-Based Biofertilizers

This case study describes an experiment conducted under greenhouse condition to evaluate if it was possible to use local arbuscular mycorrhizal fungi-based biofertilizers to decrease the transfer of trace metals between contaminated peri-urban soil and edible crops. Two metal-contaminated soils (Cd, Pb and Sb) were compared for their different trace metal origin, which was either anthropic (from Bazoche, France—BZC) or geogenic (from Nantes, France—NTE). A full description of these soils is detailed in Pierart (2016). In brief, the study examined the growing leek in these soils using a pot experimental design: half pots received a biofertilizing solution (called Biofertilization) containing arbuscular mycorrhizal fungi spores isolated from trap crops grown on a mix of these soils. The scope of this fertilization was that urban gardeners could easily prepare such biofertilizer from their own cultivated pots. The second half pots were not biofertilized (control pots). For Cd, biofertilization had a significant opposite result in trace metal concentration in the roots between both soils, with an increase in the NTE soil and a decrease in the BZC soil (Fig. 1A and 1D). On the other hand, biofertilization had no effect on Pb accumulation in the roots, and in the aerial parts of the plant (Fig. 1B and 1E). However, Pb increased significantly in the leaves of leeks grown in the BZC soil (anthropic contamination), probably because of an increase of metal translocation from root to leaves. In the case of Sb (Fig. 1C), biofertilization decreased its concentration in leaves, with significant results only on BZC soil. Sb concentration in root was found under the detection limit, which suggests a full transference from root to leaves in both soils.

For these three trace metals, the human bioaccessible fraction was estimated in edible parts of the plant to assess human health risk by food ingestion (Fig. 1F). A significant fraction of Cd in leek leaves was observed to be gastrically bioaccessible (85%), while Pb was significantly less bioaccessible (~ 60%). Sb bioaccessibility was lower, ranging between 15 and 33%. The biofertilization treatment significantly increased the bioaccessible fraction of Sb in leeks cultivated in the NTE soil (13% compared with controls). In BZC soil, a slight increase was also observed (5.5%); however, these results were not statistically significant.

This case study highlights the lack of understanding of the role of arbuscular mycorrhizal fungi in trace metal transfer from soil to plant. Furthermore, it shows that a case-by-case approach ought to be performed instead of applying a standardized bioaugmentation method, even if arbuscular mycorrhizal fungi were shown to reduce the transfer of trace metal under fully controlled conditions (see the review by Lebeau et al. 2008). Therefore, the balance between advantages and limitations needs to be assessed carefully when using these organisms.

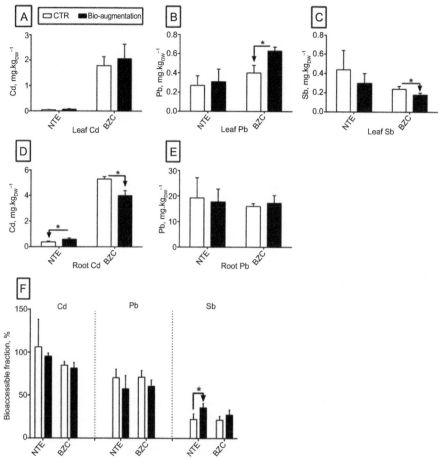

Fig. 1. Trace metal accumulation (mg/kg dry mass) in leek leaf [A-Cd, B-Pb, C-Sb] and root [D-Cd, E-Pb] in natural (control, CTR) and biofertilized soil (arbuscular mycorrhizal fungi, Bio-augmentation) from Bazoche (BZC) or Nantes (NTE). Bioaccessible fraction of each metal [F] is expressed as the ratio between the bioaccessible trace metal concentration and the total trace metal concentration. Sb was under the detection limit in leek root. Significant differences are indicated with * (T-test, $\alpha = 5$). (Results taken from Pierart et al. 2018.)

Green Manure Plants for Improving Environmental Quality and Fertility

The regeneration of brownfield sites in urban areas is a major challenge for the sustainable development of cities. In effect, these sites are generally localized in the center of the cities and could be transformed in new ecological areas with micro-farms or collective gardens used to produce consumed vegetables. Management and conversion of these large urban sites, imposed by regulations, however, require the development of tools for assessing environmental and health risks, and sustainable remediation techniques. For example, Foucault et al. (2013) developed bioavailability and ecotoxicity tests to improve the classification of contaminated soils, focusing their

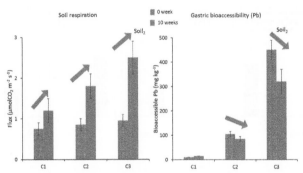

Fig. 2. Soil respiration (Lycor apparatus) and lead bioaccessibility (Barge procedure) variation during 10 weeks of phytoremediation treatment with borage. C1, C2, and C3 are the Pb concentrations in soil, which correspond to 0, 400 and 800 mg/kg dry mass of polluted soil (Data taken from Foucault et al. 2013).

study on a lead battery recycling site in Toulouse (France), which was characterized by historical Pb pollution and other metals (Uzu et al. 2011). Moreover, green manure plants (borage, mustard and phacelia), commonly used in agriculture and gardens, were tested by Foucault et al. (2013) for re-functionalization of polluted soil. It is well-known that these green manures improve biological and physicochemical properties of soils (soil respiration, microorganism biomass) with root system and large production of root exudates. The mechanisms involved in the fate of pollutants in the rhizosphere and associated microorganisms were studied in controlled experiments with the industrial soil, then the tested plants were used directly on the field in order to promote the ecosystem services. Soil respiration and human bioaccessibility of pollutants measured before and after soil remediation in order to assess the soil quality with a global environmental-health approach demonstrated the efficiency of green manure plants for soil remediation. Actually, borage (Fig. 2) improved soil respiration, reduced metal toxicity and the amount of total and gastric bioaccessible lead in soil, respectively, by phytostabilisation (borage) and storage in the roots (Pb and Sb).

Depending on metal speciation (the chemical forms in which a metal is present in the soil) as well as the type of soil and plant species, the environmental fate of metal greatly differs. Metal speciation can also be influenced by the agronomic characteristics of the soil and microbial activity in the rhizosphere. A molecular screening and meta-analysis of microbial genomics have helped to highlight the differences in bacterial communities studied by the level of metal concentration, plant species and characteristics of the soils studied (Foucault 2013).

In the context of polluted soils potentially suitable for urban agriculture, green manure plants appear an interesting tool to develop in the coming years because of their soil fertility capability, decreased metal bioavailability, and human bioaccessibility and metal ecotoxicity. After the first step of experimentation in controlled conditions, long term phytoremediation actions at the field scale are currently being performed using borage and also for comparing the vegetation which grows spontaneously on the site of the lead recycling factory. Moreover, borage plants are not perennial, they should be harvested to avoid the release of metal(loid)s in soil, and treated in a waste treatment unit (Foucault et al. 2013).

Combination of Amendments and Fertilizers in Metal-Polluted Soil

Agricultural soils may contain important levels of trace metals. For instance, the origin of Cd in agricultural soils is mainly due to the high utilization of phosphate fertilizers, which are enriched with this metal (Grant et al. 2013). Lowering metal phytoavailability, it is therefore possible to grow plants consumed by humans on soil slightly or moderately polluted by using the phenomena of adsorption (on clay minerals, wollastonite, biochars or black carbon, etc.), complexation (organic matters) or precipitation (calcareous amendments) of metals, in order to stabilize these pollutants and reduce the soil-plant transfers. Chemical speciation and compartmentalization of elements indeed alter their phytoavailability and (eco)toxicity. In China, for instance, with the aim of reducing human exposure to Cd, several materials (clay minerals, wollastonite, biochars) were tested in order to immobilize cadmium in cultivated soil. The questions asked concerned the specificity and sustainability of these added materials. The study by Wu et al. (2016) is an example on how mineral-based amendments may significantly reduce the phytoavailability of toxic metals. They applied the mineral wollastonite (calcium inosilicate, $Ca_3[Si_3O_9]$), which is a commercial product currently available in China, to Cd-polluted soils. This mineral reduced the Cd mobility in the soil, thus reducing its phytoavailability and toxicity. Therefore, wollastonite applications in cadmium-contaminated soils can reduce cadmium accumulation in plants, although the main limitation is the synchronous immobilization of micronutrients which may affect, in turn, the plant growth.

In an attempt to mitigate this disadvantage of using wollastonite, Wu et al. (2016) applied Zn- and Mn-fertilizers in wollastonite-amended soils to promote the growth and fitness of amaranth (*Amaranthus tricolor* L.). The plants were cultivated under three different treatments: Cd-contaminated soil with a micronutrient fertilizer, Cd-contaminated soil amended with wollastonite, and Cd-contamination soil amended with both wollastonite and micronutrient fertilizer. The following variables were measured: the plant biomass, photosynthesis parameters, and total Cd, Mn and Zn concentrations. This latter variable was measured also in soil samples. The results of that study demonstrated that application of wollastonite decreased the concentrations of Cd, Zn, and Mn in the plant, as well as their availability in the soil. Moreover, this mineral fertilization increased the gas exchange ability of plants. However, wollastonite treatment reduced the chlorophyll concentration in the leaves, and it had no positive influence on plant biomass. In contrast, Mn and Zn fertilization following wollastonite application corrected these two parameters, so plant biomass and photosynthetic ability significantly increased. This combination of fertilizers also reduced Cd phytoavailability more efficiently, probably because of the competition phenomenon. Therefore, this laboratory study is an example on how synergistic improvement could be tacked by combination of a mineral amendment for metal immobilization and a micronutrient fertilizer for reducing nutrient deficit in the plant.

Case Study: Cultivating Vegetable While Phytoextraction

Cropping using Low Trace Metal-Accumulating Vegetables

Firstly, it should be reminded that phytoremediation, especially phytoextraction, is one of the most common methods for the *in situ* remediation of metal-contaminated soils. Moreover, one of the main weaknesses of phytoremediation is that it takes a long time to significantly reduce the concentration of toxic metals in the soil (Pilon-Smith 2005). To partially solve this limitation, some studies have used phytoextraction together with vegetables with a low capability of metal absorption this latter is linked to human consumption. The association between phytoextraction and crops, and even setting up phytoextraction during inter-cropping seasons, are two practical options to avoid metal toxicity from soil, without taking precaution for reducing metal concentrations. An example of this functional association is the study by Bouquet et al. (Pers. Commun.) performed in an allotment garden in Nantes (France). The soil in this allotment garden is contaminated by Pb (ca. 170 mg kg^{-1} dry mass of soil), so there was a high risk of garden closure for cropping purpose. Moreover, the most workable solution was the excavation of topsoil that displayed an acceptable agronomic quality. To avoid these two measures, an *in situ* 2-years experiment was launched in July 2015. Some vegetable species and Indian mustard (*Brassica juncea*) were cultivated in rotation in this allotment garden. The Indian mustard was used as phytoextraction species. The Pb concentrations in the edible parts of tomatoes, winter cabbages, leeks and potatoes were under the EEC 466/2001 regulatory threshold set at 0.1 mg kg^{-1} of fresh matter (0.3 mg kg^{-1} for cabbage) (Fig. 3). For green beans, Pb concentrations in pods were close to the threshold (0.1 mg kg^{-1}). Biomass production was very low on those plots (7.25 times lower compared to others). Such a cropping system was already tested with success by associating the metal accumulator plant *Sedum alfredii* and a low metal-accumulating cultivar of maize (Xiaomei et al. 2005).

The phytoextraction efficiency of Pb was very low (ca. 2 mg kg^{-1} dry mass of aerial parts). The geogenic Pb availability partly explains this low phytoextraction efficiency (Bouquet et al. 2017). These same authors showed that the addition of EDTA increased significantly the Pb concentration in shoots of *B. juncea*, up to 26 times in comparison with the control sample. Bouquet et al. (2017) estimated that if the phytoextraction capability of *B. juncea* led to a Pb concentration in shoots of 45 mg kg^{-1} dry mass, then 604 years are required to remove the non-residual Pb, but much less if only the bioavailable fraction of Pb is considered. Although EDTA displays sublethal toxicity to plants at concentrations as low as micromolar (Shahid et al. 2012), phytoextraction, in the presence of this metal chelator, could be a complementary alternative (Lebeau et al. 2008).

An unexpected finding was to find a high accumulation rate of Pb in aerial parts, i.e., stems and leaves, of tomato (much more than in *B. juncea*), whereas the metal concentrations in fruits were under the EEC regulatory threshold. With the aim of examining species-specific differences of Pb accumulation in several varieties of *Brassica juncea* and *Solanum lycopersicum* (Tomato), a study by Bouquet et al. 2018 (Pers. Commun) cultivated these vegetables using a hydroponic system and exposed

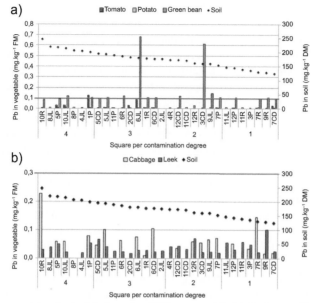

Fig. 3. Lead amount (mg kg⁻¹ of fresh matter (FM)) measured in (a) fresh vegetables and (b) leafy vegetables compared to lead content measured in soil before cropping experiment and regulatory thresholds for vegetables (red line). Note: try to correct the number in the Y-axis. Decimals indicated with point instead of comma. Also, try to define all the treatments; otherwise, this figure is not needed.

to realistic phytoavailable Pb concentrations frequently found in soil, i.e., 20 mg L⁻¹ and 40 mg L⁻¹. In these conditions, tomato plant was a "hyperaccumulator", with Pb concentrations in shoots as high as 500 mg kg⁻¹ dw and 2000 mg kg⁻¹ dw in the plants exposed to 20 and 40 mg L⁻¹, respectively. Concentrations in *S. lycopersicum* shoots were 50 to 100 times higher than those measured in *B. juncea* shoots, regardless of Indian mustard cultivar involved.

Associating Plants that Compete for Trace Metals in Soil

Another option for maintaining cropping in metal-contaminated soil with minimum risk for human exposure consists in reducing the transfer of trace metals from the soil to plants, by creating competition for the metal uptake in favor of hyperaccumulators. Some results were already published mainly regarding crops, not vegetables co-cultivated with hyperaccumulators. Intercropping with Indian Mustard (*Brassica juncea*) led to a decrease in Cd concentration by 57.1% in alfalfa cultivated in soils while it increased by 14.5% in the aboveground parts of *B. juncea* (Su et al. 2008, Xin-Bo et al. 2009). A co-cropping system with the Cd- and Zn-hyperaccumulator *Sedum alfredii* and a low-accumulating crop (*Z. mays*) was set up in a rice field historically irrigated with Pb- and Zn-enriched water from mining activities. In that study, Wu et al. (2007) showed a removal of heavy metals by *S. alfredii*, whereas safe corn for animal feed was produced allowing farmers to continue their agricultural activities. Cd-accumulating varieties of oilseed rape could also reduce Cd uptake by co-cropped cabbage. Unfortunately, final concentrations remained high (Liu et al.

2007). However, the Cd concentrations in cabbage depended on the metal speciation in the soil (De-Chun et al. 2010).

As compared to low metal-accumulating plant species, the beneficial effect of associating a hyperaccumulator species with a normal crop species was not always achieved as shown by Yu et al. (2014), who used an oilseed rape-rice rotation. Barley (*Hordeum vulgare*) co-cropped with *Noccea caerulescens* decreased Zn concentration in barley, but simultaneously increased Cd concentration (Gove et al. 2002). A similar finding was reported by Jiang et al. (2010) for the co-cropping of *N. caerulescens* and ryegrass. Comparing these results with a co-cropping using *Thlaspi arvense* (non hyperaccumulator), they suggested that the high concentration of Zn mobilized by the hyperaccumulator was used by itself, whereas it did not require Cd, explaining why the co-planted crop accumulated less Zn and more Cd.

Co-cropping of two non-hyperaccumulating plant species may alter the metal speciation in soil leading to a decrease in trace metal accumulated in harvested plants. For example, *Cunninghamia lanceolate*, as inter-crop plant, decreased the concentrations of Cu, Mn and Pb in tea leaves (Xue and Fei 2006). Similarly, Wu et al. (2003) showed that wheat/rice intercropping could reduce Cd concentration in wheat grain while it increased in rice grain. Taken together these studies suggest that this association cropping between a metal hyperaccumulator and plants intended for human consumption reduces the risk of human exposure to toxic metals.

Urban Soil Cleaning using Nature-Based Solutions: Anticipating New Regulations?

Cleaning of contaminated soils using *ex situ* procedures faces stronger regulation constraints than *on site* or *in situ* processes (soil respectively excavated or not and kept on the site itself). Indeed, according to the European legislation (European Waste Framework Directive 2006), contaminated soils are considered waste as soon as they are transported out of their original site. However, cleaning of soil is possible *ex situ* using specific treatment facilities as long as the corresponding industrial legislation and soil/waste traceability is strictly enforced to these procedures. However, phytoremediation needs both space and time, which might not be compatible with management costs.

An option could be to use the public or private space, becoming an opportunity to create green areas of urban concern. Especially, phytoremediation could be introduced when creating new quarters or in the frame of the redevelopment of existing ones, like in polluted harbor area in Amsterdam (Wilschut et al. 2013), and in the case of polluted vacant lands in Canadian municipalities (Todd et al. 2016). Developers and urban planners would thus need to anticipate such *ex situ* treatment in the planning and developing process and the regulation should evolve to allow such option. Of course, there is a need for traceability of soil movement and quality. Also, the quality of the receiving soil and local water (surficial, groundwater) should not be altered, and health risks be limited. With the rate of urbanization, green areas take a particular place in the management of the city. Thus, the use of phytomanagement, including phytoremediation, could be a solution to the need of the city dwellers for nature and vegetation in the urban environment (Boutefeu et al. 2005, Blanc 2009, Cheverry and Gascuel 2009, Bourdeau-Lepage and Vidal

2012). Therefore, the phytomanagement may be an important element to include in the regulation of the urban environment. Indeed, the vegetation should allow the decrease of the temperature during the night (Hardin and Jensen 2007, Cameron et al. 2014, Doick et al. 2014, Foissard 2015). Another function is the possibility to attenuate sound reverberations (Balaÿ 2013). In addition, it is generally accepted that urban vegetation improves air quality. For example, Setälä et al. (2013) studied the ability of urban park/forest vegetation to remove air pollutants (NO_2, anthropogenic volatile organic compounds and particle deposition) in two Finnish cities (Helsinki and Lahti), suggesting that urban vegetation is a suitable biological target to remove air pollutants. Eventually, phytomanagement can contribute to the infiltration of the rain water and thus limit the risk of floods due to the huge impermeable surface in cities (Cheverry and Gascuel 2009, Carré and Deutsch 2015).

Principles of circular economy, as the UK waste strategies WRAP—Waste & Resources Action Programme (Wrap 2010), should help making such options workable, i.e., introduce phytomanagement in the redevelopment of quarters in cities. To ensure that soil quality and use are compatible, a health risk approach can be applied as that used for polluted soil management in countries such as USA (Clay 1991), France (MEEM 2017) and Taiwan (Lai et al. 2010), or for soil reuse within land management (Bodemdecreet 2006, Vlarebo 2008, Blanc et al. 2012, Coussy et al. 2017). Indeed, such options could help to reduce the amount of excavated materials (soil) landfilled. For instance, 45 million of excavated soil (and subsoil) is expected in the Grand Paris Express Project. The objective of this project is to reuse 70% of soil (Richard 2017). Middle size cities are also concerned by the preservation of the soil resource. This is the case of the Nates city (France), where the second part of the Ile de Nantes redevelopment project is expected to produce around 100,000 tons per year of excavated soils over 15 yrs (Jeanniot et al. 2014). The redevelopment project could apply phytoextraction on green areas welcoming low to middle contaminated excavated soils.

Conclusions

The construction and the use of urban soil for cropping depend on specific rules. It is essential to determine in advance if remediation strategies, such as phytoremediation, are needed because many urban soils have high concentrations of trace metals such as Cd, Cu, Mn and Zn, compared with soils from rural areas. Yet, urban agriculture or cropping is increasing worldwide. Nature-based solutions, more especially phytotechnologies, appear to be promising solutions to manage these contaminated urban soils. However, the main limitation of such methods is the long duration that the procedure takes to decrease metal concentrations in the soil up to safety levels for human health. To avoid that, low trace metal-accumulating vegetables (secure cropping system) may be co-cultivated with metal hyperaccumulating plants (*in situ* phytoextraction). This association cropping would allow the soil to be reusable for cultivating vegetables without any regulatory constraint. When excavation of soil is the best option due to time constraints, it is expected that the regulations evolve and validate the opportunity to clean up soils—only when moderately contaminated—by means of phytotechnologies before soils are re-used for vegetable cropping.

Acknowledgments

Results is this chapter were supported by the Nature4Cities project, which is funded by the European Union's Horizon 2020 research and innovation programme under grant agreement No. 730468, and the POLLUSOLS project supported by the "Pays de la Loire" Region (France).

References

Aghababaei, F., F. Raiesi and A. Hosseinpur. 2014. The influence of earthworm and mycorrhizal co-inoculation on Cd speciation in a contaminated soil. Soil Biol. Biochem. 78: 21–29.

Amir, H., D.A. Jasper and L.K. Abbott. 2008. Tolerance and induction of tolerance to Ni of arbuscular mycorrhizal fungi from New Caledonian ultramafic soils. Mycorrhiza 19: 1–6.

Amir, H., P. Jourand, Y. Cavaloc and M. Ducousso. 2014. Role of mycorrhizal fungi in the alleviation of heavy metal toxicity in plants. pp. 241–258. *In*: Solaiman, Z.M., L.K. Abbott and A. Varma (eds.). Mycorrhizal Fungi: Use in Sustainable Agriculture and Land Restoration. Springer Berlin Heidelberg, Berlin, Heidelberg.

Angle, J.S. and N.A. Linacre. 2005. Metal phytoextraction—A survey of potential risks. Int. J. Phytoremediation 7: 241–254.

Aubry, C. and Y. Chiffoleau. 2009. Le développement des circuits courts et l'agriculture périurbaine: histoire, évolution en cours et questions actuelles. Innovations Agronomiques 5: 53–67.

Austruy, A., C. Laplanche, S. Mombo, C. Dumat, F. Deola and C. Gers. 2016. Ecological changes in historically polluted soils: Metal(loid) bioaccumulation in microarthropods and their impact on community structure. Geoderma 271: 181–190.

Badreddine, R., C. Dumat and P. Branchu. 2017. Interdisciplinary methodology for the assessment of health risks caused by pollution in urban gardens. Presented at the International Congress "Urban agricultures and ecological transition," Toulouse, France.

Baize, D., W. Deslais and N. Saby. 2007. Teneurs en huit éléments en traces (Cd, Cr, Cu, Hg, Ni, Pb, Se, Zn) dans les sols agricoles en France (Rapport final simplifié). ADEME, Angers, France.

Balaÿ, O. 2013. L'ARCHITECTE, L'HABITAT LE VEGETAL ET LA DENSITÉ.

Balian, E., H. Eggermont and X. Le Roux. 2014. Outputs of the Strategic Foresight workshop "Nature-Based Solutions in a BiodivERsA context" (BiodivERsA report). Brussels.

Baumgartl, T. 1998. Physical soil properties in specific fields of application especially in anthropogenic soils. Soil Tillage Res. 47: 51–59.

Béchet, B., F. Carré, L. Florentin, C. Leyval, L. Montanarella, J. Morel et al. 2009. Caractéristiques et fonctionnement des sols urbains. Cheverry Gascuel Éd Sous Pavés Terre Omniscience Montreuil 45–74.

Bechet, B., S. Joimel, L. Jean-Soro, A. Hursthouse, A. Agboola, T.E. Leitão et al. 2016. Spatial variability of trace elements in allotment gardens of four European cities: assessments at city, garden, and plot scale. J. Soils Sediments 1–16.

Biasioli, M., R. Barberis and F. Ajmone-Marsan. 2006. The influence of a large city on some soil properties and metals content. Sci. Total Environ. 356: 154–164.

Blanc, C., avec la participation de Lefevre, F. (MEDDTL), G. Boissard, M. Scamps and B. Hazebrouck. 2012. Guide de réutilisation hors site des terres excavées en technique routière et dans des projets d'aménagement (No. BRGM/RP-60013-FR).

Blanc, N. 2009. Vers un urbanisme écologique? URBIA Cah. Dév. Urbain Durable 8: 39–59.

Blanchart, A., G. Sere, J. Cherel, G. Warot, M. Stas, J.N. Consales et al. 2017. Contribution des sols à la production de services écosystémiques en milieu urbain—une revue. Environ. UrbainUrban Environ.

Bodemdecreet. 2006. Decreet van 27 oktober 2006 betreffende de bodemsanering en de bodembescherming—Titel VIII.

Bouquet, D., A. Braud and T. Lebeau. 2017. Brassica juncea tested on urban soils moderately contaminated by lead: Origin of contamination and effect of chelates. Int. J. Phytoremediation 19: 425–430.

Bourdeau-Lepage, L. and R. Vidal. 2012. Nature urbaine en débat: à quelle demande sociale répond la nature en ville? Déméter 20131: 195–210.

Boutefeu, E., architecture, F.P.U. construction, les transports Centre d'études sur les réseaux, l'urbanisme et les constructions publiques (France), réseaux, les transports C. d'études sur les. 2005. La demande sociale de nature en ville: enquête auprès des habitants de l'agglomération lyonnaise. Plan urbanisme construction architecture; Centre d'études sur les réseaux, les transports, l'urbanisme et les constructions.

Breure, A.M. 2004. Soil biodiversity: Measurements, indicators, threats and soil functions. In International Conference Soil and Compost Eco-biology. Spain.

Brimo, K., P. Garnier, F. Lafolie, G. Séré and S. Ouvrard. 2018. *In situ* long-term modeling of phenanthrene dynamics in an aged contaminated soil using the VSOIL platform. Sci. Total Environ. 619-620: 239–248.

Brundrett, M.C. 2009. Mycorrhizal associations and other means of nutrition of vascular plants: understanding the global diversity of host plants by resolving conflicting information and developing reliable means of diagnosis. Plant Soil 320: 37–77.

Cai, Y. 2000. Problems of farmland conservation in the rapid growth of China's economy. Resour. Sci. 22: 24–28.

Cameron, R.W., J.E. Taylor and M.R. Emmett. 2014. What's "cool" in the world of green façades? How plant choice influences the cooling properties of green walls. Build. Environ. 73: 198–207.

Campbell, M. and I. Campbell. 2010. Allotment waiting lists in England 2010. Transition Town West Kirby in conjunction with the National Society of Allotment and Leisure Gardeners Ltd.

Carré, C. and J.-C. Deutsch. 2015. L'eau dans la ville: une amie qui nous fait la guerre. Editions de l'Aube.

Chaney, R.L., A.J.M. Baker and J.L. Morel. 2018. The long road to developing agromining/phytomining. pp. 1–17. *In*: Van der Ent, A., G. Echevarria, A.J.M. Baker and J.L. Morel (eds.). Agromining: Farming for Metals: Extracting Unconventional Resources Using Plants (Cham: Springer International Publishing).

Cheverry, C. and C. Gascuel. 2009. Sous les pavés la Terre: Connaître et gérer les sols urbains. Montreuil Omnisciences Coll. Écrin.

Clay, D.R. 1991. Role of the baseline risk assessment in Superfund remedy selection decisions. OSWER Dir. 9355, 30.

Coussy, S., C. Hulot and A. Billard. 2017. Guide de valorisation hors site des terres excavées issues de sites et sols potentiellement pollués dans des projets d'aménagement. MTES.

Cundy, A., P. Bardos, M. Puschenreiter, N. Witters, M. Mench, V. Bert et al. 2015. Developing effective decision support for the application of "gentle" remediation options: The GREENLAND project. Remediat. J. 25: 101–114.

Da Silva, J.F. and R.J.P. Williams. 2001. The biological chemistry of the elements: the inorganic chemistry of life. Oxford University Press.

Danh, L.T., P. Truong, R. Mammucari and N. Foster. 2014. A critical review of the arsenic uptake mechanisms and phytoremediation potential of pteris vittata. Int. J. Phytoremediation 16: 429–453.

Daniel, A.-C. 2017. Fonctionnement et durabilité des microfermes urbaines, une observation participative sur le cas des fermes franciliennes. Chaire Eco-conception, AgroParisTech, INRA, UMR SADAPT, France.

De Kimpe, C.R. and J.-L. Morel. 2000. Urban soil management: A growing concern. Soil Sci. 165: 31–40.

De-Chun, S., J. Wei-Ping, Z. Man and C. Xia. 2010. Can cadmium uptake by Chinese cabbage be reduced after growing Cd-accumulating rapeseed? Pedosphere 20: 90–95.

Denys, S., J. Caboche, K. Tack and P. Delalain. 2007. Bioaccessibility of lead in high carbonate soils. J. Environ. Sci. Health Part A 42: 1331–1339.

Doick, K.J., A. Peace and T.R. Hutchings. 2014. The role of one large greenspace in mitigating London's nocturnal urban heat island. Sci. Total Environ. 493: 662–671.

Dubbeling, M., H. de Zeeuw and R. van Veenhuizen. 2010. Cities, poverty and food: multi-stakeholder policy and planning in urban agriculture. Practical Action, 192.

Dudka, S., M. Piotrowska and H. Terelak. 1996. Transfer of cadmium, lead, and zinc from industrially contaminated soil to crop plants: a field study. Environ. Pollut. 94: 181–188.

Dumat, C., A. Pierart, M. Shahid and S. Khalid. 2017. Pollutants in urban agriculture: sources, health risk assessment and sustainable management. *In*: Bioremediation of Agricultural Soils. CRC Press/Taylor & Francis Group.

Dyckhoff, H. and K. Allen. 2001. Measuring ecological efficiency with data envelopment analysis (DEA). European Journal of Operational Research 132(2): 312–325.

El Khalil, H., C. Schwartz, O. Elhamiani, J. Kubiniok, J.L. Morel and A. Boularbah. 2008. Contribution of technic materials to the mobile fraction of metals in urban soils in Marrakech (Morocco). J. Soils Sediments 8: 17–22.

European Commission. 2015. Towards an EU research and innovation policy agenda for nature-based solutions & re-naturing cities (No. Final report of the Horizon 2020 expert group on "Nature-based solutions and re-naturing cities."). Brussels.

European Waste Framework Directive. 2006. Directive 2006/12/EC of the European Parliament and of the Council of 5 April 2006 on waste (Text with EEA relevance).

Foissard, X. 2015. L'îlot de chaleur urbain et le changement climatique: application à l'agglomération rennaise. Rennes 2.

Foucault, Y. 2013. Réhabilitation écologique et gestion durable d'un site industriel urbain : cas d'une pollution historique en éléments inorganiques potentiellement toxiques (Pb, Cd, Zn, Cu, Sb et As). Université de Toulouse, Toulouse, France.

Foucault, Y., T. Lévêque, T. Xiong, E. Schreck, A. Austruy, M. Shahid et al. 2013. Green manure plants for remediation of soils polluted by metals and metalloids: Ecotoxicity and human bioavailability assessment. Chemosphere 93: 1430–1435.

Garbisu, C. and I. Alkorta. 2001. Phytoextraction: a cost-effective plant-based technology for the removal of metals from the environment. Bioresour. Technol. 77: 229–236.

Gerhardt, A. 2002. Bioindicator species and their use in biomonitoring. Environmental monitoring I. Encyclopedia of life support systems. UNESCO ed. Oxford (UK): Eolss Publisher.

Gerhardt, K.E., X.-D. Huang, B.R. Glick and B.M. Greenberg. 2009. Phytoremediation and rhizoremediation of organic soil contaminants: Potential and challenges. Plant Sci. 176: 20–30.

Goddard, M.A., A.J. Dougill and T.G. Benton. 2010. Scaling up from gardens: biodiversity conservation in urban environments. Trends Ecol. Evol. 25: 90–98.

Gove, B., J.J. Hutchinson, S.D. Young, J. Craigon and S.P. McGrath. 2002. Uptake of metals by plants sharing a rhizosphere with the hyperaccumulator Thlaspi caerulescens. Int. J. Phytoremediation 4: 267–281.

Grant, C., D. Flaten, M. Tenuta, S. Malhi and W. Akinremi. 2013. The effect of rate and Cd concentration of repeated phosphate fertilizer applications on seed Cd concentration varies with crop type and environment. Plant Soil 372: 221–233.

Haase, D., N. Larondelle, E. Andersson, M. Artmann, S. Borgström, J. Breuste et al. 2014. A Quantitative Review of Urban Ecosystem Service Assessments: Concepts, Models, and Implementation. AMBIO 43: 413–433.

Hardin, P.J. and R.R. Jensen. 2007. The effect of urban leaf area on summertime urban surface kinetic temperatures: a Terre Haute case study. Urban For. Urban Green. 6: 63–72.

Heink, U. and I. Kowarik. 2010. What are indicators? On the definition of indicators in ecology and environmental planning. Ecol. Indic. 10: 584–593.

Holt, E.A. and S.W. Miller. 2010. Bioindicators: using organisms to measure environmental impacts. Nature 3(10): 8–13.

Jacobs, A., T. Drouet, T. Sterckeman and N. Noret. 2017. Phytoremediation of urban soils contaminated with trace metals using Noccaea caerulescens: comparing non-metallicolous populations to the metallicolous "Ganges" in field trials. Environ. Sci. Pollut. Res. 24: 8176–8188.

Jeanniot, E., M. Carreau, C. Le Guern, V. Baudouin, P. Bâlon, C. Blanc et al. 2014. La gestion des terres excavées sur les zones d'aménagement de l'Ile de Nantes. Presented at the Journées Techniques Nationales "Reconversion des friches urbaines polluées", Paris, France.

Jézéquel, K. and T. Lebeau. 2008. Soil bioaugmentation by free and immobilized bacteria to reduce potentially phytoavailable cadmium. Bioresour. Technol. 99: 690–698.

Jézéquel, K., J. Perrin and T. Lebeau. 2005. Bioaugmentation with a *Bacillus* sp. to reduce the phytoavailable Cd of an agricultural soil: comparison of free and immobilized microbial inocula. Chemosphere 59: 1323–1331.

Jiang, C., Q.-T. Wu, T. Sterckeman, C. Schwartz, C. Sirguey, S. Ouvrard et al. 2010. Co-planting can phytoextract similar amounts of cadmium and zinc to mono-cropping from contaminated soils. Ecol. Eng. 36: 391–395.

Jim, C.Y. 1998. Impacts of intensive urbanization on trees in Hong Kong. Environ. Conserv. 25: 146–159.

Joimel, S., J. Cortet, C.C. Jolivet, N.P.A. Saby, E.D. Chenot, P. Branchu et al. 2016. Physico-chemical characteristics of topsoil for contrasted forest, agricultural, urban and industrial land uses in France. Sci. Total Environ. 545: 40–47.

Joner, E.J., R. Briones and C. Leyval. 2000. Metal-binding capacity of arbuscular mycorrhizal mycelium. Plant Soil 226: 227–234.

Kabata-Pendias, A. 2010. Trace Elements in Soils and Plants, Fourth Edition [WWW Document]. CRC Press. URL https://www.crcpress.com/Trace-Elements-in-Soils-and-Plants-Fourth-Edition/Kabata-Pendias/p/book/9781420093681 (accessed 9.4.17).

Kabisch, N., S. Qureshi and D. Haase. 2015. Human–environment interactions in urban green spaces—A systematic review of contemporary issues and prospects for future research. Environ. Impact Assess. Rev. 50: 25–34.

Kabisch, N., N. Frantzeskaki, S. Pauleit, S. Naumann, M. Davis, M. Artmann et al. 2016. Nature-based solutions to climate change mitigation and adaptation in urban areas: perspectives on indicators, knowledge gaps, barriers, and opportunities for action. Ecol. Soc. 21.

Kalkhajeh, Y.K., B. Huang, W. Hu, P.E. Holm and H.C. Bruun Hansen. 2017. Phosphorus saturation and mobilization in two typical Chinese greenhouse vegetable soils. Chemosphere 172: 316–324.

Karagiannidis, N. and N. Nikolaou. 2000. Influence of Arbuscular Mycorrhizae on Heavy Metal (Pb and Cd) Uptake, Growth, and Chemical Composition of Vitis vinifera L. (cv. Razaki). Am. J. Enol. Vitic. 51: 269–275.

Khan, AG. 2003. Vetiver grass as an ideal phytosymbiont for Glomalian fungi for ecological restoration of derelict land. pp. 466–74. *In*: Truong, P, X. Hanping (eds.). Proceedings of the third International Conference on Vetiver and Exhibition: Vetiver and Water, Guangzou, China, October 2003. Beijing: China Agricultural Press.

Khan, A.G. 2005. Role of soil microbes in the rhizospheres of plants growing on trace metal contaminated soils in phytoremediation. J. Trace Elem. Med. Biol. 18: 355–364.

Knecht, M.F. and A. Göransson. 2004. Terrestrial plants require nutrients in similar proportions. Tree Physiol. 24: 447–460.

Lai, H.-Y., Z.-Y. Hseu, T.-C. Chen, B.-C. Chen, H.-Y. Guo and Z.-S. Chen. 2010. Health risk-based assessment and management of heavy metals-contaminated soil sites in Taiwan. Int. J. Environ. Res. Public. Health 7: 3595–3614.

Lambert, M., B. Leven and R. Green. 2000. New methods of cleaning up heavy metal in soils and water. Environ. Sci. Technol. Briefs Citiz. Kans. State Univ. Manhattan KS.

Lebeau, T., A. Braud and K. Jézéquel. 2008. Performance of bioaugmentation-assisted phytoextraction applied to metal contaminated soils: A review. Environ. Pollut. 153: 497–522.

Lebeau, T. 2011. Bioaugmentation for *in situ* soil remediation: how to ensure the success of such a process. *In*: Singh, A, N. Parmar and R.C. Kuhad (eds.). Bioaugmentation, Biostimulation and Biocontrol for Soil Biology. Springer-Verlag, Berlin-Heidelberg, Germany 28: 129–186.

Li, J., Z. Zhang, L. Ma, Q. Gu, K. Wang and Z. Xu. 2016. Assessment on the Impact of Arable Land Protection Policies in a Rapidly Developing Region. ISPRS Int. J. Geo-Inf. 5: 69.

Li, Z., X. Feng, G. Li, X. Bi, J. Zhu, H. Qin et al. 2013. Distributions, sources and pollution status of 17 trace metal/metalloids in the street dust of a heavily industrialized city of central China. Environ. Pollut. Barking Essex 1987 182: 408–416.

Linger, P., J. Müssig, H. Fischer and J. Kobert. 2002. Industrial hemp (*Cannabis sativa* L.) growing on heavy metal contaminated soil: fibre quality and phytoremediation potential. Ind. Crops Prod. 16: 33–42.

Liu, J., M. Liu, D. Zhuang, Z. Zhang and X. Deng. 2003. Study on spatial pattern of land-use change in China during 1995–2000. Sci. China Ser. Earth Sci. 46: 373–384.

Liu, Y.-G., Y. Fei, G. Zeng, F. Ting, M. Lei and H. Yuan. 2007. Effects of added Cd on Cd uptake by oilseed rape and pai-tsai co-cropping. Trans. Nonferrous Met. Soc. China 17: 846–852.

Lovely, D. and J. Lloyd. 2000. Microbes with a metal for bioremediation. Nat. Biotechnol. 18: 600–601.

Maes, J. and S. Jacobs. 2017. Nature-based solutions for Europe's sustainable development. Conservation Letters 10(1): 121–124.

McGrath, S.P. and F.-J. Zhao. 2003. Phytoextraction of metals and metalloids from contaminated soils. Curr. Opin. Biotechnol. 14: 277–282.

MEEM. 2017. Méthodologie nationale de gestion des sites set sols pollués - French national methodology to manage polluted soils and sites. Ministère de l'Environnement, de l'Energie et de la Mer.

Mench, M. and E. Martin. 1991. Mobilization of cadmium and other metals from two soils by root exudates of Zea mays L., Nicotiana tabacum L. and Nicotiana rustica L. Plant Soil 132: 187–196.

Mench, M., J. Vangronsveld, H. Clijsters, N.W. Lepp and R. Edwards. 1999. *In situ* metal immobilisation and phytostabilisation of contaminated soils. pp. 323–358. *In*: Terry, N. and G. Banuelos (eds.). Phytoremediation of Contaminated Soil and Water. Lewis, Boca Raton, USA.

Mench, M., A. Manceau, J. Vangronsveld, H. Clijsters and B. Mocquot. 2000. Capacity of soil amendments in lowering the phytoavailability of sludge-borne zinc. Agronomie 20: 383–397.

Meuser, H. 2010. Contaminated Urban Soils. Springer Science & Business Media, 340pp.

Mombo, S., Y. Foucault, F. Deola, I. Gaillard, S. Goix, M. Shahid et al. 2016. Management of human health risk in the context of kitchen gardens polluted by lead and cadmium near a lead recycling company. J. Soils Sediments 16: 1214–1224.

Morel, J., C. Schwartz, L. Florentin and C. De Kimpe. 2005. Urban soils. Encycl. Soils Environ. 4: 202–208.

Morel, K. and F. Léger. 2016. A conceptual framework for alternative farmers' strategic choices: the case of French organic market gardening microfarms. Agroecol. Sustain. Food Syst. 40: 466–492.

Nehls, T., S. Rokia, B. Mekiffer, C. Schwartz and G. Wessolek. 2013. Contribution of bricks to urban soil properties. J. Soils Sediments 13: 575–584.

Parraga-Aguado, I., M.N. Gonzalez-Alcaraz, J. Alvarez-Rogel, F.J. Jimenez-Carceles and H.M. Conesa. 2013. The importance of edaphic niches and pioneer plant species succession for the phytomanagement of mine tailings. Environ. Pollut. 176: 134–143.

Pascaud, G., T. Leveque, M. Soubrand, S. Boussen, E. Joussein and C. Dumat. 2014a. Environmental and health risk assessment of Pb, Zn, As and Sb in soccer field soils and sediments from mine tailings: solid speciation and bioaccessibility. Environ. Sci. Pollut. Res. 21: 4254–4264.

Pascaud, G., T. Leveque, M. Soubrand, S. Boussen, E. Joussein and C. Dumat. 2014b. Environmental and health risk assessment of Pb, Zn, As and Sb in soccer field soils and sediments from mine tailings: solid speciation and bioaccessibility. Environ. Sci. Pollut. Res. 21: 4254–4264.

Pérez de Mora, A., P. Madejón, P. Burgos, F. Cabrera, N.W. Lepp and E. Madejón. 2011. Phytostabilization of semiarid soils residually contaminated with trace elements using by-products: Sustainability and risks. Environ. Pollut., Nitrogen Deposition, Critical Loads and Biodiversity 159: 3018–3027.

Pierart, A. 2016. Role of arbuscular mycorrhizal fungi and bioamendments in the transfer and human bioaccessibility of Cd, Pb, and Sb contaminant in vegetables cultivated in urban areas. Université Paul Sabatier, Toulouse.

Pierart, A., C. Dumat, A.Q. Maes, C. Roux and N. Sejalon-Delmas. 2018. Opportunities and risks of biofertilization for leek production in urban areas: Influence on both fungal diversity and human bioaccessibility of inorganic pollutants. Environ. Res. under review.

Pilon-Smits, E. 2005. Phytoremediation. Annual Review of Plant Biology 56: 15–39.

Pourias, J., C. Aubry and E. Duchemin. 2016. Is food a motivation for urban gardeners? Multifunctionality and the relative importance of the food function in urban collective gardens of Paris and Montreal. Agric. Hum. Values 33: 257–273.

Rangel, W. de M., Schneider, J., Costa, E.T. de S., Soares, C.R.F.S., Guilherme, L.R.G., Moreira, F.M. de S., 2014. Phytoprotective effect of arbuscular mycorrhizal fungi species against arsenic toxicity in tropical Leguminous species. Int. J. Phytoremediation 16: 840–858.

Richard, J. 2017. Valtex: how to offer and industrial platform of excavated soil management in the context of waste legislation regarding circular economy purposes? Presented at the Suez Environment, Aquaconsoil, Lyon, France.

Ross, S.M. 1994. Toxic metals in soil-plant systems. John Wiley and Sons Ltd.

Scharenbroch, B.C., J.E. Lloyd and J.L. Johnson-Maynard. 2005. Distinguishing urban soils with physical, chemical, and biological properties. Pedobiologia 49: 283–296.

Selosse, M.-A. and F.L. Tacon. 1998. The land flora: a phototroph-fungus partnership? Trends Ecol. Evol. 13: 15–20.

Setälä, H., V. Viippola, A.L. Rantalainen, A. Pennanen and V. Yli-Pelkonen. 2013. Does urban vegetation mitigate air pollution in northern conditions? Environmental Pollution 183: 104–112.

Shahid, M., E. Pinelli and C. Dumat. 2012. Review of Pb availability and toxicity to plants in relation with metal speciation; role of synthetic and natural organic ligands. J. Hazard. Mater 219-220: 1–12.

Smith, S.E. and F.A. Smith. 2011. Roles of arbuscular mycorrhizas in plant nutrition and growth: new paradigms from cellular to ecosystem scales. Annu. Rev. Plant Biol. 62: 227–250.

Su, D., X. Lu and J. Wong. 2008. Could co-cropping or successive cropping with Cd accumulator oilseed rape reduce Cd uptake of sensitive Chinese Cabbage? Pract. Period. Hazard. Toxic Radioact. Waste Manag. 12: 224–228.

Tan, M., X. Li, H. Xie and C. Lu. 2005. Urban land expansion and arable land loss in China—a case study of Beijing–Tianjin–Hebei region. Land Use Policy 22: 187–196.

Tang, Y.-T., T.-H.-B. Deng, Q.-H. Wu, S.-Z. Wang, R.-L. Qiu, Z.-B. Wei et al. 2012. Designing Cropping Systems for Metal-Contaminated Sites: A Review. Pedosphere 22: 470–488.

Tedoldi, D., G. Chebbo, D. Pierlot, P. Branchu, Y. Kovacs and M.-C. Gromaire. 2017. Spatial distribution of heavy metals in the surface soil of source-control stormwater infiltration devices – inter-site comparison. Sci. Total Environ. 579: 881–892.

Tisdale, S.L., W.L. Nelson, J.D. Beaton and J.L. Havlin. 1985. Soil and fertilizer potassium. Soil Fertility and Fertilizers 4: 249–291.

Todd, L.F., K. Landman and S. Kelly. 2016. Phytoremediation: An interim landscape architecture strategy to improve accessibility of contaminated vacant lands in Canadian municipalities. Urban For. Urban Green. 18: 242–256.

Tonin, C., P. Vandenkoornhuyse, E.J. Joner, J. Straczek and C. Leyval. 2001. Assessment of arbuscular mycorrhizal fungi diversity in the rhizosphere of Viola calaminaria and effect of these fungi on heavy metal uptake by clover. Mycorrhiza 10: 161–168.

Uzu, G., J.-J. Sauvain, A. Baeza-Squiban, M. Riediker, M. Sánchez Sandoval Hohl, S. Val et al. 2011. *In vitro* assessment of the pulmonary toxicity and gastric availability of lead-rich particles from a lead recycling plant. Environ. Sci. Technol. 45: 7888–7895.

Vlarebo. 2008. Besluit van de Vlaamse Regering houdende vaststelling van het Vlaams reglement betreffende de bodemsanering en de bodembescherming—Title XIII.14.

Volesky, B. and Z.R. Holan. 1995. Biosorption of heavy metals. Biotechnol. Prog. 11: 235–250.

Whitford, V., A.R. Ennos and J.F. Handley. 2001. City form and natural process—indicators for the ecological performance of urban areas and their application to Merseyside, UK. Landscape and Urban Planning 57(2): 91–103.

Wilschut, M., P. Theuws and I. Duchhart. 2013. Phytoremediative urban design: Transforming a derelict and polluted harbour area into a green and productive neighbourhood. Environ. Pollut. 183: 81–88.

Wrap. 2010. Guidelines for measuring and reporting construction, demolition and excavation waste.

Wu, H., L. Li and F. Zhang. 2003. The influence of interspecific interactions on Cd uptake by rice and wheat intercropping. Rev. China Agric. Sci. Technol. 5: 43–46.

Wu, J., C. Dumat, H. Lu, Y. Li, H. Li, Y. Xiao et al. 2016. Synergistic improvement of crop physiological status by combination of cadmium immobilization and micronutrient fertilization. Environ. Sci. Pollut. Res. 23: 6661–6670.

Wu, Q., Z. Wei and Y. Ouyang. 2007. Phytoextraction of metal-contaminated soil by Sedum alfredii H: effects of chelator and co-planting. Water Air. Soil Pollut. 180: 131–139.

Xiaomei, L., W. Qitang and M.K. Banks. 2005. Effect of simultaneous establishment of Sedum Alfredii and Zea Mays on heavy metal accumulation in plants. Int. J. Phytoremediation 7: 43–53.

Xin-Bo, L., X. Jian-Zhi, L. Bo-Wen and W. Wei. 2009. Ecological responses of Brassica juncea-alfalfa intercropping to cadmium stress. Yingyong Shengtai Xuebao 20.

Xue, J. and Y. Fei. 2006. Effects of intercropping Cunninghamia lanceolata in tea garden on contents and distribution of heavy metals in soil and tea leaves. J. Ecol. Rural Environ. 22: 71–73.

Yu, L., J. Zhu, Q. Huang, D. Su, R. Jiang and H. Li. 2014. Application of a rotation system to oilseed rape and rice fields in Cd-contaminated agricultural land to ensure food safety. Ecotoxicol. Environ. Saf. 108: 287–293.

Zhao, F.-J., Y. Ma, Y.-G. Zhu, Z. Tang and S.P. McGrath. 2015. Soil contamination in china: current status and mitigation strategies. Environ. Sci. Technol. 49: 750–759.

Chapter 6

Phytoremediation of Agricultural Soils Polluted with Metals

A. Navazas,[1] *L.E. Hernández,*[2] *F. Martínez,*[2]
C. Ortega-Villasante,[2] *A. Bertrand*[1] and
A. González[1,*]

INTRODUCTION

Traditional agricultural practices are based mainly on maintaining sufficient soil fertility to obtain appropriate crop yield (Wei and Zhou 2008, Memon et al. 2008). However, soils are not only a source of nutrients for plants, but also a potential source of metals, which decrease crop productivity and may pose a risk to human health (Kabata-Pendias 2010). This is not only due to inherent toxicity of these substances but to their bioaccumulation, biomagnification and long residence times in organisms (Kabata-Pendias 2010). Over the last decades, recurrent cases of soil pollution by metals are appearing in diverse geographical locations throughout the world, mostly associated with inefficient management of industrial wastewater effluents, poor containment of mining leachates and residues, and as a consequence of natural geological sources (i.e., weathering of bead rock) (Nicholson et al. 2003, Ruttens et

[1] Departamento de Biología de Organismos y Sistemas, Universidad de Oviedo, Oviedo, Spain.
 Email: Alejandro.navazas@gmail.com
[2] Department of Biology, Universidad Autónoma de Madrid, Madrid, Spain.
 Email: luise.hernandez@uam.es
* Corresponding author: aidag@uniovi.es

al. 2006, Gallego et al. 2016, Sahito et al. 2016, Azzi et al. 2017). Finally, the use of groundwater for irrigation purposes may also deliver metals such as arsenic, which eventually accumulate in watered crops, even if cultivation takes place far from the point source of pollution (Neilson and Rajakaruna 2012).

Among the eco-friendly cleanup technologies, *in situ* contaminant stabilization ("inactivation") and plant-based approaches ("phytoremediation") are proposed to cope with the above mentioned contamination challenge (Kidd et al. 2015). These green technologies, already described more than two decades ago by Raskin et al. (1994), exploit the ability of certain plants species to accumulate metals in their tissues, thus reducing their concentrations in soil (Pilon-Smits 2005).

The magnitude of metal dispersal into the environment increases with increasing contaminated soil area, where cleanup strategies become more complex. For example, brownfields affect soil and cause serious environmental and health risks, as well as economic and social costs (Morio et al. 2013). The large extension of these areas makes them an illustrative example of study, since soils and subsoils may contain complex mixtures of metals, which appear frequently as multi-contaminant situations. In addition, specific weathering conditions and different edaphic factors modify chemical speciation and toxicity of these contaminants (Adriano 2001, Wenzel et al. 2003). In this chapter, we will describe different approaches focused on improving the efficiency of phytoremediation in brownfields and agricultural soils. Considering those varied characteristics, we will compare recent studies from the literature, including ours, to formulate effective strategies in the phytoremediation of agricultural soils. These methodological approaches vary in function of the soil contaminant concentration, and include the use of metal hyperaccumulators and/ or excluder plants, while considering certain agronomic practices (plant density, pruning, pH correction, fertilization and addition of amendments) (Nicholson et al. 2003, Sirbu et al. 2009, Souza et al. 2014, Chen et al. 2015, Greger and Landberg 2015, Liu et al. 2017). As in Chapter 11, herein we opted to use the term metal (or "metalloid" in the case of As, At, B, Ge, Po, Sb, Se, Si, and Te), as suggested by Chapman and Holzmann (2007), because of inaccurate definition of the term "heavy metal" (Pourret and Bollinger 2018).

Plant Selection for a Phytoremediation Process

One of the most complex problems in the optimization procedure for a phytoremediation program is the selection of suitable plant species. It is well-known that toxic metal(loid)s induce loss of plant biomass, among other deleterious effects, associated mainly with growth inhibition (Gill et al. 2015). Nevertheless, some plant species and soil biota populations, usually endemic of polluted soils, are able to colonize and thrive in such highly polluted environments, even when high concentrations of metals are found in their cells and tissues. Although plants growing in these areas are tolerant to the predominant natural occurring metals, in most of these cases, they show a slow growth rate and have a limited biomass (Maywald and Weigel 1997). This is a fundamental drawback when hyperaccumulator plants are used in phytoremediation, because of their limited growth rate and shallow-rooting traits, which hamper effective phytoremediation strategies of metal-polluted soils (Fernández-Fernández et al. 2010).

Since the beginning of 20th century, more than 500 hyperaccumulator plant species have been identified. Some examples are the As-hyperaccumulator *Pteris vittata* (Ma et al. 2001), the Zn- and Cd-hyperaccumulators *Sedum alfredii* (Long et al. 2002) and *Noccaea* species (Tlusto et al. 2016), the Mn-hyperaccumulator Australian *Gossia* species (Fernando et al. 2009), and the Cu-hyperaccumulator *Crassula helmsii* (Kupper et al. 2009). Due to the ability of these plants to translocate large amounts of toxic metal(loid)s to shoots, the time required to phytoremediate is substantially shortened by the use of these hyperaccumulator species (Niazi et al. 2012, Pandey 2012, Tripathi et al. 2012). However, this approach is a matter of current debate, since most of metal-hyperaccumulator species produce very low biomass, which may pose serious limitation to this cleaning procedure. Furthermore, many studies dealing with hyperaccumulators have been performed under lab conditions, with controlled supply of both water and nutrients, with no routine agronomic practices and management implementation (Robinson et al. 2006, McGrath et al. 2006, Fernández et al. 2008, 2012). To overcome this limitation, a great effort has been performed to obtain new higher yield metal-hyperaccumulating cultivars from indigenous species, so high biomass yield may be achieved. This is the case of *Siegesbeckia orientalis*, a high-biomass species that shows notable tolerance and accumulation of Cd in field studies. These plants also display Cd bioconcentration and translocation factors higher than one, suggesting that this species has a significant Cd-hyperaccumulating capability. Furthermore, this species is easily cultivable, and has strong competitive ability and wide geographical distribution; characteristics that make *S. orientalis* a good candidate for phytoremediation of Cd-contaminated soils (Zhang et al. 2013).

Although most authors agree that native plant species are the best option for an optimized phytoremediation procedure, those species adapted to pedo-climatic and environmental conditions similar to those reigning in the polluted site may also constitute a suitable and affordable possibility (Kazakou et al. 2008). Nonetheless, the introduction of non-native hyperaccumulating species may incur in side-effect environmental problems and increases the costs of the phytoremediation process. Such strategy may need additional environmental monitoring to prevent unforeseen spreading of contaminants or competition with natural flora and microbiota, and may require prevention of unintentional cross-fertilization and hybridization with local plants, which may reduce phytoremediation effectiveness (Che-Castaldo and Inouye 2015).

Since most of hyperaccumulating plant species are specific for one metal, co-planting of different hyperaccumulating species can also be an interesting option. Though there might be a competition for nutrients, there is also evidence that only weak competitive effects between the different metal-tolerant species may occur (Che-Castaldo and Inouye 2015).

On the other hand, some studies have dealt with high and moderate metal accumulating species that produce high biomass and, in turn, display a better adaptation to field conditions than hyperaccumulator species. In this sense, Cd concentration in rice grains is a serious health problem in different Asian polluted sites, and recent large-scale experiments were carried out in paddy fields containing low to moderate levels of Cd, where two rice crops were compared: one was cultivated for phytoremediation (Chokoukoku) and the other for human consumption (Yumesayaka) (Murakami et al. 2009). The authors of this field study observed that Chokoukoku cultivar was able to accumulate high levels of Cd, reducing 38% of initial Cd concentration in soil after two years, which resulted in a relatively low accumulation of this metal in grains of the edible cultivar (Yumesayaka) in the following crop seasons with no significant decrease in crop yield. Furthermore, this decrease of Cd concentrations in soil was higher than the reduction obtained after three-year cropping with the hyperaccumulator *Noccaea caerulescens* (by 15% of total Cd concentration). The results of this study showed that some highly Cd tolerant plants can display greater Cd uptake rates than those of hyperaccumulator species. In addition, recent studies explored the feasibility of other high biomass plants to extract metals from polluted soils such as willow (*Salix viminalis*). In this case, the high biomass compensated for the moderate metal concentrations found in the aboveground tissues (Hammer et al. 2003).

Another alternative to hyperaccumulating species is the use of excluder or stabilizing species that accumulate low levels of metals, even under high metal concentrations in soil. In this case, it is important to ensure that edible organs of plants are free of metals and may not represent risks for putative consumers. If this is guaranteed, cultivation of these plants could help to meet the growing demands for food, forage, and industrial crops in the nowadays situation of continuously receding arable land areas of fertile soils around the world (Gramss and Voigt 2015). Some of the excluder species that may be used for such purpose include *Hyparrhenia hirta* for Cu accumulation (Poschenrieder et al. 2008), *Silene vulgaris* for Ni (Wenzel et al. 2003); *Armeria maritima* for Co (Brewin et al. 2003); *Oenothera biennis* for Cd (Wei et al. 2005) or *Acacia pycnantha* for Cu, Zn, Cd, and Pb (Niola et al. 2016). Some examples of excluder plants are different Japonica rice cultivars able to limit Cd translocation to grains. Out of 39 cultivars assayed in Shenyang region of China, only two (named Fuhe 90, and Yanfeng 47) showed Cd-excluding behavior and safe crop use (Zhan et al. 2013).

The ideal strategy for selecting potential candidates for phytoremediation is the characterization of native species in terms of their high biomass yield. Besides their capability for metal accumulation, selection of hyperaccumulator species should consider other features such as biomass production, percentage of cover/ aggregation, and distribution in polluted areas. Therefore, it is important to identify the characteristics that make these plants more prone to hyperaccumulate metals, such as to be a nitrophilous, ruderal and resistant species to other abiotic or biotic stressors (Fernández et al. 2010). In the studies performed in our laboratory, some of the selected plants to phytoremediate polluted soils in Asturias (North Spain) were: *Dittrichia viscosa* and *Betula pubescens* for Cd accumulation, *Melilotus alba* for Pb accumulation, and *Eupatorium cannabinum* and *Salix atrocinerea* for As accumulation. Likewise, the most promising metal-accumulator species propagated from seeds was selected and cloned *in vitro* to be used later in phytoremediation actions in order to reduce genetic variability (Fernández et al. 2008, 2012).

A complementary approach to search plant species with high capability for metal extraction from soils is chemical mutagenesis programs to generate genetic variations within a particular genotype. Such approach was followed in *Helianthus annuus* treated with ethyl methanesulfonate, where screening of the second progeny generation showed some individuals with shoot metal concentration 2–3 times higher than parental wild-type congeners (Nehnevajova et al. 2007).

Metal uptake from soils is a complex physicochemical process influenced by a variety of edaphic factors (Wenzei et al. 2003). Among them, microorganisms associated to rhizosphere are of particular concern. In Cd-contaminated soils, it has been reported that the rhizomicrobial activity associated to the hyperaccumulator *S. alfredii* ecotype resulted in slight soil acidification, higher metal bioavailability and a more developed root system than for the non-hyperaccumulator ecotype (Hou et al. 2017). Therefore, the bioavailability and tolerance to Cd (and other metals) of *S. alfredii* was significantly correlated with the bacterial community found in their rhizosphere. This rhizosphere granted the hyperaccumulating trait to one plant over the other. In other related study, the mobilization of As from the roots to the fronds of ferns growing in As-contaminated soils was found to be facilitated by the *ars* genes expressed in the endophytic bacteria living in the roots and leaves (Chang et al. 2010). Taken together these studies suggests that isolation of indigenous bacteria from rhizosphere may provide a better understanding of metal mobility in the phytoremediation of contaminated sites. However, future studies are still required to explore the genetic basis for metal tolerance and accumulation modulated by microflora interaction in many of the phytoremediation examples reported in the literature.

Agronomic Techniques to Increase Phytoremediation Efficiency

Most experiments setting the basis for phytoremediation are normally carried out in lab or greenhouse conditions, with plants grown in hydroponic cultures or in pots filled with convenient soils or substrates, in which the medium is spiked with metals and chelating agents to mimic field conditions. However, extrapolation from greenhouse to field demands requires knowledge of agronomical practices that may affect the remediation process, similar to those needed to increase crop yield. After proper plant species selection, agronomic practices can be implemented to improve plant biomass, nutrient absorption, and facilitate metal availability, which consequently, would increase their accumulation in plant tissues. Some works highlighted the importance of convenient crop managing to maximize the efficiency of metal removal from soils (Ji et al. 2011). These agronomic practices are based on principles of simplicity, cost effectiveness and an environmentally friendly approach for application in *in situ* phytoremediation.

A summary of the different agronomic techniques and practices aimed to increase metal accumulation by plants, leading simultaneously to an enhanced biomass, root growth and rhizosphere stabilization, are discussed hereafter. Among these practices, we will put particular attention to pruning, plant density, correction of soil pH and fertilization.

Pruning

Phytoremediation is, at first instance, a low-cost technique because plant cultivation in the field is an inexpensive method to absorb metals from soils. However, there is certain management associated to the maintenance of crop that must also be considered to enhance the efficiency of phytoremediation. In this regard, pruning is used routinely to increase the biomass yield and, consequently, may help to improve the phytoextraction of metals. Some studies proved that regular pruning of *Atriplex nummularia* was efficient for cropping of this species in salt-affected soils, mainly because of its stimulated regrowth of less lignified material, promoting Na and Cl phytoextraction in salty soils (Souza et al. 2014). Pruning of *Vetiveria zizanioides* roots before transplanting in Cd polluted soil improved Cd uptake and accelerated plant growth (Chen et al. 2015). In addition, soil amended with pruning residues may constitute a major source of carbon, among other benefits, to maintain sufficient rhizospheric microbial activity. For example, Santo et al. (2011) showed that pruning debris increased the phytoremediation capability of *A. nummularia* Lindl in several polluted soils.

In our experience, we found that the benefits of pruning in the field are strongly influenced by both soil pH and the plant species. For example, pruning enhanced *Dittrichia viscosa* and *Eupatorium cannabium* biomass at pH values between 2.9 and 6.7, but this acid environment did not affect the *Melilotus alba* biomass (Fig. 1). Nonetheless, As, Cd, Pb or Zn tissue accumulation decreased or remained the same after pruning (Table 1). These results resemble those obtained in the phytoremediation of polychlorinated biphenyls using *Cucurbita pepo*, where concentration of these contaminants did not increase by pruning in the primary

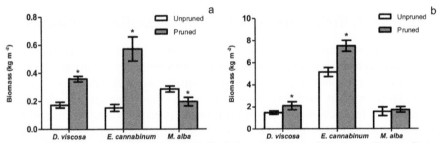

Fig. 1. Effect of pruning on the biomass (kg/m²) of aboveground part of *D. viscosa*, *E. cannabinum* and *M. alba* after 7 months growing in a polluted soil at different pH values: (a) pH = 2.9 – 5 and (b) pH = 5.8 – 6.7. Asterisks denote significant differences between groups (mean ± SE, n = 6, $p < 0.05$).

Table 1. Effect of pruning on the metal(loid)s accumulation (mg/kg) in leaves of *D. viscosa*, *E. cannabinum* and *M. alba* after 7 months growing in a polluted soil at pH = 6.5. Different letters indicate significant differences between groups (mean ± SE, n = 6, $p < 0.05$).

	D. viscosa		*E. cannabinum*		*M. alba*	
Metal	**Unpruned**	**Pruned**	**Unpruned**	**Pruned**	**Unpruned**	**Pruned**
Fe	530 ± 40 a	470 ± 34 a	210 ± 15 b	214 ± 18 b	80 ± 6 c	89 ± 8 c
Zn	420 ± 14 a	413 ± 19 a	174 ± 12 c	150 ± 10 d	210 ± 17 b	150 ± 14 cd
Cu	23 ± 3 a	17 ± 2 b	10 ± 2 c	6 ± 2 d	12 ± 2 c	12 ± 1 c
As	5 ± 0.2 b	4 ± 0.8 bc	7 ± 0.5 a	3 ± 0.2 c	2 ± 0.3 e	2 ± 0.1 e
Cd	14 ± 1 b	13 ± 2 b	4 ± 0.2 c	3 ± 0.1 d	18 ± 3 a	4 ± 0.2 c
Pb	13 ± 2 b	12 ± 2 b	4 ± 0.1 c	3 ± 0.2 d	17 ± 2 a	5 ± 1 c

stem (Low et al. 2011). Therefore, it was generally observed that, while biomass is increased by pruning, metal accumulation decreased. In our studies, pruning is not a necessary technique since it requires an additional cost with no significant improvement of metal phytoextraction.

Pruning residue may be used as a soil amendment, therefore acting as an added-value agronomic product. For example, Fellet et al. (2014) transformed pruning residues in biochar (carbonaceous material elaborated from biomass by pyrolysis), which reduced the accumulation of Cd and Pb by different plant species cultivated in experimental pots. A field study using maize plants demonstrated that the bioavailable fraction of Cu in soil as well as Cu uptake was significantly lower in biochar-amended soils compared to biochar-free soils. By contrary, an opposite finding was achieved in soils contaminated with As (Brennan et al. 2014). Amendment of an As-contaminated soil with biochar generated from orchard pruning residues resulted in increased solubility and mobility of this metalloid, although plant uptake and toxicity-transfer risk was reduced. Therefore, addition of biochar to As-contaminated soils seemed to improve leaching of As, and consequently may attenuate metalloid transfer through food web (Beesley et al. 2013). Thus, biochar may contribute to produce safer crops because it is possible to prevent metal incorporation into terrestrial food webs, although with a higher environmental cost associated mainly to biochar production and application. Therefore, further work is required to ameliorate undesirable side effects coming from biochar in the bioremediation of metal-contaminated soils. Chapter 10 in this book provides a more detailed description on the potential of biochar in the remediation of contaminated soils, while protects soil metabolic activity.

Plant Density

Pot experiments are of great value to set the basic knowledge for phytoremediation. However, field trials are necessary to obtain practical information on a variety of interrelated processes including soil physiochemical properties and nutrient absorption, root and shoot enlargement, water availability and fluctuating light intensity (shadding effect), maintenance and replacement of soil mineral elements, the survival to stress and seed production, among others. In the field, planting density can be modified by varying the distance between the plants, so an immediate question arises: What is the appropriate number of plants for an optimal rate of soil decontamination?

The minimum population density depends on different factors such as plant species and edaphic characteristics of the polluted soil. At a first glance, one may hypothesize that higher plant density results in shorter time required to clean the polluted soil, due to a higher biomass able to absorb metals. Therefore, higher density of individuals would facilitate faster and effective phytoremediation. In this context, it may also be thought that higher plant density would reduce the toxic effect of metal on plant physiology, since the toxic dose received (i.e., plant uptake) by each plant would be inversely related to plant density. This assumption was demonstrated by Hansi et al. (2014) who found that at higher plant density, the dose of metal per individual became diluted. Likewise, it was also observed that high density of

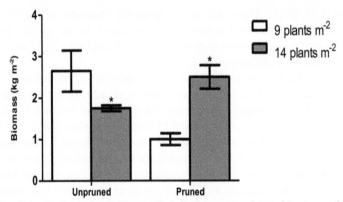

Fig. 2. Effect of planting density on the biomass (kg/m²) of aboveground part of *D. viscosa* after 7 months growing in a polluted soil at pH = 6.5. Asterisks denote significant differences between groups (mean ± SE, n = 6, $p < 0.05$).

giant reed (*Arundo donax*) had a beneficial impact on the accumulation of As, Cd, and Pb, demonstrating to be a workable approach to promote phytoremediation of polluted sites (Liu et al. 2017). In this line, other studies also showed *Salix viminalis* cultivated at high density removed Cd from agricultural soils up to 27% of original concentrations, leading to a Cd reduction in wheat grains up to 33% (Greger and Landberg 2015).

However, cropping at high plant density may lead to a reduction in biomass production because of competition for limited resources. The study by Liu et al. (2009) is an illustrative example of this drawback. These researchers observed that, despite higher phytoremediation efficiency, cropping of *Sedum plumbizincicola* at high density prevented further increases in biomass and metal extraction ability. Similarly, an increase in the individual density of *Potamogeton crispus* L. to remediate sediments contaminated by polycyclic aromatic hydrocarbons (PAHs) did not reflect a greater contaminant uptake (Meng and Chi 2015). However, *P. crispus* could eventually improve contaminant removal by increasing biological activity at the rhizosphere which helped to enhance the dissipation of PAHs in sediments. Likewise, bioremediation of PAH-contaminated soils using *Vallisneria spiralis* showed that there were no significant differences in the concentration of PAHs removed from soil under different planting density treatments. Accordingly, lower plant density should be a better selection for phytoremediation of PAHs since it will be associated to a lower cost (Liu et al. 2014).

From our field experience, we observed that unpruned *D. viscosa* planted at the lowest density (9 plants/m²) showed higher dry biomass than that obtained at the highest planting density (14 plants/m²) (Fig. 2). However, if this plant species is pruned, the total biomass of the plot with the highest planting density was increased, although the toxic metal(loid)s accumulation was lower at the highest planting density (Table 2). Taken together these results indicate that higher planting density is not a useful strategy to achieve improved phytoremediation efficiency with *D. viscosa*. These findings agree with those from other field trials (Whitfield et al. 2008), where increased planting density of *Cucurbita pepo* ssp. *pepo* led to a significant decrease in both plant biomass and phytoextraction capability of

Table 2. Effect of planting density on the metal(loid)s accumulation (mg/kg) in leaves of *D. viscosa* after 7 months growing in a polluted soil at pH = 6.5. Asterisks denote significant differences between groups (mean ± SE, n = 6, $p < 0.05$).

Metal(loid)s	Individuals/m^2	
	9	14
Fe	950 ± 70	765 ± 34*
Zn	385 ± 7	380 ± 19
Cu	28 ± 4	21 ± 2*
As	9 ± 0.7	10 ± 0.3
Cd	10 ± 0.1	5 ± 2*
Pb	26 ± 2	12 ± 1*

persistent organic pollutants. Similar results were also observed with different crops of *Miscanthus x giganteus* grown in soils highly contaminated with Cd, Pb, and Zn, where planting at high densities was counterproductive to optimize plant yield and phytoextraction effectiveness (Nsanganwimana et al. 2015).

In consequence, both the planting density and the total biomass per unit area should be considered to improve phytoremediation since this agronomic practice is clearly dependent on edapho-climatic factors (White 2009). This is the case of a field trial using the Cd hyper-accumulator plant *Solanum nigrum* L. cultivated in a Cd-contaminated soil, where the dry biomass per single plant at the lowest planting density was higher than that at the highest density, as a probable plant competition for nutrients and water. Therefore, high plant density could have a negative effect on plant biomass and thus on overall Cd accumulation. Nonetheless, Ji et al. (2011) found a positive correlation between the total biomass per plot and planting densities due to a larger number of individuals per plot at high density. For this reason, an equilibrium must be found to reach the maximum extraction capacity while competition between individuals does not prevent proper plant growth and phytoremediation effectiveness.

Soil pH

Soil pH determines the bioavailability fraction of metals in soil, so it is a significant variable in plant growth and phytoremediation efficiency (Marschner 1995). Changes in soil pH as well as in root exudates, and rhizospheric microorganisms may cause precipitation of metals onto root surfaces (Prasad and Freitas 2003). Likewise, under acic conditions, H$^+$ ions displace metal cations from the cation exchange complex (CEC) of colloidal components of soil, favoring metal uptaking by plants (McBride 1994). Furthermore, the capability of soil organic matter to retain and sequester metals (e.g., Cd, Cu, Hg, Ni, Pb, and Zn) is weaker at low pH (< 5.5), resulting in a more available metal fraction in the soil solution suitable for root absorption (McBride 1994). Some studies dealing with the effect of soil pH on growth and metal accumulation in plants showed that at pH lower than 6.0, maize accumulated Cd with low phytotoxicity symptoms, and *Noccaea caerulescens* had even higher

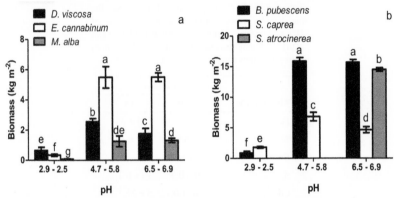

Fig. 3. Effect of pH on the biomass (kg/m²) of aboveground part of (a) herbaceous species: *D. viscosa*, *E. cannabinum* and *M. alba* and (b) woody species: *B. pubescens*, *S. caprea* and *S. atrocinerea*; after 7 months growing in a polluted soil at pH = 6.5. Different letters indicate significant differences between groups (mean ± SE, n = 6, $p < 0.05$).

Cd concentrations than maize without exhibiting phytotoxicity symptoms at any pH assayed (Broadhurst et al. 2015).

Moreno-Jimenez et al. (2011) used Mediterranean native plants to clean up an As-polluted site and concluded that total As concentrations were higher in soils with pH below 5, but As quantity extracted by the plant was higher when soil pH was above 5. This apparent disagreement was probably due to higher available fraction of As (up to 60%) in soils with pH above 5 than that at lower pH. Other studies showed that different plant species, under As exposure, presented a pH-independent metal accumulation (Wei et al. 2006, Moreno-Jiménez et al. 2010). This suggests that some plants have a fixed As-accumulation pattern, so soil As concentrations and soil properties have little effect on the accumulation rate of this metalloid by plants.

Field studies performed by our research group revealed that metal accumulation in plant tissues was highly influenced by soil pH. When the pH was fairly acic (< 4.5), both herbaceous and woody plants showed a minimum growth (Fig. 3), but high accumulation of toxic metal(loid)s were found in the leaves (Table 3). Moreover, the species *D. viscosa* accumulated more Fe, Zn, Cu and Cd than other plant species tested in that field study, and *M. alba* accumulated the highest Pb concentration. Both *D. viscosa* and *E. cannabinum* showed the highest As accumulation (Table 3). The results of this field study also showed that woody plants had the highest metal accumulation in their roots; except for Zn in *B. pubescens* probably due to a high translocation to leaves (Ferández-Fuego et al. 2017). In addition, metal(loid) concentration was always higher in the older leaves compared to younger leaves, which would imply that metals had a low mobility in these plant species. On the other hand, these studies suggest that soil pH as well as metal the concentration of metals in soil exert a high pressure on the composition of soil microbial communities (Sullivan et al. 2013), so a deeper knowledge on soil properties at local scale is highly recommended to propose an efficient phytoremediation program.

Table 3. Effect of pH on the metal(loid)s accumulation (mg/kg) in leaves of *D. viscosa*, *E. cannabinum* and *M. alba* after 7 months growing in a polluted soil at pH = 6.5. Different letters indicate statistically significant differences between groups (mean ± SE, n = 6, $p < 0.05$).

Metal(loid)	pH values								
	D. viscosa			*E. cannabinum*			*M. alba*		
	2.5 – 2.9	4.7 – 5.8	6.5 – 6.9	2.5 – 2.9	4.7 – 5.8	6.5 – 6.9	2.5 – 2.9	4.7 – 5.8	6.5 – 6.9
Fe	1000 ± 50 a	1010 ± 45 a	514 ± 18 b	450 ± 20 c	310 ± 15 d	270 ± 17 e	394 ± 19 c	80 ± 6 f	92 ± 12 g
Zn	520 ± 14 a	434 ± 12 b	450 ± 10 b	290 ± 17 c	187 ± 9 d	174 ± 12 d	310 ± 10 c	210 ± 17 d	274 ± 13 c
Cu	23 ± 3 ab	28 ± 2 a	19 ± 2 b	23 ± 2 ab	10 ± 1 d	11 ± 2 cd	23 ± 2 ab	12 ± 2 cd	14 ± 2 c
As	9 ± 0.2 b	15 ± 1 a	3 ± 0.2 d	13 ± 0.3 a	10 ± 0.5 b	5 ± 0.5 c	6 ± 0.8 c	7 ± 1 c	3 ± 0.5 d
Cd	7 ± 1 b	13 ± 0.2 a	8 ± 0.1 b	4 ± 0.1 c	7 ± 0.2 b	4 ± 0.2 c	3 ± 0.1 d	2 ± 0.1 e	1 ± 0.01 f
Pb	29 ± 2 b	27 ± 3 b	15 ± 2 d	22 ± 2 c	4 ± 0.1 f	7 ± 0.3 e	60 ± 7 a	14 ± 2 d	19 ± 1 c

Fertilization

Growth of plants in contaminated lands take up the essential mineral elements for their own growth and development, but they are also exposed to non-essential metals. Taking this into account the question that arises is: what will be the effects of fertilization or soil amendments on plant metal accumulation? Frequently, contaminated soils not only have high concentrations of metal, but also nutritional deficiency, poor soil structure, low organic matter content and low water retention. Thus, fertilization regimes can be designed with the aim of improving plant growth and establishment or increasing metal plant uptake (Kidd et al. 2015). It is generally hypothesized that soils fertilized with various types of nutrients would quicken the metal phytoremediation process, partially because plants with better growth would be more able to withstand soil contaminants (Delorme et al. 2000). Furthermore, fertilization together with soil amendments could increase mobility of metals in soil, increasing thereby their bioavailability to plants which, in turn, would improve phytoremediation efficiency (Nicholson et al. 2003). However, certain metal may display antagonistic interactions with several macro- and micro-nutrients, leading to a deceased phytoremediation efficiency (Zhou et al. 2015).

In the same way farmers use fertilizers or soil amendments to obtain a better crop yield and to compensate for productivity loss, this agronomical practice could be implemented in contaminated sites. Nonetheless, excessive application of fertilizers may turn into greater accumulation of As, Cd, Zn, Fe and Pb in crops cultivated in polluted soil with a consequent metal transfer to food webs (McLaughlin et al. 1996, Singh et al. 2011). Therefore, fertilizer implementation in a metal contaminated soil must be evaluated with great caution, and strategies need to be tailored to site-specific conditions in order to achieve the desired outcome and to avoid negative impacts on soil ecosystem.

On the other side, it has also been observed that addition of acidifying agents such as NH_4-containing fertilizers (Grant et al. 1999, Prasad 2008) and citric acid (Mao et al. 2016) to contaminated soils improved the phytoextraction process (Amoakwah et al. 2014). This effect may be due to the lower retention of metals to soil organic matter under acidic conditions, resulting in more available metal in the soil solution for root absorption. However, these effects are highly species-specific and, in some cases, may actually decrease remediation potential (White et al. 2006).

Results from field studies performed in our group showed that fertilization just after planting increased the biomass of *D. viscosa*, *E. cannabinum* and *M. alba* during the first two months (Fig. 4a), but metal accumulation decreased (Table 4). These results can be explained because the added fertilizer (an organic fertilizer, named NPK: 6-8-15) may interfere with other metals such as Pb or Zn, reducing their bioavailability, solubility and mobility, and consequently inhibiting their accumulation. However, seven months later, fertilization did not increase the plant growth when compared with unfertilized plants (Fig. 4b), although a higher accumulation of Fe, Zn, Cu, As and Pb was observed in *D. viscosa* leaves (Table 4). This trend was also observed one year later, which suggests that fertilization can improve metal accumulation particularly in industrial polluted soils. Nevertheless, the use of fertilizers to improve phytoremediation of metal-contaminated lands shows contradictory results. For

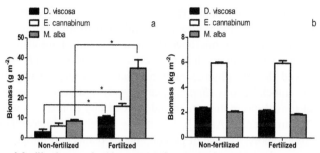

Fig. 4. Effect of fertilization on the biomass (g/m² and kg/m²) of aboveground part of *D. viscosa, E. cannabinum* and *M. alba*; after (a) 2 months and (b) 7 months growing in a polluted soil at pH = 6.5. Asterisks denote significant differences between groups (mean ± SE, n = 6, $p < 0.05$).

Table 4. Effect of fertilization on the metal(loid)s accumulation (mg/kg) in leaves of *D. viscosa* after 2 and 7 months growing in a polluted soil at pH = 6.5. Different letters indicate statistically significant differences between groups (mean ± SE, n = 6, $p < 0.05$).

	Fertilizer (NPK: 6-8-15)			
	2 months		**7 months**	
Metal(loid)	**Unfertilized**	**Fertilized**	**Unfertilized**	**Fertilized**
Fe	401 ± 12 c	402 ± 15 c	810 ± 16 b	1605 ± 28 a
Zn	293 ± 10 c	260 ± 8 d	382 ± 14 b	820 ± 14 a
Cu	16 ± 2 c	15 ± 1 c	26 ± 1 b	45 ± 4 a
As	5 ± 0.4 c	2 ± 0.1 d	8 ± 0.6 b	16 ± 1 a
Cd	9 ± 1 a	4 ± 0.3 c	10 ± 1 a	6 ± 0.3 b
Pb	8 ± 0.3 c	5 ± 0.1 d	18 ± 1 b	61 ± 6 a

example, it has been demonstrated that adding P-fertilizer to As-contaminated soils may affect phytoremediation efficiency due to the physicochemical similarities between phosphate and arsenate ions (Adriano 2001). Replacement of arsenate by phosphate from the soil binding sites as a competitive anion exchange could, result in enhanced As mobility and higher As accumulation in such phosphate-amended soils (Smith et al. 2002, Jankong et al. 2007). However, Fitz and Wenzel (2002) reported that the effects of phosphorus fertilization on the uptake and toxicity of As in plants is unpredictable because it depends on plant species, chemical form of available As and the edaphic-climatic conditions. In the case of *Pteris vittata*, Caille et al. (2004) found that the addition of phosphate had no effect on As uptake in plants growing in As-contaminated soils coming from mining and smelting activities. However, Cao et al. (2003) showed that *P. vittata* growing in soils amended with phosphate-rock fertilizer increased the As uptake rate. A plausible explanation to this contradictory result could be related to the different concentrations of phosphate used in these two studies, although the bioavailable fraction of As as well as its concentration ratio with phosphate ions should not be excluded. These examples make clear that application of a certain technique is not equally effective in altering bioavailable TEs in different soils, and pilot studies should be performed for each specific contaminated site before to initiate a larger scale phytoremediation action (Neunhauserer et al. 2001,

Meeinkuirt et al. 2012). In our study, fertilization caused a significant increase of As accumulation in *D. viscosa* leaves after a 7-months fertilization program (Table 4).

In addition, rhizosphere's microbes (bacteria and fungi) are also an important component to improve phytoremediation. From our experience on As-contaminated sites, we observed that As phytoextraction efficiency could be improved by the inoculation of autochthonous *Betula* plants with indigenous bacteria, which exhibited plant growth-promoting traits and the capacity to resist as well as reduce arsenate (As[V]) to arsenite (As[III]), facilitating thereby its bioavailability (Mesa et al. 2017). Additionally, the use of autochthonous plants and indigenous bacteria alleviated the synecological requirement of both partners, ensuring a successful plant establishment at the remediation site. In this trial, the addition of fertilizer increased shoot biomass, but not As accumulation by plants, probably due to different soil specific conditions (pH, metal bioavailability, and so on). Gullap et al. (2014) concluded that the time required for a significant removal of Pb, Ni, B, Mn, and Zn could be further decreased to approximately 4, 6, 3, 3, and 1 growing seasons, respectively, if a mixture of 33 kg/ha of phosphorus pentoxide (P_2O_5) and 10^8 cfu/mL of plant growth promotion factor *Bacillus megaterium* var. *phosphaticum* were applied at a rate of 250 mL/10 m^2 in the spring season.

Compost is a promising natural organic fertilizer that may improve phytoremediation. Furthermore, metal stabilization using soil amendments such as waste materials derived from industrial activities has been extensively applied as an economical and environmentally friendly remediation method (Onibon and Fagbola 2013). Nonetheless, the use of poultry waste (without proper treatment) prior to cultivation of vegetables was related to high levels of organic and inorganic toxicants, including metals, in plant tissues (Sahito et al. 2016). Therefore, possible risk of contamination should be assessed first. It has been proved that both, compost and chemical fertilizers, can be used to enhance phytoremediation of metal contaminated soils; however, compost addition is recommended since it enables greater vegetative coverage than fertilizers (Palmroth et al. 2006).

On the other hand, some agricultural practices to enhance crop productivity rely on the combined use of chelating substances use added to fertilizers to facilitate uptake of micronutrients important for plant metabolism of the plants (Sirbu et al. 2009). Therefore, phytoremediation could also benefit from the use of such chelators to enhance the decontamination rate of toxic metal(loid)s. Several experiments proved that addition of chelating substances to soil generally accelerate the soil decontamination rate due to increased metal absorption capability. However, supply of chelating agents may result also in certain environmental risks and side-effects, particularly affecting the viability of soil microorganisms (Grcman et al. 2001).

The chelator EDTA is widely used to increase metal solubility, particularly at low soil pH (Luo et al. 2016). EDTA combined with ethyl lactate significantly enhanced the efficiency of willow in removing Cd from soils located at the lower course of the Yangtze River (Li et al. 2010). However, excessive use of EDTA during phytoremediation processes can cause undesirable side effects on soil fauna such as earthworms, with the subsequent negative ecological impact (Jones et al. 2007). In this sense, a field experiment was conducted to evaluate impacts of a mixture of chelators upon the growth and phytoextraction of metals by the hyperaccumulator

Sedum alfredii, which increased significantly the accumulation of Zn, Cd, and Pb by this plant species (Wu et al. 2007). Similarly, the application of chloride (NH$_4$Cl as N fertilizer and Ca$_2$Cl) approximately doubled the amount of Cd taken up by sunflower and kenaf (*Hibiscus cannabinus*) leaves compared with the control, but markedly decreased Cd uptake in sorghum plants because of root damage. The effect of soil acidification on Cd uptake differed therefore between the plant species owing to differences in metal tolerance under low pH. The results showed that the application of Cl$^-$ increased Cd uptake by decreasing soil pH, so it could be a promising method for phytoremediation of Cd depending on the tolerance of the plant species to low pH (Hattori et al. 2006). Another common chelating agent is the citric acid, which has less side effects on the environment than EDTA. Citric acid can be very effective in desorbing metal(loid)s from the soil (Mao et al. 2016) as well as into shaping the rhizosphere communities (Abbas et al. 2017). Huang et al. (1997) reported that the addition of citric acid increased uranium accumulation in *Brassica juncea* tissues more than nitric or sulfuric acid, although all these acids decreased soil pH. These authors speculated that citric acid not only reduced the pH, but also chelated U in the soil, enhancing thereby its solubility and availability in the soil solution and causing a 1000-fold increase of U concentration in the shoots of *B. juncea*.

Occasionally, immobilization of metals in soil is the most recommended strategy, especially in low contaminated agricultural lands. Thomas Basic Slag (TBS) is an alkaline by-product coming from the steel industry that is commercially available and used as a fertilizer by farmers to increase soil pH. The beneficial effects of TBS on an agriculture soil are reached by an increase of soil pH that consequently decreases metal mobility, in a short-term scale, but further chemical reactions (substitution, precipitation) stabilize the metals at a long-term scale (Panfili et al. 2005, Bert et al. 2012).

Conclusion Remarks and Future Directions

Phytoremediation is an economic technology to restore metal-contaminated lands, but it is a slow process, requiring several years to reduce pollutants levels to meet soil environmental quality standards. Nonetheless, when soil decontamination is not priority, phytoremediation is an attractive and environmentally friendly technique with a plethora of possible strategies to increase its effectiveness. In this chapter, we attempted to summarize and discuss the available information on different phytoremediation aspects, highlighting the need for getting a greater knowledge not only on metal concentration in soil, but also on all other different soil variables that might affect plant growth. Understanding the intertwining of all these processes will certainly improve the phytoremediation of contaminated soils.

Most phytoremediation studies have been focused on industrial contaminated soils, where phytoremediation appears as the most suitable strategy to decrease metal concentration and toxicity. However, in agricultural contaminated soils, edible parts of plants must guarantee maximum safety for human consumption (see Chapters 4 and 5). Therefore, if contamination of agricultural soil is too high, a first phytoremediation step is required to mobilize and accumulate metals in plants that will be not destined to human consumption followed by cropping of plant species of concern to human consumption. On the other hand, if soil contamination is low, metal mobility could be reduced by addition of organic amendments such as biochar, that significantly increase metal immobilization, therefore reducing their accumulation in plants. For all the progress done in finding the best phytoremediation approach and the presence in the literature of field trials with a positive outcome, much remains to be learned. In this sense, field trials are still needed to obtain a deeper understanding on the processes that govern phytoremediation.

Acknowledgements

This research was supported by the projects CTM2011-29972, LIFE11/ENV/ES/000547 and AGL2014-53771-R. A. Navazas was funded by fellowship Education, Culture and Sport Ministry (FPU13/05809).

References

Abbas, T., M. Rizwan, S. Ali, M. Zia-ur-Rehman, M. Farooq Qayyum, F. Abbas et al. 2017. Effect of biochar on cadmium bioavailability and uptake in wheat (*Triticum aestivum* L.) grown in a soil with aged contamination. Ecotoxicology and Environmental Safety 140: 37–47.

Adriano, D.C. 2001. Trace Elements in Terrestrial Environments: Biogeochemistry, Bioavailability and Risks of Metals. 2nd Edition, Springer, New York, 867.

Amoakwah, E., S. Van Slycken and D.K. Essumang. 2014. Comparison of the solubilizing efficiencies of some pH lowering (Sulphur and $(NH_4)_2SO_4$) amendments on Cd and Zn mobility in soils. Bulletin of Environmental Contamination and Toxicology 93(2): 187–191.

Azzi, V., V. Kazpard, B. Lartiges, A. Kobeissi, A. Kanso and A.G. El Samrani. 2017. Trace metals in phosphate fertilizers used in eastern mediterranean countries. CLEAN—Soil, Air, Water 45(1): 1–8.

Beesley, L., M. Marmiroli, L. Pagano, V. Pigoni, G. Fellet, T. Fresno et al. 2013. Biochar addition to an arsenic contaminated soil increases arsenic concentrations in the pore water but reduces uptake to tomato plants (*Solanum lycopersicum* L.). Sci. Total Environ. 454-455: 598–603.

Bert, V., C. Lors, J.-F. Ponge, L. Caron, A. Biaz, M. Dazy et al. 2012. Metal immobilization and soil amendment efficiency at a contaminated sediment landfill site: A field study focusing on plants, springtails, and bacteria. Environ. Pollut. 169: 1–11.

Brennan, A., E.M. Jiménez, M. Puschenreiter, J.A. Alburquerque and C. Switzer. 2014. Effects of biochar amendment on root traits and contaminant availability of maize plants in a copper and arsenic impacted soil. Plant and Soil 379(1): 351–360.

Brewin, L.E., A. Mehra, P.T. Lynch and M.E. Farago. 2003. Mechanisms of copper tolerance by Armeria maritima in dolfrwynog bog, North Wales—Initial Studies. Environmental Geochemistry and Health 25(1): 147–156.

Broadhurst, C.L., R.L. Chaney, A.P. Davis, A. Cox, K. Kumar, R.D. Reeves et al. 2015. Growth and Cadmium Phytoextraction by Swiss Chard, Maize, Rice, *Noccaea caerulescens*, and *Alyssum murale* in Ph Adjusted Biosolids Amended Soils. International Journal of Phytoremediation 17(1): 25–39.

Caille, N., S. Swanwick, F.J. Zhao and S.P. McGrath. 2004. Arsenic hyperaccumulation by Pteris vittata from arsenic contaminated soils and the effect of liming and phosphate fertilisation. Environ. Pollut. 132(1): 113–120.

Cao, X., L.Q. Ma and A. Shiralipour. 2003. Effects of compost and phosphate amendments on arsenic mobility in soils and arsenic uptake by the hyperaccumulator, *Pteris vittata* L. Environ. Pollut. 126(2): 157–167.

Chapman, P.M. and M. Holzmann. 2007. Heavy metal—music, not science. Environmental Science and Technology 41: 6C.

Chang, J.S., S.Y. Lee and K.W. Kim. 2010. Arsenic in an As-contaminated abandoned mine was mobilized from fern-rhizobium to frond-bacteria via the ars Gene. Biotechnol. Bioprocess Eng. 15(5): 862–873.

Che-Castaldo, J.P. and D.W. Inouye. 2015. Interspecific competition between a non-native metal-hyperaccumulating plant (*Noccaea caerulescens*, Brassicaceae) and a native congener across a soil-metal gradient. Aust. J. Bot. 63(1-2): 141–151.

Chen, X.-C., Y.-G. Liu, G.-M. Zeng, G.-F. Duan, X.-J. Hu, X. Hu et al. 2015. The optimal root length for *Vetiveria zizanioides* when transplanted to Cd polluted soil. International Journal of Phytoremediation 17(6): 563–567.

Delorme, T.A., J.S. Angle, F.J. Coale and R.L. Chaney. 2000. Phytoremediation of phosphorus-enriched soils. International Journal of Phytoremediation 2(2): 173–181.

Fellet, G., M. Marmiroli and L. Marchiol. 2014. Elements uptake by metal accumulator species grown on mine tailings amended with three types of biochar. Sci. Total Environ. 468-469: 598–608.

Fernández, R., I. Carballo, H. Nava, R. Sánchez-Tamés, A. Bertrand and A. González. 2010. Looking for Native Hyperacumulator species useful in Phytoremediation. pp. 297–330. *In*: Golubev, I.A. (ed.). Handbook of Phytoremediation. Nova Science Publishers Inc. New York.

Fernandez, R., A. Bertrand, A. Casares, R. Garcia, A. Gonzalez and R.S. Tames. 2008. Cadmium accumulation and its effect on the *in vitro* growth of woody fleabane and mycorrhized white birch. Environ. Pollut. 152(3): 522–529.

Fernández, R., A. Bertrand, J.I. García, R.S. Tamés and A. González. 2012. Lead accumulation and synthesis of non-protein thiolic peptides in selected clones of Melilotus alba and Melilotus officinalis. Environmental and Experimental Botany 78: 18–24.

Fernandez-Fuego, D., A. Bertrand and A. Gonzalez. 2017. Metal accumulation and detoxification mechanisms in mycorrhizal Betula pubescens. Environ. Pollut. 231(1): 1153–1162.

Fernando, D.R., G. Guymer, R.D. Reeves, I.E. Woodrow, A.J. Baker and G.N. Batianoff. 2009. Foliar Mn accumulation in eastern Australian herbarium specimens: prospecting for 'new' Mn hyperaccumulators and potential applications in taxonomy. Annals of Botany 103(6): 931–939.

Fitz, W.J. and W.W. Wenzel. 2002. Arsenic transformations in the soil–rhizosphere–plant system: fundamentals and potential application to phytoremediation. Journal of Biotechnology 99(3): 259–278.

Gallego, J.R., E. Rodríguez-Valdés, N. Esquinas, A. Fernández-Braña and E. Afif. 2016. Insights into a 20-ha multi-contaminated brownfield megasite: An environmental forensics approach. Science of The Total Environment 563: 683–692.

Gill, R.A., L. Zang, B. Ali, M.A. Farooq, P. Cui, S. Yang et al. 2015. Chromium-induced physio-chemical and ultrastructural changes in four cultivars of *Brassica napus* L. Chemosphere 120: 154–164.

Gramss, G. and K.-D. Voigt. 2015. Regulation of the mineral concentrations in pea seeds from uranium mine and reference soils diverging extremely in their heavy metal load. Scientia Horticulturae 194: 255–266.

Grant, C.A., L.D. Bailey, M.J. Mclaughlin and B.R. Singh. 1999. Management factors which influence cadmium concentrations in crops. pp. 151–198. *In*: McLaughlin, M.J. and B.R. Singh (eds.). Cadmium in Soils and Plants. Dordrecht: Springer Netherlands.

Grčman, H., Š. Velikonja-Bolta, D. Vodnik, B. Kos and D. Leštan. 2001. EDTA enhanced heavy metal phytoextraction: metal accumulation, leaching and toxicity. Plant and Soil 235(1): 105–114.

Greger, M. and T. Landberg. 2015. Novel field data on phytoextraction: Pre-cultivation with salix reduces cadmium in wheat grains. International Journal of Phytoremediation 17(10): 917–924.

Gullap, M.K., M. Dasci, H.I. Erkovan, A. Koc and M. Turan. 2014. Plant growth-promoting rhizobacteria (PGPR) and phosphorus fertilizer-assisted phytoextraction of toxic heavy metals from contaminated soils. Commun. Soil Sci. Plant Anal. 45(19): 2593–2606.

Hammer, D., A. Kayser and C. Keller. 2003. Phytoextraction of Cd and Zn with *Salix viminalis* in field trials. Soil Use Manage 19(3): 187–192.

Hansi, M., J.D. Weidenhamer and A. Sinkkonen. 2014. Plant growth responses to inorganic environmental contaminants are density-dependent: Experiments with copper sulfate, barley and lettuce. Environ. Pollut. 184: 443–448.

Hattori, H., K. Kuniyasu, K. Chiba and M. Chino. 2006. Effect of chloride application and low soil pH on cadmium uptake from soil by plants. Soil Sci. Plant Nutr. 52(1): 89–94.

Hou, D.D., K. Wang, T. Liu, H.X. Wang, Z. Lin, J. Qian et al. 2017. Unique Rhizosphere micro-characteristics facilitate phytoextraction of multiple metals in soil by the hyperaccumulating plant sedum alfredii. Environ. Sci. Technol. 51(10): 5675–5684.

Huang, J.W., J. Chen, W.R. Berti and S.D. Cunningham. 1997. Phytoremediation of Lead-Contaminated Soils: Role of Synthetic Chelates in Lead Phytoextraction. Environ. Sci. Technol. 31(3): 800–805.

Jankong, P., P. Visoottiviseth and S. Khokiattiwong. 2007. Enhanced phytoremediation of arsenic contaminated land. Chemosphere 68(10): 1906–1912.

Ji, P.H., T.H. Sun, Y.F. Song, M.L. Ackland and Y. Liu. 2011. Strategies for enhancing the phytoremediation of cadmium-contaminated agricultural soils by *Solanum nigrum* L. Environ. Pollut. 159(3): 762–768.

Jones, L., M. O'Reilly and A.J. Morgan. 2007. Responses of a non-target organism to metalliferous field soils amended by a phytoremediation-promoting chelator (EDTA): The earthworm, *Eisenia fetida*. Eur. J. Soil Biol. 43: S289–S296.

Kabata-Pendias, A. 2010. *In*: Trace Elements in Soils and Plants, 4th Edition. CRC Press, Boca Raton, Florida, 407–505.

Kazakou, E., P.G. Dimitrakopoulos, A.J.M. Baker, R.D. Reeves and A.Y. Troumbis. 2008. Hypotheses, mechanisms and trade-offs of tolerance and adaptation to serpentine soils: from species to ecosystem level. Biological Reviews 83(4): 495–508.

Kidd, P., M. Mench, V. Álvarez-López, V. Bert, I. Dimitriou, W. Friesl-Hanl et al. 2015. Agronomic practices for improving gentle remediation of trace element-contaminated soils. International Journal of Phytoremediation 17(11): 1005–1037.

Kupper, H., B. Gotz, A. Mijovilovich, F.C. Kupper and W. Meyer-Klaucke. 2009. Complexation and toxicity of copper in higher plants. I. Characterization of copper accumulation, speciation, and toxicity in Crassula helmsii as a new copper accumulator. Plant Physiology 151(2): 702–714.

Li, J.H., Y.Y. Sun, Y. Yin, R. Ji, J.C. Wu, X.R. Wang et al. 2010. Ethyl lactate-EDTA composite system enhances the remediation of the cadmium-contaminated soil by Autochthonous Willow (Salix x aureo-pendula CL 'J1011') in the lower reaches of the Yangtze River. J. Hazard Mater 181(1-3): 673–678.

Liu, H.Y., F.B. Meng, Y.D. Tong and J. Chi. 2014. Effect of plant density on phytoremediation of polycyclic aromatic hydrocarbons contaminated sediments with Vallisneria spiralis. Ecol. Eng. 73: 380–385.

Liu, L., L.H. Wu, N. Li, L.Q. Cui, Z. Li, J.P. Jiang et al. 2009. Effect of planting densities on yields and zinc and cadmium uptake by *Sedum plumbizincicola*. Huan jing ke xue = Huanjing kexue 30(11): 3422–3426.

Liu, Y.-N., Z.-H. Guo, X.-Y. Xiao, S. Wang, Z.-C. Jiang and P. Zeng. 2017. Phytostabilisation potential of giant reed for metals contaminated soil modified with complex organic fertiliser and fly ash: A field experiment. Science of the Total Environment 576: 292–302.

Long, X., E. Yang Xiao, Z. Ye, W. Ni and W. Shi. 2002. Differences of uptake and accumulation of zinc in four species of sedum. Acta Botanica Sinica 44(2): 152–157.

Low, J.E., M.L. Whitfield Åslund, A. Rutter and B.A. Zeeb. 2011. The effects of pruning and nodal adventitious roots on polychlorinated biphenyl uptake by Cucurbita pepo grown in field conditions. Environ. Pollut. 159(3): 769–775.

Luo, J., S.H. Qi, X.W.S. Gu, T. Hou and L.H. Lin. 2016. Ecological risk assessment of EDTA-assisted phytoremediation of Cd under different cultivation systems. Bulletin of Environmental Contamination and Toxicology 96(2): 259–264.

Ma, L.Q., K.M. Komar, C. Tu, W. Zhang, Y. Cai and E.D. Kennelley. 2001. A fern that hyperaccumulates arsenic. Nature 409(6820): 579.

Mao, X.Y., F.X.X. Han, X.H. Shao, Z. Arslan, J. McComb, T.T. Chang et al. 2016. Remediation of lead-, arsenic-, and cesium-contaminated soil using consecutive washing enhanced with electro-kinetic field. Journal of Soils and Sediments 16(10): 2344–2353.

Marschner, H. 1995. Nutrient availability in soils. pp. 483–507. *In*: Mineral Nutrition of Higher Plants (2nd Edition). London, Academic Press.

Maywald, F. and H.J. Weigel. 1997. Biochemistry and molecular biology of heavy metal accumulation in higher plants. Landbauforsch Volk 47(3): 103–126.

McBride, M.B. 1994. Environmental chemistry of soils. Oxford University Press, Inc., New York.

McGrath, S.P., E. Lombi, C.W. Gray, N. Caille, S.J. Dunham and F.J. Zhao. 2006. Field evaluation of Cd and Zn phytoextraction potential by the hyperaccumulators Thlaspi caerulescens and Arabidopsis halleri. Environ. Pollut. 141(1): 115–125.

McLaughlin, M.J., K.G. Tiller, R. Naidu and D.P. Stevens. 1996. Review: The Behaviour and Environmental Impact of Contaminants in Fertilizers. Aust. J. Soil Res. 34: 1–54.

Meeinkuirt, W., P. Pokethitiyook, M. Kruatrachue, P. Tanhan and R. Chaiyarat. 2012. Phytostabilization of a Pb-contaminated mine tailing by various tree species in pot and field trial experiments. Int. J. Phytoremediation 14(9): 925–938.

Memon, A.R., Y. Yildizhan and E. Kaplan. 2008. Metal accumulation in crops—human health issues. pp. 81–97. *In*: Prasad, M.N.V. (ed.). Trace Elements as Contaminants and Nutrients. New York: John Wiley & Sons, Inc.

Meng, F. and J. Chi. 2015. Phytoremediation of PAH-contaminated sediments by *Potamogeton crispus* L. with four plant densities. Transactions of Tianjin University 21(5): 440–445.

Mesa, V., A. Navazas, R. Gonzalez-Gil, A. Gonzalez, N. Weyens, B. Lauga et al. 2017. Use of endophytic and Rhizosphere bacteria to improve phytoremediation of arsenic-contaminated industrial soils by autochthonous *Betula celtiberica*. Applied and Environmental Microbiology 83(8).

Moreno-Jimenez, E., R. Manzano, E. Esteban and J. Penalosa. 2010. The fate of arsenic in soils adjacent to an old mine site (Bustarviejo, Spain): mobility and transfer to native flora. Journal of Soils and Sediments 10(2): 301–312.

Moreno-Jimenez, E., S. Vazquez, R.O. Carpena-Ruiz, E. Esteban and J.M. Penalosa. 2011. Using Mediterranean shrubs for the phytoremediation of a soil impacted by pyritic wastes in Southern Spain: A field experiment. J. Environ. Manage 92(6): 1584–1590.

Morio, M., S. Schädler and M. Finkel. 2013. Applying a multi-criteria genetic algorithm framework for brownfield reuse optimization: Improving redevelopment options based on stakeholder preferences. J. Environ. Manage 130: 331–346.

Murakami, M., F. Nakagawa, N. Ae, M. Ito and T. Arao. 2009. Phytoextraction by rice capable of accumulating Cd at high levels: Reduction of Cd content of rice grain. Environ. Sci. Technol. 43(15): 5878–5883.

Nehnevajova, E., R. Herzig, G. Federer, K.H. Erismann and J.P. Schwitzguebel. 2007. Chemical mutagenesis—A promising technique to increase metal concentration and extraction in sunflowers. International Journal of Phytoremediation 9(1-3): 149–165.

Neilson, S. and N. Rajakaruna. 2012. Roles of Rhizospheric processes and plant physiology in applied phytoremediation of contaminated soils using brassica oilseeds. pp. 313–330. *In*: Anjum, N.A., I. Ahmad, M.E. Pereira, A.C. Duarte, S. Umar and N.A. Khan (eds.). The Plant Family Brassicaceae: Contribution Towards Phytoremediation. Dordrecht: Springer Netherlands.

Neunhauserer, C., M. Berreck and H. Insam. 2001. Remediation of soils contaminated with molybdenum using soil amendments and phytoremediation. Water Air Soil Poll. 128(1-2): 85–96.

Niazi, N.K., B. Singh, L. Van Zwieten and A.G. Kachenko. 2012. Phytoremediation of an arsenic-contaminated site using *Pteris vittata* L. and *Pityrogramma calomelanos* var. austroamericana: a long-term study. Environmental Science and Pollution Research 19(8): 3506–3515.

Nicholson, F.A., S.R. Smith, B.J. Alloway, C. Carlton-Smith and B.J. Chambers. 2003. An inventory of heavy metals inputs to agricultural soils in England and Wales. Science of the Total Environment 311(1–3): 205–219.

Nirola, R., M. Megharaj, R. Aryal and R. Naidu. 2016. Screening of metal uptake by plant colonizers growing on abandoned copper mine in Kapunda, South Australia. International Journal of Phytoremediation 18(4): 399–405.

Nsanganwimana, F., B. Pourrut, C. Waterlot, B. Louvel, G. Bidar, S. Labidi et al. 2015. Metal accumulation and shoot yield of Miscanthus × giganteus growing in contaminated agricultural soils: Insights into agronomic practices. Agriculture, Ecosystems & Environment 213: 61–71.

Onibon, V.O. and O. Fagbola. 2013. Evaluation of bioremediation efficiency of crude oil-polluted soils as influenced by application of composts and NPK fertilizer. Fresenius Environ. Bull. 22(1): 61–66.

Palmroth, M.R.T., P.E.P. Koskinen, J. Pichtel, K. Vaajasaari, A. Joutti, T.A. Tuhkanen et al. 2006. Field-scale assessment of phytotreatment of soil contaminated with weathered hydrocarbons and heavy metals. Journal of Soils and Sediments 6(3): 128–136.

Pandey, V.C. 2012. Phytoremediation of heavy metals from fly ash pond by *Azolla caroliniana*. Ecotoxicology and Environmental Safety 82: 8–12.

Panfili, F., A. Manceau, G. Sarret, L. Spadini, T. Kirpichtchikova, V. Bert et al. 2005. The effect of phytostabilization on Zn speciation in a dredged contaminated sediment using scanning electron microscopy, X-ray fluorescence, EXAFS spectroscopy, and principal components analysis. Geochimica et Cosmochimica Acta 69(9): 2265–2284.

Pilon-Smits, E. 2005. Phytoremediation. Annual Review of Plant Biology 56: 15–39.

Poschenrieder, C., J. Allué, R. Tolrà, M. Llugany and J. Barceló. 2008. Trace elements and plant secondary metabolism: Quality and efficacy of herbal products. pp. 99–119. *In*: Prasad, M.N.V. (ed.). Trace Elements as Contaminants and Nutrients. New York: John Wiley & Sons, Inc.

Pourret, O. and J.C. Bollinger. 2018. Heavy metal—What to do now: To use or not to use. Sci. Total Environ. 610-611: 419–420.

Prasad, M.N.V. and H.M.D. Freitas. 2003. Metal hyperaccumulation in plants—Biodiversity prospecting for phytoremediation technology. Electron. J. Biotechnol. 6(3): 285–321.

Prasad, M.N.V. 2008. Biofortification: Nutritional security and relevance to human health. pp. 161–181. *In*: Prasad, M.N.V. (ed.). Trace Elements as Contaminants and Nutrients. New York: John Wiley & Sons, Inc.

Raskin, I., P.B.A.N. Kumar, S. Dushenkov and D.E. Salt. 1994. Bioconcentration of heavy metals by plants. Current Opinion in Biotechnology 5(3): 285–290.

Robinson, B., R. Schulin, B. Nowack, S. Roulier, M. Menon, B. Clothier et al. 2006. Phytoremediation for the management of metal flux in contaminated sites. Forest Snow and Landscape Research 80(2): 221–234.

Ruttens, A., M. Mench, J.V. Colpaert, J. Boisson, R. Carleer and J. Vangronsveld. 2006. Phytostabilization of a metal contaminated sandy soil. I: Influence of compost and/or inorganic metal immobilizing soil amendments on phytotoxicity and plant availability of metals. Environ. Pollut. 144(2): 524–532.

Sahito, O.M., T.G. Kazi, H.I. Afridi, J.A. Baig, F.N. Talpur, S. Baloch et al. 2016. Assessment of toxic metal uptake by different vegetables grown on soils amended with poultry waste: Risk assessment. Water Air Soil Poll. 227(11): 423.

Santos, K., M. Steffane Lopes da Silva, L. Eduardo da Silva, M. Alves Miranda and M.B.G.D.S. Freire. 2011. Biological activity in saline sodic soil saturated by water under cultivation of *Atriplex nummularia*. Revista Ciencia Agronomica 42(3): 619–627.

Singh, B.R., S.K. Gupta, H. Azaizeh, S. Shilev, D. Sudre, W.Y. Song et al. 2011. Safety of food crops on land contaminated with trace elements. Journal of the Science of Food and Agriculture 91(8): 1349–1366.

Sirbu, C., T.M. Cioroianu, I. Cojocaru, V. Trandafir and M.G. Albu. 2009. Fertilizers with protein chelated structures with biostimulator role. Rev. Chim. 60(11): 1135–1140.

Smith, E., R. Naidu and A.M. Alston. 2002. Chemistry of inorganic arsenic in soils: II. Effect of phosphorus, sodium, and calcium on arsenic sorption. J. Environ. Qual. 31(2): 557–563.

Souza, E., M.B. Freire, D. Vandeval Maranhão de Melo and A. de Antônio Assunção Montenegro. 2014. Management of *Atriplex nummularia* Lindl. in a Salt Affected Soil in a Semi Arid Region of Brazil 16.

Sullivan, T.S., M.B. McBride and J.E. Thies. 2013. Soil bacterial and archaeal community composition reflects high spatial heterogeneity of pH, bioavailable Zn, and Cu in a metalliferous peat soil. Soil Biol. Biochem. 66: 102–109.

Tlustos, P., K. Brendova, J. Szakova, J. Najmanova and K. Koubova. 2016. The long-term variation of Cd and Zn hyperaccumulation by *Noccaea* spp. and Arabidopsis halleri plants in both pot and field conditions. International Journal of Phytoremediation 18(2): 110–115.

Tripathi, P., S. Dwivedi, A. Mishra, A. Kumar, R. Dave, S. Srivastava et al. 2012. Arsenic accumulation in native plants of West Bengal, India: Prospects for phytoremediation but concerns with the use of medicinal plants. Environmental Monitoring and Assessment 184(5): 2617–2631.

Wei, C.Y., X. Sun, C. Wang and W.Y. Wang. 2006. Factors influencing arsenic accumulation by Pteris vittata: A comparative field study at two sites. Environ. Pollut. 141(3): 488–493.

Wei, S., Q. Zhou and X. Wang. 2005. Identification of weed plants excluding the uptake of heavy metals. Environment International 31(6): 829–834.

Wei, S. and Q. Zhou. 2008. Trace elements in agro-ecosystems. pp. 55–79. *In*: Prasad, M.N.V. (ed.). Trace Elements as Contaminants and Nutrients. New York: John Wiley & Sons, Inc.

Wenzel, W.W., M. Bunkowski, M. Puschenreiter and O. Horak. 2003. Rhizosphere characteristics of indigenously growing nickel hyperaccumulator and excluder plants on serpentine soil. Environ. Pollut. 123(1): 131–138.

White, J.C. 2009. Optimizing planting density for p,p '-DDE Phytoextraction by *Cucurbita pepo*. Environ. Eng. Sci. 26(2): 369–375.

White, P.M., D.C. Wolf, G.J. Thoma and C.M. Reynolds. 2006. Phytoremediation of alkylated polycyclic aromatic hydrocarbons in a crude oil-contaminated soil. Water Air Soil Poll. 169(1-4): 207–220.

Whitfield Åslund, M.L., A. Rutter, K.J. Reimer and B.A. Zeeb. 2008. The effects of repeated planting, planting density, and specific transfer pathways on PCB uptake by *Cucurbita pepo* grown in field conditions. Science of the Total Environment 405(1): 14–25.

Wu, Q.T., Z.B. Wei and Y. Ouyang. 2007. Phytoextraction of metal-contaminated soil by Sedum alfredii H: Effects of chelator and co-planting. Water Air Soil Poll. 180(1-4): 131–139.

Zhan, J., S. Wei, R. Niu, Y. Li, S. Wang and J. Zhu. 2013. Identification of rice cultivar with exclusive characteristic to Cd using a field-polluted soil and its foreground application. Environmental Science and Pollution Research 20(4): 2645–2650.

Zhang, S.R., H.C. Lin, L.J. Deng, G.S. Gong, Y.X. Jia, X.X. Xu et al. 2013. Cadmium tolerance and accumulation characteristics of *Siegesbeckia orientalis* L. Ecol. Eng. 51: 133–139.

Zhou, S., J. Liu, M. Xu, J. Lv and N. Sun. 2015. Accumulation, availability, and uptake of heavy metals in a red soil after 22-years fertilization and cropping. Environmental Science and Pollution Research 22(19): 15154–15163.

Chapter 7

Bioremediation of Pesticides and Metals
A Discussion about Different Strategies using Actinobacteria

Alvarez Analía,[1,2,]* *Saez Juliana Maria,*[1]
Dávila Costa José Sebastián,[1] *Polti Marta
Alejandra*[1,2] and *Benimeli Claudia Susana*[1]

INTRODUCTION

Highly toxic compounds have been released into the environment by direct or indirect inputs over a long time. With the scope of reducing the concentration and toxicity of these chemicals, eco-friendly techniques have emerged for cleaning up polluted sites using plants and/or microbial species. This approach, known as bioremediation, is considered less invasive compared to conventional physicochemical techniques (Kidd et al. 2009).

Pesticides are probably the most widely used and released chemicals in the world. The disposal of obsolete pesticide stocks has resulted in many long-term contaminated sites in Latin America, many of which are currently illegal.

[1] Planta Piloto de Procesos Industriales Microbiológicos (PROIMI-CONICET), Avenida Belgrano y Pasaje Caseros, Tucumán 4000, Argentina.
  Emails: jsaez@proimi.org.ar; jsdavilacosta@gmail.com; cbenimeli@yahoo.com.ar
[2] Facultad de Ciencias Naturales e Instituto Miguel Lillo, Universidad Nacional de Tucumán (UNT), Miguel Lillo 205, Tucumán 4000, Argentina.
  Email: mpolti@proimi.org.ar
* Corresponding author: alvanalia@gmail.com

For instance, the most important known illegal disposal of organochlorine (OC) pesticides (lindane, chlordane, methoxychlor, among others) and metals (Cr[VI], Cu[II], Cd[II], among others) has been found in the northwest of Argentina (Chaile et al. 1999, Fuentes et al. 2010). Moreover, pollution arising from agrochemical inputs is distributed over large areas and at low concentrations. This non-point or diffusive contamination leads to pesticide residues often detected in samples of air (Lammel et al. 2007), water (Kumari et al. 2007), soil (Fuentes et al. 2010), fishes (Malik et al. 2007), and humans (Ridolfi et al. 2014).

Metal contamination is another environmental issue of current concern. Polluted areas can be decontaminated using multiple *in situ* and *ex situ* approaches; however, in case of metal contamination, only a few techniques can be used because of the immutable and generally immobile nature of metals (Dávila Costa et al. 2011a,b). A wide range of human activities such as industrial, agriculture, municipal landfill, and sewage disposal, significantly contribute to increase metal concentrations at levels higher than those established as background (Fernández et al. 2014).

However, the general scenario is to find a mixture of pollutants of both organic and inorganic nature, which are typical in industrial, urban and agricultural areas (Mansour 2012). For example, metals and OC pesticides were found in water and silt samples from a major river basin from Northern Argentina ("The Salí Basin") at concentrations up to 10 times higher than allowed by national regulatory (Polti et al. 2007). Environments contaminated by a mixture chemical including metals and organic contaminants are difficult to remediate because of the complex nature of these pollutants; however, this is the common scenario in polluted sites (Mansour 2012).

Among the organisms used in the bioremediation of polluted sites, actinobacteria are one of the most popular biological vectors for *in situ* bioremediation. These bacteria play an important role in recycling substances in natural world because they metabolize complex organic molecules and polymers (Kieser et al. 2000). The ecological role performed by actinobacteria is demonstrated by their capability to breakdown and transform pesticides and metals, among other chemicals (Benimeli et al. 2003, 2006, 2007, Polti et al. 2009, 2011a,b, 2014, Alvarez et al. 2017). Furthermore, the physiological diversity of actinobacteria allows the production of a large number of metabolites with biotechnological importance such as antibiotics (Whitman et al. 2012). This chapter focuses on how different techniques using actinobacteria can contribute to improve the bioremediation of pesticide and metal-polluted environments.

Bioremediation: A Sustainable Technology for Recovering Polluted Environments

Bioremediation is defined as the "use of living organisms to clean up pollutants from soil, water, or wastewater" (EPA 2016). Bioremediation of toxic organic compounds is often less controversial than bioremediation of metals because organic compounds may be completely degraded to carbon dioxide and water by a process known as mineralization. However, metals cannot be obviously mineralized, so bioremediation actions are based on the conversion of metal into less toxic forms and/ or immobilization, both goals aimed to reduce metal bioavailability, and subsequent toxicity.

Areas polluted with toxic organic and inorganic compounds have been detected worldwide. This simultaneous contamination (co-contamination) represents the real challenge of grey biotechnology. Bioremediation of organic compounds and metals has been shown to be successful, although each process has generally been performed singly. In fact, a multifunctional biological process is needed for bioremediation of sites polluted by multiple environmental contaminants. In this context, actinobacteria show great potential because they have been demonstrated to be efficient biological entities for bioremediation of both organic and inorganic toxic compounds (Polti et al. 2014, Aparicio et al. 2015).

The Phylum Actinobacteria

The phylum Actinobacteria constitutes one of the main and most diverse group within the domain Bacteria. This phylum encompasses presently six classes, 19 orders, 50 families, and 221 genera (Whitman et al. 2012). Actinobacteria are Gram-positive or Gram-variable aerobes, facultative anaerobes or anaerobes, which have a rigid cell wall that contains muramic acid. The phylum includes microorganisms with a wide variety of morphologies that range from cocci to highly differentiated mycelia and spore production. Actinobacteria exhibits a cosmopolitan distribution, so members of the group are widely distributed in aquatic and terrestrial ecosystems (Whitman et al. 2012), those organisms involved in the formation of soil organic matter being especially important. Nowadays, they are considered among the most successful colonizers of all environments, in opposition to the traditional perception of actinobacteria as autochthonous soil and freshwater organisms.

Actinobacteria exhibit diverse physiological and metabolic properties, such as the production of extracellular enzymes or exoenzymes, and the formation of a wide variety of secondary metabolites (Whitman et al. 2012). For instance, the *Streptomyces* genus remains the richest source of antibiotics, antimetabolites, and antitumor agents (Bérdy 2005, Olano et al. 2009). Actinobacteria also produce non-antibiotic molecules with biological activity such as phytotoxins, biopesticides, biosurfactants, nanoparticles, probiotics and enzymes involved in the degradation of complex molecules such as lignin or cellulose (Manivasagan et al. 2013). This versatility in metabolite production makes actinobacteria important vectors in many biotechnological applications such as bioremediation.

Bioremediation using Actinobacteria

Actinobacteria Strains for Application in Phytoremediation

In the last decade, gentle soil remediation options have been developed for phyto-management of contaminated soils, with the aim of disrupting the links between pollutants and ecosystems (Vangronsveld et al. 2009, Mench et al. 2010, Kidd et al. 2015). Phyto-management includes the management of degraded sites and their restoration through phytoremediation techniques that promote biodiversity, improve the integral functionality of ecosystem and allow the sustainable use of resources.

Phytoremediation of soils uses plants and their associated microorganisms to remove, stabilize or detoxify pollutants (Kidd et al. 2009). Microbes profit from plants because of the enhanced availability of nutrients, whereas plants can receive benefits from some bacterial associates by growth enhancement or stress reduction (Weyens et al. 2009, Compant et al. 2010). Exploiting the plant-microbial partnerships in phytoremediation is generally based on the capacity of the bacteria, on one hand, to improve establishment, growth and plant survival (plant-growth promotion) and, on the other hand, to act directly on contaminants. This complementary action between plants and their associated microorganisms to degrade environmental contaminants has been studied in detail in the rhizosphere zone.

Microorganism-assisted phytoremediation is mainly effective for organic pollutants (Gerhardt et al. 2009). Several studies have demonstrated an enhanced dissipation of xenobiotics in the rhizosphere due to the release of plant rhizodeposits containing root exudates (Gerhardt et al. 2009, Kidd et al. 2009, Becerra-Castro et al. 2013). These root exudates increase bacterial metabolic biodiversity in the rhizosphere and can additionally stimulate contaminant degradation through co-metabolism. Indeed, Álvarez et al. (2015) found that *Streptomyces* sp. M7, A11, A2 and A5 were able to grow in a liquid system with maize plants as the carbon source, suggesting that the maize plant, and/or the root exudates, were a suitable carbon and energy source for these bacteria strains. In previous studies, Alvarez et al. (2012) evaluated the effect of root exudates isolated from maize on the growth of three *Streptomyces* strains and, simultaneously, on their ability to remove the OC pesticide lindane. These authors found that *Streptomyces* sp. A5 showed maximum biomass and the highest pesticide dissipation rate (55%) in presence of root exudates. Following these findings, Álvarez et al. (2015) cultivated a mixed culture of four *Streptomyces* strains with maize plants on soils artificially contaminated with lindane. A comparable trend in the lindane removal was observed between inoculated and non-inoculated planted pots, suggesting that pesticide dissipation was not significantly affected by the *Streptomyces* strains. However, the vigor index of the maize plants grown in the presence of the pesticide was enhanced, so the inoculation of the actinobacteria consortium could protect plants against the lindane (or metabolites) toxicity. In a phytoremediation framework, mixed cultures can provide several benefits to plants, including the synthesis of protective compounds before they can negatively impact the plant (Gerhardt et al. 2009). Similar results were found by Becerra-Castro et al. (2013), who inoculated substrates seeded with *Cytisus striatus* with the endophytic *Rhodococcus erythropolis* ET54b and *Sphingomonas* sp. D4. These authors reported that planted pots had a higher dissipation of the hexachlorocyclohexane isomers (including lindane), and that the bacteria consortia protected the plants against the effects of these contaminants.

The metabolic capacity of endophytic bacterial communities plays an important role in contaminant dissipation. Thus, the plant-endophyte association can be exploited for the treatment of contaminated systems (Weyens et al. 2009). Furthermore, plant inoculation with degrading strains can lead to the transfer of the degradation pathway among the endogenous members of the community, via horizontal gene transfer (Weyens et al. 2009). Recently, Mesquini et al. (2015) isolated the endophytic *Streptomyces* sp. atz2 from sugarcane, which was able to reduce 98% of the initial concentration of atrazine in culture medium. Degradation of this herbicide was also achieved when *Arthrobacter* sp. DNS10 was inoculated in polluted soil and in the presence of *Pennisetum*. The authors reported that the *Pennisetum-Arthrobacter* interaction was more efficient in atrazine dissipation (98% of atrazine removal) compared with single strain and single-plant effects (Zhang et al. 2014).

Actinobacteria Immobilization for Bioremediation

Recently, the immobilization of whole cells has been developed as biocatalysts for environmental applications. The use of immobilized cells emerged as an alternative to enzyme immobilization since the heat stability and functional stability of immobilized cells are greater than those of the immobilized extracellular enzymes. Moreover, the processes for the extraction, separation, and purification of extracellular enzymes are not necessary in the case of immobilized cells (Martins et al. 2013, Bayat et al. 2015).

Cell immobilization has been defined as the physical confinement of cells to a certain defined space as limiting their mobility with a simultaneous preservation of their viability and catalytic function, and providing hydrodynamic characteristics, which differ from those of the surrounding environment (Martins et al. 2013, Dzionek et al. 2016). However, bioremediation with immobilized cells not only uses viable cells, but also dead, living, and growing cells. The selection of the suitable cell state will depend, however, on the application purpose and the environmental features (Bayat et al. 2015).

Immobilized cell systems provide many advantages over the use of suspended free cells for bioremediation purpose. Among these advantages, a high biomass, greater cell viability (weeks or months), elimination of cell washout at high dilution rates, improving genetic stability, and easier solid-liquid separation are the most remarkable. Moreover, immobilized cells often present a higher resistance to toxic chemicals, and fluctuating pH and temperature (Bayat et al. 2015). Likewise, the use of immobilized cells provides, in general, high efficiency and good operational stability, with the possibility of cell reuse, hence reducing the costs of the catalytic process (Ahamad and Kunhi 2011, León-Santiestebán et al. 2011, Bayat et al. 2015).

The use of immobilized cells has been recently investigated as a complementary technology for environmental applications, such as the bioremediation of organic and inorganic compounds (Martins et al. 2013, Bayat et al. 2015). In fact, this approach brings many benefits to bioremediation, such as higher efficiency of pollutant degradation, multiple use of biocatalysts, reduced costs, ensuring a stable microenvironment for cells, reduced risks of genetic mutations, ensured resistance to shear forces present in bioreactors, increased resistance of biocatalysts to adverse environmental conditions, increased biocatalyst survival during storage and increased tolerance to high pollutant concentrations. It is assumed that the carrier protects the cells and hinders the spread of the pollutants, thus reducing the surface contaminants concentration on the immobilized microorganisms. However, changes in microenvironment after immobilization may lead to changes in cell morphology, physiology and metabolic activity, which could significantly impact contaminant degradation (Dzionek et al. 2016).

Apart from the method itself, the selection of the physical support is critical during the preparation of the immobilization process. For bioremediation purposes, carriers must not be toxic and biodegradable. Ideally, cell-immobilized carriers should meet the following criteria: a high loading capacity to hold a high cell mass, high mechanical, biological and chemical stability, long shelf life, optimum diffusion capacity, easy separation of the cell-bound carrier from media, and easy to handle and regenerate (Bayat et al. 2015). Besides, in the case of bioremediation, it is important that carriers are cost-effective and easily accessible on a large scale (Dzionek et al. 2016). For example, Poopal and Laxman (2009) revealed that polyvinyl alcohol-alginate was the most effective matrix, among others such as agar, agarose, and polyacrylamide-alginate, for the immobilization of a *Streptomyces griseus* strain, regarding beads integrity and chromate reduction. The polyvinyl alcohol-alginate immobilized cells could be used up to five cycles without compromising on its performance. In contrast, Humphries et al. (2005) found that agar and agarose were the best immobilization matrices for *Microbacterium* sp. NCIMB 13776 for Cr(VI) reduction. Therefore, the selection of the immobilization technique would depend on the microorganism.

Immobilization supports are divided into organic and inorganic carriers according to their chemical structure; in turn, these carriers can be naturals and synthetics (Dzionek et al. 2016). Organic carriers are more abundant than inorganic ones; however, the latter generally have a major advantage over other materials, namely, their toughness. Most of inorganic carriers are inert, resistant to temperature, pH, chemicals, microbial degradation, and crushing or abrasion (Bayat et al. 2015). On the other hand, natural carriers are preferred for the immobilization of microorganisms to synthetic ones because they cause less environmental disposal problems compared to synthetic carriers. However, most of the natural polymers are non-mechanically resistant. Therefore, the selection of an ideal carrier to bioremediate polluted soils is currently one of the primary challenges in the immobilization cell technology (Dzionek et al. 2016).

Particularly, the immobilization of actinobacteria in different carriers has been described as a promising technology for the bioremediation of a wide range of pollutants such as pesticides, metals, hydrocarbons and phenolic compounds (Kitova et al. 2004, Bazot and Lebeau 2009, Poopal and Laxman 2009, Zhang et al. 2016, Alessandrello et al. 2017). For instance, Tallur et al. (2015) studied the degradation of the pyrethroid insecticide cypermethrin by the actinobacterium *Micrococcus* sp. strain CPN 1, which was kept either free or immobilized in various matrices such as polyurethane foam (PUF), polyacrylamide, sodium alginate and agar. They demonstrated that the PUF-immobilized cells degraded the insecticide more efficiently than free suspended cells and cells immobilized in other matrices. Moreover, the PUF-immobilized *Micrococcus* sp. strain CPN 1 retained their degradation capacity and could be reused for more than 32 cycles, without significantly losing their degradation capability. Similarly, Briceño et al. (2013) found that *Streptomyces* sp.

AC1-6 immobilized in alginate beads degraded the organophosphorus pesticide diazinon more efficiently (almost 20% higher) than the degradation rate by free cells. They also confirmed the reusability of the encapsulated cells obtaining more than 50% of pesticide removal after the third batch cycle.

Besides the use of physical support to retain functional cells, efforts have been also addressed to develop viable mixed cultures for bioremediation and wastewater treatment, which attempts to exploit synergistic effects (Alvarez et al. 2017). In this sense, Bazot et al. (2007) evaluated the mineralization of the herbicide diuron by a co-culture made with the actinobacterium *Arthrobacter* sp. N4 and *Delftia acidovorans* W34. These strains were individually unable to mineralize the pesticide, but they could individually either degrade diuron to 3,4-DCA or mineralize 3,4-DCA, respectively. The immobilization of both microbes together into alginate beads increased the degradation rate of diuron almost three-fold compared with that achieved by free cells. These authors attributed their finding to an optimum expression of both bacteria due to a better concentration of both oxygen and substrates into the microenvironment of the beads. Saez et al. (2012) evaluated the removal of lindane by pure and mixed cultures of *Streptomyces* immobilized in agar, PVA-alginate, silicone tubes and clothes sachets. They proved that the removal efficiency achieved by the microorganisms immobilized in all the tested supports was significantly higher than by the free cells, although cells immobilized in cloth sachets showed the best behavior. Besides, a *Streptomyces* consortium immobilized in this support could be reused during three cycles, obtaining a maximum of 71.5% of lindane removal efficiency, although they did not perform further cycles of lindane degradation. This result emphasizes that one important advantage of cell immobilization consists of the extended or repeated use of the cells. Afterward, this *Streptomyces* consortium immobilized in cloth sachets carried out a successful lindane removal in soil slurry reactors contaminated with 50 mg kg^{-1} of lindane (Saez et al. 2014, 2015). Immobilized actinobacteria have been also used to remove mixture of pesticides. For instance, a mixture of chlorpyrifos and pentachlorophenol was removed by using a mixed culture of *Streptomyces* strains immobilized in alginate beads. The degradation rates increased more than 30% for chlorpyrifos and 10% for pentachlorophenol through the use of this immobilized culture, with respect to the degradation rates obtained with free cells (Fuentes et al. 2013).

In the case of metal bioremediation, Jézéquel and Lebeau (2008) compared the inoculation technique, i.e., free or immobilized cells, of *Bacillus* sp. ZAN-044 or *Streptomyces* sp. R25 for the reduction of the potentially phytoavailable Cd in soil. They found that the immobilization technique did not improve the cell survival in the bioaugmented soil; however, the entrapment of the bacteria into alginate beads allowed the adaptation of the bacterial cells to their new environment, before the cells were released from the beads to the soil. Altogether, these studies confirm that the use of immobilization technique for actinobacteria cultures is being increasingly considered as an alternative for the bioremediation of polluted matrices. However, specific studies are needed on the selection of the suitable physical support depending on each microorganism and pollutant, the longevity and reusability of the immobilized cells, among other factors.

Microemulsions as a Novel Technology to Enhance Bioremediation

Microemulsions are macroscopically homogeneous colloidal dispersions of two immiscible liquids, generally oil and water, stabilized by the presence of a surfactant, either alone or in combination with a co-surfactant (Zhang et al. 2011, Sanchez-Dominguez et al. 2012). They are transparent isotropic mixtures with a micellar size from 20 to 200 nm, which exhibit good wettability, low to moderate viscosity, low interfacial tension and high solubilization capacity for both hydrophilic and hydrophobic compounds (Zheng et al. 2011). Unlike conventional emulsions, whose microstructure is static, microemulsions are dynamic systems with an interface continuously and spontaneously fluctuating. In addition, microemulsions are thermodynamically stable and spontaneously forming, making them easy to prepare. In contrast, emulsions require a large input of mechanical energy for their formation; therefore, the cost of preparation is higher (Talegaonkar et al. 2008).

Microemulsions can be classified as oil-in-water, water-in-oil or bicontinuous systems depending on their structure. Oil in water microemulsions, also called direct microemulsions, consist of oil droplets dispersed in the aqueous phase. The polar portion of the surfactant is oriented towards the outside of the micelle, while the hydrophobic extremes are oriented inwardly, resulting in a globular structure. Generally, they are formed when the oil concentration is low (< 30%). Water-in-oil or reverse microemulsions consist of water droplets dispersed in the oil phase. The orientation of the surfactant molecules is the opposite of the above mentioned and they are formed when the aqueous concentration is low (Muñoz Hernández et al. 2005, Flanagan and Singh 2006). Finally, the bicontinuous microemulsions are formed when the amounts of water and oil are similar (Talegaonkar et al. 2008).

A large number of oils, surfactants, and co-surfactants may be used in the preparation of microemulsions. However, they must be biocompatible, non-toxic, and clinically acceptable compounds. Also, their concentrations should be appropriate so that they do not result in an aggressive microemulsion. This point, therefore, makes special emphasis on the use of generally regarded as safe substances (Talegaonkar et al. 2008).

The main component of microemulsions is the surfactant. This is a molecule with amphiphilic properties, i.e., it contains a hydrophilic and a hydrophobic portion. The surfactant chosen must be able to reduce the interfacial tension between the medium in which is dissolved and any other fluid in contact, which facilitates the process of dispersion during the preparation of the microemulsion (Muñoz Hernández et al. 2005, Talegaonkar et al. 2008).

In most cases, surfactants alone are unable to reduce the interfacial tension between oil/water enough to enable the formation of the microemulsion, while the presence

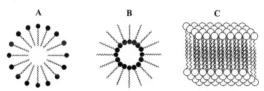

Fig. 1. Schematic representation of the types of microemulsions. (A) Oil in water or direct microemulsion, (B) water in oil or reverse microemulsion, and (C) bicontinuous microemulsion (Adapted from Muñoz Hernández et al. 2005).

of a co-surfactant allows that the interfacial film has sufficient flexibility to assume the curvature required to form a microemulsion. The co-surfactant inserts between the surfactant molecules of the interfacial film. Short-, intermediate- and long-chain alcohols are suitable to be used as co-surfactants. The short-chain alcohols (ethanol, propanol and isopropanol) are more hydrophilic and slightly increase the affinity of the surfactant for the aqueous phase, whereas the longer chain alcohols (pentanol, hexanol) are located mainly towards the oil phase (Talegaonkar et al. 2008).

The oil has the ability to penetrate the microemulsions and, therefore, increase the region of the surfactant monolayer. Short-chain oils penetrate the tail group region to a greater degree than the long-chain oils and, therefore, reduce the hydrophilic-lipophilic balance. Saturated and unsaturated fatty acids and esters of fatty acids have been employed as oil phase for microemulsions preparation (Talegaonkar et al. 2008).

Microemulsions have attracted attention in various fields of application, such as dry cleaning, cosmetics industry and the production of biodiesel (Worakitkanchanakul et al. 2008). Microemulsions are also applied in the pharmaceutical industry, improving the oral supply of soluble drugs and for the controlled release of drugs (Talegaonkar et al. 2008, Fanun 2012). Microemulsions have been also used in the manufacture of nanoparticles, food industry, oil recovery, among other new applications that are constantly being developed (Talegaonkar et al. 2008, Sanchez-Dominguez et al. 2012). In the last decade, a special emphasis has been put in the use of microemulsions for remediation and bioremediation processes of organic and inorganic contaminants (Bragato and El Seoud 2003, Castro Dantas et al. 2009, Vargas-Ruiz et al. 2016). In the case of hydrophobic organic contaminants, such as organochlorine pesticides, due to their poor solubility in water, they tend to adhere strongly to soil organic carbon by adsorption, electrostatic interaction and covalent bonding (Zheng and Wong 2010). For this reason, the biodegradation of these pesticides in soils is often slow and unsatisfactory. In this sense, synthetic surfactants are extensively utilized in bioremediation of polluted water and soil to increase the bioaccessibility of these lipophilic contaminants (Bustamante et al. 2012). Moreover, microemulsions can enhance the degradation of organochlorine pesticides. For soil remediation, non-ionic surfactants are more suitable for the preparation of the microemulsions than anionic or cationic surfactants; the anionic surfactants can complex with divalent cations in soil, while cationic surfactants, due to their positive charge, tend to significantly sorb onto the soil (Wang and Keller 2008). On the other hand, environmental considerations dictate that components of microemulsions should be biodegradable, which justifies why vegetable oils are preferred microemulsion ingredients to mineral ones (Bragato and El Seoud 2003).

Microemulsions are attractive tools for *in situ* and *ex situ* soil remediation because of the following features:

(1) Efficiency. The microemulsion reduces the interfacial tension between the contaminant and soil to very low values; then the desorbed contaminants are readily dissolved in the microemulsions. This is particularly important in the case of highly lipophilic chemicals.

(2) Convenience. Handled volumes are much smaller than those produced from soil washings with either water or aqueous micellar solutions, which is a desirable option in the case of *ex situ* remediation.

(3) Recycling. Separation of microemulsions can be achieved between a recycling phase (aqueous and surfactant-rich phase) and a disposal phase (the organic phase containing the pollutants), simply by changing the temperature of the system (Bragato and El Seoud 2003).

The use of microemulsions for the extraction of metals has also proved to be a good alternative because metals have some properties that make the microemulsion-assisted extraction a cost-effective approach to more conventional solvent extractions. The increase of the metal extraction rate in microemulsion systems is explained by an enormous rise of the microinterfacial surface area in the microemulsion phase, and the participation of the microemulsion to transport metal ions from the aqueous phase to the organic phase. Castro Dantas et al. (2003) and Dantas Neto et al. (2004) applied microemulsion systems to remove heavy metals from aqueous solutions. They evaluated the efficiency of these systems by liquid–liquid extractions, and by a surface modifier agent of natural clay. The extraction yields were higher than 98%, demonstrating the potential of microemulsion systems to remediate metal-contaminated waters. Similarly, Castro Dantas et al. (2009) also used microemulsion systems to remove chromium from leather tannery sediments, obtaining percentages of chromium extraction up to 93.4% with only one extraction stage. These results demonstrated the high potential of microemulsions to treat tannery sediments. In this sense, the use of actinobacteria, resistant to high concentrations of heavy metals combined with microemulsions, would represent a promising approach to bioremediate metal-polluted soils.

Although the application of microemulsions to enhance the efficiency of a bioremediation process is still in its preliminary stages, the results seen so far are promising (Zheng et al. 2012, Salam and Das 2013). In this way, Saez et al. (2017) reported for the first time the enhancement of lindane removal by an actinobacterium using direct microemulsions as bioremediation tools. In that study, the authors evaluated the use of different non-ionic synthetic surfactants and vegetable oils as well as different proportions of the components for the preparation of microemulsions. Based on their results, they selected a stable microemulsion prepared with Tween 80, 1-pentanol and soybean oil with a co-surfactant/surfactant ratio of 1:1 and an oil/surfactant ratio of 1/10, therefore obtaining a greater capacity of lindane solubilization at a surfactant concentration of 1000 mg L^{-1}, compared with the other microemulsions and with the surfactant solution alone. The results of that study demonstrated that the actinobacterium *Streptomyces* sp. M7 was able to grow in 100 mg L^{-1} of lindane, achieving a biomass almost three times greater than that obtained in the absence of solubilizing agents. Moreover, the microemulsion increased the solubility of lindane in the liquid medium, and therefore its bioavailability, which favored the removal of the pesticide by *Streptomyces* sp. M7. This removal rate was almost two times with respect to the one obtained with the surfactant Tween 80 alone. In a loam soil system, the microemulsion allowed an 87% of lindane removal by *Streptomyces* sp. M7, although it did not differ significantly from that obtained with the surfactant solution.

This microemulsion could be used as a potential tool in soil washing technologies or *ex situ* bioremediation processes of wastewaters containing hydrophobic organic compounds such as organochlorine pesticides. However, the results by Saez et al. (2017) corresponded to a preliminary laboratory study, so further research is still needed to validate and expand the usage of microemulsions for bioremediation purposes in a field scale, which assesses its impact on indigenous microorganisms.

Challenges Associated with Mixed Pollution Remediation

Mixed contamination by both organic and inorganic chemicals, so called co-contamination, is an environmental problem of concern in the scientific community, and the restoration of the affected environments poses a challenge for the current technologies (Arjoon et al. 2013, Tariq et al. 2016).

Co-contamination is generally studied using an unrealistic approach since, for simplicity, studies generally involve a single class of pollutant or type of contamination. The presence of multiple contaminants in the environment makes difficult not only the detection and measurement of contaminants, but also the evaluation of their ecotoxicity and environmental impact. Furthermore, this scenario makes the choice of the suitable remediation technologies difficult because they are developed and validated for specific classes of pollutants (Olaniran et al. 2013). In this context, bioremediation is a more flexible technology because living systems can perform complex reactions that include the degradation of organic pollutants and the conversion of inorganic compounds into non-toxic by-products, simultaneously or sequentially (Chen et al. 2016, Aparicio et al. 2017).

In co-contaminated environments, the type of metals and their concentrations determine the effectiveness of bioremediation because metals are more reactive than organic compounds, and subsequently more toxic. Therefore, in order to select an efficient biological system in bioremediation, the tolerance of the organism to metal toxicity should first be evaluated. Even, some studies have shown that, although the organism tolerance to metals is high, it decreases significantly in the presence of a mixture of contaminants (Olaniran et al. 2011). For example, Polti et al. (2014) evaluated the tolerance to Cr(VI) and lindane in actinobacteria. They determined that although the strain *Streptomyces* sp. A2 was slightly tolerant to Cr(VI) and lindane individually, this actinobacteria strain was completely sensitive to the mixture of both pollutants.

Metal toxicity can affect not only the microbial growth but also the biodegradation of organic pollutants by inhibiting the enzymatic systems involved in the microbial metabolism or those enzymes directly implicated in pollutant degradation. The metals affect the microbial activity through different mechanisms, which depend on the properties of the metal such as bioavailability, speciation, and redox potential, and the physicochemical properties of the soil such as texture, pH, among others (Arjoon et al. 2013). Therefore, treatment of a co-contaminated environment may have, as the primary goal, to decrease the metal bioavailability by physical (heating) or chemical (chelating agents) treatments, or by biostimulation or bioaugmentation with metal-resistant microbiota.

Bioaugmentation using metal-resistant actinobacteria, able to degrade organic pollutants, appears to be a promising remediation approach. For instance, Sprocati et al. (2012) applied a bioaugmentation approach by introducing a microbial formula composed of 12 allochthonous strains, including 4 actinobacteria (*Arthrobacter* and *Rhodococcus*). These strains were previously isolated from a chronically polluted site and were both actively hydrocarbon degraders and, in turn, metal resistant. Bioaugmentation tests demonstrated that soil components and microbial processes did not affect metal immobilization. However, biodegradation of diesel oil was similar or even higher than the biodegradation observed in absence of metals. In conclusion, this study evidenced that the microbial formula introduced as a bioaugmentation agent was able to promote hydrocarbon biodegradation under the double chemical stress posed by diesel oil and heavy metals.

Similarly, Polti et al. (2014) assessed the capability of actinobacteria to degrade lindane and detoxify Cr(VI). The authors evaluated single and mixed cultures of four actinobacteria strains in soil samples contaminated with Cr(VI) and lindane. All actinobacteria strains were able to grow and remove both contaminants as single cultures; however, the consortium formed by four bacteria (*Streptomyces* spp. A5, M7, MC1, and *Amycolatopsis tucumanensis*) showed the highest Cr(VI) removal rate. Interestingly, the single culture of *Streptomyces* sp. M7 produced the maximum lindane removal, although it was expected that the consortium could be more efficient in removing the pollutants than individual strains. However, it is necessary to consider the chemical interactions that take place in a complex matrix such as soil. Aparicio et al. (2015) evaluated the influence of physicochemical properties, including moisture, temperature, initial Cr(VI) and lindane concentrations, on the remediation of lindane and Cr(VI) by *Streptomyces* sp. M7. They found that all physicochemical variables had a significant impact on the removal rate of both lindane and Cr(VI). For example, at high concentrations of both pollutants, the major mechanism implicated in Cr(VI) removal was the high adsorption of this metal to soil particles, thereby reducing its mobility. In contrast, at low contaminant concentrations, microbial activity could account for Cr(VI) removal. Likewise, low soil moisture increased lindane removal, which is explained by its hydrophobic nature. Moreover, the resistance of *Streptomyces* sp. M7 to Cr(VI) exposure could save its metabolic activity that, in turn, would be involved in the degradation of lindane.

In a related study, Aparicio et al. (2017) used the same actinobacteria strains for vinasse bioremediation. These strains, as single or mixed cultures, were able to remove the metals present in the vinasse and reduce its biochemical oxygen demand, which is one of the main problems of this residue generated during the alcoholic distillation of sugarcane. The microbial biomass produced using vinasse as a carbon and energy source was used to bioremediate soil samples contaminated with Cr(VI) and lindane, which resulted in an good solution to two important environmental problems: the highly toxic vinasse residue and the environmental contaminants.

The bioremediation of environments contaminated with organic compounds and metals is a subject still little studied. In particular, bioremediation of complex systems such as soil requires the study of interactions between different biological, physical and chemical factors in order to understand the activity of the bacteria involved in bioremediation (Aparicio et al. 2015). This is critical to extrapolate the results obtained in laboratory conditions at a field scale.

Further studies and the support of new methodologies, such as metagenomics and metabolomics, are needed to achieve satisfactory results in complex and large co-contaminated areas. In the following section, we provide an overview on the potential applications of these emerging techniques of molecular biology in bioremediation.

Omics Technologies for Bioremediation

Over the last few years, "omics" technologies have arisen as reliable tools that strengthen biological results and are useful tools to increase the understanding of many biological processes. Omics technologies include genomics, transcriptomics, proteomics and metabolomics. Although they may be applied individually, a combination of them is often used to get a more detailed picture of a particular physiological cellular state. The more omics technologies are applied, the more precise and reliable the results will be. Molecules that can be analyzed by omics approaches are schematized in Fig. 2.

Generally, "omics" allow accurately identifying and quantifying several kinds of biomolecules including proteins, carbohydrates and lipids. Proteomics and metabolomics are the most recently developed "omics" and the continuous advances in liquid chromatography and mass spectrometry contribute to the reliability of these technologies.

Mass spectrometry-based proteomics, in addition to the available software, made possible the identification and quantification of proteins regulated in response to physiological alterations by an increase or decrease in their abundances. Besides the detection of unmodified proteins by label-free proteomic, the study of oxidative modification of proteins is important to understand, for instance, the regulation of degradative pathways. Although oxidative modifications often cause irreparable damage, not all protein oxidations are irreversible. Particularly, oxidation of the thiol-group of cysteines are reversed by antioxidant *in vivo*, such as the thioredoxin or glutaredoxin systems. It has been demonstrated that oxidative thiol modification is an effective and elegant mechanism for activity modulation of regulatory proteins. For example, OxICAT methodology was successfully applied in the identification

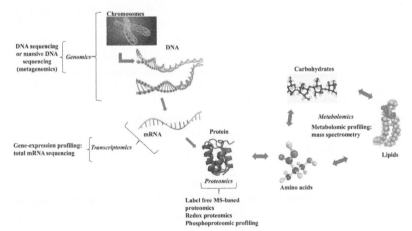

Fig. 2. Omics technologies and their molecular level of application.

of redox-regulated proteins (Dávila Costa et al. 2015, Leichert et al. 2008). As for metabolomics, it was developed to elucidate the cellular metabolome. The nature and abundance of each metabolite present in a particular physiological cellular state represents the metabolome. These metabolites may be detected by mass spectrometry, infrared spectroscopy or ultraviolet spectroscopy, among other techniques (Han et al. 2009, Zhang et al. 2012).

Integrated Omics to Understand Bioremediation Processes

According to the central dogma of molecular genetics, a direct link between mRNA expression levels and protein abundance could be established. Nevertheless, many studies demonstrated that to assume this correlation may lead to an incorrect interpretation of a physiological state. Thus, a combination of transcriptomic and proteomic techniques is strongly recommended. Similarly, the combination of transcriptomics and metabolomics has become a powerful approach to establish a cause-effect relationship between transcripts and metabolites (functional elements) in cells.

Integrated omics approaches have been successfully applied to study the potential of bacteria for bioremediation. For instance, multi-omics data were integrated to investigate the metabolic network of *Mycobacterium vanbaalenii* PYR-1 to degrade polycyclic aromatic hydrocarbons. Proteomic data and metabolic analysis allowed the identification of 160 proteins implicated in the degradation of the polyaromatic hydrocarbons (Kweon et al. 2011). *Polaromonas* sp. strain JS666 was reported as the only isolated bacterium able to use cis-dichloroethene as a sole carbon and energy source (Jennings et al. 2009). Proteomic and transcriptomic approaches revealed the up-regulation of genes encoding glutathione *S*-transferase, cyclohexanone monooxygenase, and haloacid dehalogenase during the degradation of cis-dichloroethene in this novel bacterium. Based on these results, a degradation pathway of this contaminant was proposed (Jennings et al. 2009). Independently, or in parallel with the use of bacteria, phytoremediation is an interesting alternative to traditional soil remediation. In spite of the potential of phytoremediation, it still has to gain reliability as a viable remediation alternative because of its variable effectiveness. Interestingly, Bell et al. (2014) postulated how the application of integrated omics methodologies will help to understand the interaction between plant and microorganisms in order to improve and design reliable phytoremediation processes.

Concluding Remarks

Actinobacteria have presented a big potential as tools for bioremediation of pesticides and metals, among other contaminants, based on their physiological and metabolic versatility. Moreover, recent works demonstrated that actinobacteria strains are able to remove both pesticides and metals, either simultaneously or sequentially, being excellent candidates for the bioremediation of environments co-contaminated with both inorganic and organic environmental contaminants, which reflect in fact the real challenge of the biotechnology.

In the last 15 years, different strategies were developed to enhance the capabilities of actinobacteria in bioremediation. Among them, the use of immobilized actinobacteria for this purpose resulted in high performances, greater degradative enzymes production, and extended biotransformation reaction time. Moreover, it enables the re-utilization of the immobilized cells.

Phytoremediation techniques, involving the interactions between plants and microorganisms, have been also proposed as an eco-friendly method for cleaning up polluted soils. In this relation, actinobacteria would improve pollutant mobilization and their uptake by plants.

The application of microemulsions combined with actinobacteria represents a potential tool in soil washing technologies or *ex situ* bioremediation processes of wastewaters containing hydrophobic organic compounds such as organochlorine pesticides.

The use of integrated omic methodologies applied to bioremediation would improve the performance of the processes through engineering techniques and would allow the design of reliable processes.

In spite of the advances reached in bioremediation, the information available on bioremediation using actinobacteria is related to laboratory scale and hence there is still a lack of information available on bioremediation technologies applicable for actual field-scale. For this reason, further and deeper research is needed to expand the usage of the mentioned different techniques for bioremediation purposes in field scale using actinobacteria.

Acknowledgements

This work was supported by Consejo Nacional de Investigaciones Científicas y Técnicas (CONICET, PIP 0372, PIO-CONICET-YPF), Agencia Nacional de Promoción Científica y Tecnológica (ANPCyT, PICT 0480, 0141 and 0229) and Secretaría de Ciencia, Arte e Innovación Tecnológica (SCAIT, PIUNT D504).

References

Ahamad, P.Y.A. and A.A.M. Kunhi. 2011. Enhanced degradation of phenol by *Pseudomonas* sp. CP4 entrapped in agar and calcium alginate beads in batch and continuous processes. Biodegradation 22: 253–265.

Alessandrello, M.J., M.S. Juárez Tomás, E.E. Raimondo, D.L. Vullo and M.A. Ferrero. 2017. Petroleum oil removal by immobilized bacterial cells on polyurethane foam under different temperature conditions. Mar. Pollut. Bull. DOI: 10.1016/j.marpolbul.2017.06.040.

Álvarez, A., C.S. Benimeli, J.M. Saez, A. Giuliano and M.J. Amoroso. 2015. Lindane removal using *Streptomyces* strains and maize plants: a biological system for reducing pesticides in soils. Plant Soil 395: 401–413.

Alvarez, A., C.S. Benimeli, J.M. Saez, M.S. Fuentes, S.A. Cuozzo, M.A. Polti et al. 2012. Bacterial bio-resources for remediation of hexachlorocyclohexane. Int. J. Mol. Sci. 13: 15086–15106.

Alvarez, A., J.M. Saez, J.S. Davila Costa, V.L. Colin, M.S. Fuentes, S.A. Cuozzo et al. 2017. Actinobacteria: Current research and perspectives for bioremediation of pesticides and heavy metals. Chemosphere 166: 41–62.

Aparicio, J.D., M.Z. Simón Solá, C.S. Benimeli, M.J. Amoroso and M.A. Polti. 2015. Versatility of *Streptomyces* sp. M7 to bioremediate soils co-contaminated with Cr(VI) and lindane. Ecotoxicol. Environ. Saf. 116: 34–39.

Aparicio, J.D., C.S. Benimeli, C.A. Almeida, M.A. Polti and V.L. Colin. 2017. Integral use of sugarcane vinasse for biomass production of actinobacteria: Potential application in soil remediation. Chemosphere 181: 478–484.

Arjoon, A., A.O. Olaniran and B. Pillay. 2013. Co-contamination of water with chlorinated hydrocarbons and heavy metals: Challenges and current bioremediation strategies. Int. J. Environ. Sci. Technol. 10: 395–412.

Bayat, Z., M. Hassanshahian and S. Cappello. 2015. Immobilization of microbes for bioremediation of crude oil polluted environments: A mini review. Open Microbiol. J. 9: 48–54.

Bazot, S., P. Bois, C. Joyeux and T. Lebeau. 2007. Mineralization of diuron [3-(3,4-dichlorophenyl)-1, 1-dimethylurea] by co-immobilized *Arthrobacter* sp. and *Delftia acidovorans*. Biotechnol. Lett. 29: 749–754.

Bazot, S. and T. Lebeau. 2009. Effect of immobilization of a bacterial consortium on diuron dissipation and community dynamics. Bioresour. Technol. 100: 4257–4261.

Becerra-Castro, C., A. Prieto-Fernández, P.S. Kidd, N. Weyens, B. Rodríguez-Garrido, M. Touceda-González et al. 2013. Improving performance of *Cytisus striatus* on substrates contaminated with hexachlorocyclohexane (HCH) isomers using bacterial inoculants: developing a phytoremediation strategy. Plant Soil 362: 247–260.

Bell, T.H., S. Joly, F.E. Pitre and E. Yergeau. 2014. Increasing phytoremediation efficiency and reliability using novel omics approaches. Trends Biotechnol. 32: 271–280.

Benimeli, C.S., M.J. Amoroso, A.P. Chaile and G.R. Castro. 2003. Isolation of four aquatic streptomycetes strains capable of growth on organochlorine pesticides. Bioresour. Technol. 89: 133–138.

Benimeli, C.S., G.R. Castro, A.P. Chaile and M.J. Amoroso. 2006. Lindane removal induction by *Streptomyces* sp. M7. J. Basic Microbiol. 46: 348–357.

Benimeli, C.S., G.R. Castro, A.P. Chaile and M.J. Amoroso. 2007. Lindane uptake and degradation by aquatic *Streptomyces* sp. strain M7. Int. Biodeterior. Biodegradation 59: 148–155.

Bérdy, J. 2005. Bioactive microbial metabolites. J. Antibiot. 58: 1–26.

Bragato, M. and O.A. El Seoud. 2003. Formation, properties, and "*ex situ*" soil decontamination by vegetable oil-based microemulsions. J. Surfactants Deterg. 6: 143–150.

Briceño, G., M.S. Fuentes, O. Rubilar, M. Jorquera, G. Tortella, G. Palma et al. 2013. Removal of the insecticide diazinon from liquid media by free and immobilized *Streptomyces* sp. isolated from agricultural soil. J. Basic Microbiol. 53: 1–10.

Bustamante, M., N. Durán and M.C. Diez. 2012. Biosurfactants are useful tools for the bioremediation of contaminated soil: A review. J. Soil Sci. Plant Nutr. 12: 667–687.

Castro Dantas, T.N., A.A. Dantas Neto, M.C.P.A. Moura, E.L. Barros Neto, K.R. Forte and R.H.L. Leite. 2003. Heavy metals extraction by microemulsions. Water Res. 37: 2709–2717.

Castro Dantas, T.N., K.R. Oliveira, A.A. Dantas Neto and M.C.P.A. Moura. 2009. The use of microemulsions to remove chromium from industrial sludge. Water Res. 43: 1464–1470.

Chaile, A.P., N. Romero, M.J. Amoroso, M.V. Hidalgo and M.C. Apella. 1999. Organochlorine pesticides in Salí River. Tucumán-Argentina. Rev. Bol. Ecol. 6: 203–209.

Chen, F., M. Tan, J. Ma, S. Zhang, G. Li and J. Qu. 2016. Efficient remediation of PAH-metal co-contaminated soil using microbial-plant combination: A greenhouse study. J. Hazard. Mater. 302: 250–261.

Compant, S., C. Clément and A. Sessitsch. 2010. Plant growth-promoting bacteria in the rhizo- and endosphere of plants: their role, colonization, mechanisms involved and prospects for utilization. Soil Biol. Biochem. 42: 669–678.

Dantas Neto, A.A., T.N. Castro Dantas and M.C.P.A. Moura. 2004. Evaluation and optimization of chromium removal from tannery effluent by microemulsion in the Morris extractor. J. Hazard. Mater. 114: 115–122.

Dávila Costa, J.S., V.H. Albarracín and C.M. Abate. 2011a. Cupric reductase activity in copper-resistant *Amycolatopsis tucumanensis*. Water Air Soil Pollut. 216: 527–535.

Dávila Costa, J.S., V.H. Albarracín and C.M. Abate. 2011b. Responses of environmental *Amycolatopsis* strains to copper stress. Ecotoxicol. Environ. Saf. 74: 2020–2028.

Dávila Costa, J.S., O.M. Herrero, H.M. Alvarez and L. Leichert. 2015. Label-free and redox proteomic analyses of the triacylglycerol-accumulating *Rhodococcus jostii* RHA1. Microbiology 161: 593–610.

Dzionek, A., D. Wojcieszyńska and U. Guzik. 2016. Natural carriers in bioremediation: A review. Electron. J. Biotechnol. 23: 28–36.

EPA. 2016. United States Environmental Protection Agency. https://www3.epa.gov/ (Accessed May 2016).

Fanun, M. 2012. Microemulsions as delivery systems. Curr. Opin. Colloid Interface Sci. 17: 306–313.

Fernández, D.S, M.E. Puchulu and S.M. Georgieff. 2014. Identification and assessment of water pollution as a consequence of a leachate plume migration from a municipal landfill site (Tucumán, Argentina). Environ. Geochem. Health 36: 489–503.

Flanagan, J. and H. Singh. 2006. Microemulsions: a potential delivery system for bioactives in food. Crit. Rev. Food Sci. Nutr. 46: 221–237.

Fuentes, M.S., C.S. Benimeli, S.A. Cuozzo and M.J. Amoroso. 2010. Isolation of pesticide-degrading actinomycetes from a contaminated site: Bacterial growth, removal and dechlorination of organochlorine pesticides. Int. Biodeterior. Biodegradation 64: 434–441.

Fuentes, M.S., G.E. Briceño, J.M. Saez, C.S. Benimeli, M.C. Diez and M.J. Amoroso. 2013. Enhanced removal of a pesticides mixture by single cultures and consortia of free and immobilized *Streptomyces* strains. Biomed. Res. Int. ID 392573.

Gerhardt, K.E., X.D. Huang, B.R. Glick and B.M. Greenberg. 2009. Phytoremediation and rhizoremediation of organic soil contaminants: potential and challenges. Plant Sci. 176: 20–30.

Han, J., R. Datla, S. Chan and C.H. Borchers. 2009. Mass spectrometry-based technologies for high-throughput metabolomics. Bioanalysis 1: 1665–1684.

Humphries, A.C., K.P. Nott, L.D. Hall and L.E. Macaskie. 2005. Reduction of Cr(VI) by immobilized cells of *Desulfovibrio vulgaris* NCIMB 8303 and *Microbacterium* sp. NCIMB 13776. Biotechnol. Bioeng. 90: 589–596.

Jennings, L.K., M.M. Chartrand, G. Lacrampe-Couloume, B.S. Lollar, J.C. Spain and J.M. Gossett. 2009. Proteomic and transcriptomic analyses reveal genes upregulated by cis-dichloroethene in *Polaromonas* sp. strain JS666. Appl. Environ. Microbiol. 75: 3733–3744.

Jézéquel, K. and T. Lebeau. 2008. Soil bioaugmentation by free and immobilized bacteria to reduce potentially phytoavailable cadmium. Bioresour. Technol. 99: 690–698.

Kidd, P.S., J. Barceló, M.P. Bernal, F. Navari-Izzo, C. Poschenrieder, S. Shilev et al. 2009. Trace element behaviour at the root–soil interface: Implications in phytoremediation. Environ. Exp. Bot. 67: 243–259.

Kidd, P., M. Mench, V. Álvarez-López, V. Bert, I. Dimitriou, W. Friesl-Hanl et al. 2015. Agronomic practices for improving gentle remediation of trace element-contaminated soils. Int. J. Phytoremediation 17: 1005–1037.

Kieser, T., M.J. Bibb, M.J. Buttner, K.F. Chater and D.A. Hopwood. 2000. Practical *Streptomyces* Genetics. John Innes Foundation, Colney, Norwich NR4 7UH, England.

Kitova, A.E., T.N. Kuvichkina, A.Y. Arinbasarova and A.N. Reshetilov. 2004. Degradation of 2,4-dinitrophenol by free and immobilized cells of *Rhodococcus erythropolis* HL PM-1. Appl. Biochem. Microbiol. 40: 258–261.

Kumari, B., V.K. Madan and T.S. Kathpal. 2007. Status of insecticide contamination of soil and water in Haryana, India. Environ. Monit. Assess. 136: 239–244.

Kweon, O., S.J. Kim, R.D. Holland, H. Chen, D.W. Kim, Y. Gao et al. 2011. Polycyclic aromatic hydrocarbon metabolic network in *Mycobacterium vanbaalenii* PYR-1. J. Bacteriol. 193: 4326–4337.

Lammel, G., Y.S. Ghimb, A. Grados, H. Gao, H. Hühnerfuss and R. Lohmann. 2007. Levels of persistent organic pollutants in air in China and over the Yellow Sea. Atmos. Environ. 41: 452–464.

Leichert, L.I., F. Gehrke, H.V. Gudiseva, T. Blackwell, M. Ilbert, A.K. Walker et al. 2008. Quantifying changes in the thiol redox proteome upon oxidative stress *in vivo*. Proc. Natl. Acad. Sci. USA 105: 8197–8202.

León-Santiestebán, H., M. Meraz, K. Wrobel and A. Tomasini. 2011. Pentachlorophenol sorption in nylon fiber and removal by immobilized *Rhizopus oryzae* ENHE. J. Hazard. Mater. 190: 707–712.

Malik, A., K.P. Singh and P. Ojha. 2007. Residues of organochlorine pesticides in fish from the Gomti River, India. Bull. Environ. Contam. Toxicol. 78: 335–340.

Manivasagan, P., J. Venkatesan, K. Sivakumar and S.K. Kim. 2013. Marine actinobacterial metabolites: Current status and future perspectives. Microbiol. Res. 168: 311–332.

Mansour, S. 2012. Evaluation of residual pesticides and heavy metals levels in conventionally and organically farmed potato tubers in Egypt. pp. 493–506. *In:* He, Z., R. Larkin and W. Honeycutt (eds.). Sustainable Potato Production: Global Case Studies. Springer, Dordrecht, Netherlands.

Martins, S.C.S., C.M. Martins and S.T. Santaella. 2013. Immobilization of microbial cells: A promising tool for treatment of toxic pollutants in industrial wastewater. Afr. J. Biotechnol. 12: 4412–4418.

Mench, M., N. Lepp, V. Bert, J.-P. Schwitzguébel, S.W. Gawronski, P. Schröder et al. 2010. Successes and limitations of phytotechnologies at field scale: outcomes, assessment and outlook from COST Action 859. J. Soils Sediments 10: 1039–1070.

Mesquini, J.A., A.C.H.F. Sawaya, B.G.C. López, V.M. Oliveira and N.R.S. Miyasaka. 2015. Detoxification of atrazine by endophytic *Streptomyces* sp. isolated from sugarcane and detection of nontoxic metabolite. Bull. Environ. Contam. Toxicol. 95: 803–809.

Muñoz Hernández, M., J.R. Ochoa Gómez and C. Fernández Sánchez. 2005. Formación de microemulsiones inversas de acrilamida. Revista Tecnología y Desarrollo 3: 1696-8085.

Olaniran, A.O., A. Balgobind and B. Pillay. 2011. Quantitative assessment of the toxic effects of heavy metals on 1,2-dichloroethane biodegradation in co-contaminated soil under aerobic condition. Chemosphere 85: 839–847.

Olaniran, A.O., A. Balgobind and B. Pillay. 2013. Bioavailability of heavy metals in soil: Impact on microbial biodegradation of organic compounds and possible improvement strategies. Int. J. Mol. Sci. 14: 10197–10228.

Olano, C., C. Méndez and J.A. Salas. 2009. Antitumor compounds from marine actinomycetes. Mar. Drugs 7: 210–248.

Polti, M.A., M.J. Amoroso and C.M. Abate. 2007. Chromium(VI) resistance and removal by actinomycete strains isolated from sediments. Chemosphere 67: 660–667.

Polti, M.A., R.O. García, M.J. Amoroso and C.M. Abate. 2009. Bioremediation of chromium(VI) contaminated soil by *Streptomyces* sp. MC1. J. Basic Microbiol. 49: 285–292.

Polti, M., M.J. Amoroso and C.M. Abate. 2011a. Intracellular chromium accumulation by *Streptomyces* sp. MC1. Water Air Soil Pollut. 214: 49–57.

Polti, M.A., M.C. Atjian, M.J. Amoroso and C.M. Abate. 2011b. Soil chromium bioremediation: synergic activity of actinobacteria and plants. Int. Biodeterior. Biodegradation 65: 1175–1181.

Polti, M.A, J.D. Aparicio, C.S. Benimeli and M.J. Amoroso. 2014. Simultaneous bioremediation of Cr(VI) and lindane in soil by actinobacteria. Int. Biodeterior. Biodegradation 88: 48–55.

Poopal, A.C. and R.S. Laxman. 2009. Chromate reduction by PVA-alginate immobilized *Streptomyces griseus* in a bioreactor. Biotechnol. Lett. 31: 71–76.

Ridolfi, A.S., G.B. Álvarez and M.E. Rodríguez Giraul. 2014. Organochlorinated contaminants in general population of Argentina and other Latin American Countries. pp. 17–40. *In:* Alvarez, A. and M.A. Polti (eds.). Bioremediation in Latin America. Current Research and Perspectives. Springer, New York, NY, USA.

Saez, J.M., C.S. Benimeli and M.J. Amoroso. 2012. Lindane removal by pure and mixed cultures of immobilized actinobacteria. Chemosphere 89: 982–987.

Saez, J.M., A. Alvarez, C.S. Benimeli and M.J. Amoroso. 2014. Enhanced lindane removal from soil slurry by immobilized *Streptomyces* consortium. Int. Biodeterior. Biodegrad. 93: 63–69.

Saez, J.M., J.D. Aparicio, M.J. Amoroso and C.S. Benimeli. 2015. Effect of the acclimation of a *Streptomyces* consortium on lindane biodegradation by free and immobilized cells. Process Biochem. 50: 1923–1933.

Saez, J.M., V. Casillas García and C.S. Benimeli. 2017. Improvement of lindane removal by *Streptomyces* sp. M7 by using stable microemulsions. Ecotoxicol. Environ. Safe. 144: 351–359.

Salam, J.A. and N. Das. 2013. Enhanced biodegradation of lindane using oil-in-water biomicroemulsion stabilized by biosurfactant produced by a new yeast strain, *Pseudozyma* VITJzN01. J. Microbiol. Biotechnol. 23: 1598–1609.

Sanchez-Dominguez, M., K. Pemartin and M. Boutonnet. 2012. Preparation of inorganic nanoparticles in oil-in-water microemulsions: a soft and versatile approach. Curr. Opin. Colloid Interface Sci. 17: 297–305.

Sprocati, A.R., C. Alisi, F. Tasso, P. Marconi, A. Sciullo, V. Pinto et al. 2012. Effectiveness of a microbial formula, as a bioaugmentation agent, tailored for bioremediation of diesel oil and heavy metal co-contaminated soil. Process Biochem. 47: 1649–1655.

Talegaonkar, S., A. Azeem, J. Ahmad Farhan, K. Khar Roop and A. Shadab. 2008. Microemulsions: a novel approach to enhanced drug delivery. Recent Pa. Drug Deliv. Formul. 2: 238–257.

Tallur, P.N., S.I. Mulla, V.B. Megadi, M.P. Talwar and H.Z. Ninnekar. 2015. Biodegradation of cypermethrin by immobilized cells of *Micrococcus* sp. strain CPN 1. Braz. J. Microbiol. 46: 667–672.

Tariq, S.R., M. Shafiq and G.A. Chotana. 2016. Distribution of heavy metals in the soils associated with the commonly used pesticides in cotton fields. Scientifica ID 7575239.

Vangronsveld, J., R. Herzig, N. Weyens, J. Boulet, K. Adriaensen, A. Ruttens et al. 2009. Phytoremediation of contaminated soils and groundwater: lessons from the field. Environ. Sci. Pollut. Res. 16: 765–794.

Vargas-Ruiz, S., C. Schulreich, A. Kostevic, B. Tiersch, J. Koetz, S. Kakorin et al. 2016. Extraction of model contaminants from solid surfaces by environmentally compatible microemulsions. J. Colloid. Interface Sci. 471: 118–126.

Wang, P. and A.A. Keller. 2008. Particle-size dependent sorption and desorption of pesticides within a water–soil–nonionic surfactant system. Environ. Sci. Technol. 42: 3381–3387.

Weyens, N., D. van der Lelie, S. Taghavi, L. Newman and J. Vangronsveld. 2009. Exploiting plant–microbe partnerships to improve biomass production and remediation. Trends Biotechnol. 27: 591–598.

Whitman, W., M. Goodfellow, P. Kämpfer, H.-J. Busse, M.E. Trujillo, W. Ludwig et al. 2012. Bergey's Manual of Systematic Bacteriology. Springer-Verlag, New York, USA.

Worakitkanchanakul, W., T. Imura, T. Fukuoka, H. Morita, H. Sakai, M. Abe et al. 2008. Aqueous-phase behavior and vesicle formation of natural glycolipid biosurfactant, mannosylerythritol lipid-B. Colloids Surf. B 65: 106–112.

Zhang, A., H. Sun, P. Wang, Y. Han and X. Wang. 2012. Modern analytical techniques in metabolomics analysis. Analyst 137: 293–300.

Zhang, H., J. Tang, L. Wang, J. Liu, R.G. Gurav and K. Sun. 2016. A novel bioremediation strategy for petroleum hydrocarbon pollutants using salt tolerant *Corynebacterium variabile* HRJ4 and biochar. J. Environ. Sci. 47: 7–13.

Zhang, Y., J.W.C. Wong, Z. Zhao and A. Selvam. 2011. Microemulsion-enhanced remediation of soils contaminated with organochlorine pesticides. Environ. Technol. 32: 1915–1922.

Zhang, Y., S. Ge, M. Jiang, Z. Jiang, Z. Wang and B. Ma. 2014. Combined bioremediation of atrazine-contaminated soil by *Pennisetum* and *Arthrobacter* sp. strain DNS10. Environ. Sci. Pollut. Res. 21: 6234–6238.

Zheng, G. and J.W.C. Wong. 2010. Application of microemulsion to remediate organochlorine pesticides contaminated soils. *In*: Proceedings of the 26th Annual International Conference on Soils, Sediments, Water and Energy. USA 15: 4.

Zheng, G., Z. Zhao and J.W.C. Wong. 2011. Role of non-ionic surfactants and plant oils on the solubilization of organochlorine pesticides by oil-in-water microemulsions. Environ. Technol. 32: 269–279.

Zheng, G., A. Selvam and J.W.C. Wong. 2012. Oil-in-water microemulsions enhance the biodegradation of DDT by *Phanerochaete chrysosporium*. Bioresour. Technol. 126: 397–403.

Chapter 8

Improving Agricultural Soils with Organic Wastes from Wastewater Plants, Salmon Aquaculture and Paper-Pulp Industry

José E. Celis[1,]* and *Marco A. Sandoval*[2]

Agricultural Practices and Soil Degradation

Agricultural soils constitute natural, fragile and very complex ecosystems where physical, chemical and biological phenomena occur. Soils are formed gradually and slowly by rock disintegration and decomposition of organic materials. For this reason, soils are built up of solid, liquid and gaseous elements. The solid phase is formed by mineral material and organic material. The liquid phase consists of water, which is absorbed by the solid particles of the soil, forming what is known as the soil solution. The gaseous phase is the air that uses the porous spaces that leave solid particles between them, when these spaces are not saturated with water.

As a result of industrial development, many anthropogenic activities have affected the physical, chemical and biological properties of the soil. At present, 12% of the Earth's land surface (13.2 billion ha) is used in crop production, whereas

---

[1] Department of Animal Science, Faculty of Veterinary Sciences, Universidad de Concepción, Av. Vicente Méndez 595, Chillán, Chile.
[2] Department of Soil and Natural Resources, Faculty of Agronomy, Universidad de Concepción, Av. Vicente Méndez 595, Chillán, Chile.
  Email: masandov@udec.cl
* Corresponding author: jcelis@udec.cl

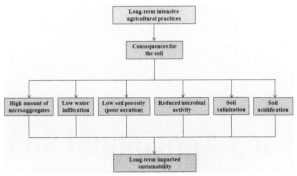

Fig. 1. Soil properties affected by anthropogenic agricultural activities.

degraded lands reach 25% (FAO 2011). Excessive land cultivation, pesticide use, soil compaction, monoculture, stubble burning, and contamination with organic and inorganic compounds are among the most aggressive threats to soil quality. In fact, current land management practices and the existence of natural slopes in agricultural land have led to the deterioration of soil (Sandoval-Estrada et al. 2008), seriously affecting the sustainability of terrestrial ecosystems (Fig. 1). Accordingly, several soil conservation measures or bioremediation practices have been implemented in order to reduce the impact of unsuitable land management practices and recover degraded soils. Soil amendment with organic wastes such as cattle manure and sludge from sewage treatment plants is a common agricultural practice to combat the loss of soil organic matter and stimulate microbial communities (biostimulation). The next sections will deal with the importance of organic carbon to soil function and the beneficial effects of several organic wastes from municipal wastewater treatment plants, salmon plants and paper-pulp industry on soil fertility.

The Role of Organic Carbon in the Soil

From a physical point of view, structural stability of soil may vary according to organic matter (OM) content of soil as well as management practices (Six et al. 2000). In this sense, Oades and Waters (1991) described an aggregate hierarchy in the structure of soils, where there are macroaggregates ($\geq$ 0.25 mm) and microaggregates (< 0.25 mm). Soil macroaggregate stability also varies with OM and management practices. There is evidence that, when agricultural production is intensified, the carbon present in soil macroaggregates decreases (Six et al. 2000). In addition, immature and more unstable OM is mostly found in macroaggregates (Elliott 1986, Jastrow et al. 1996) and, consequently, stability is more sensitive to management practices and soil OM content (Tisdall and Oades 1982). Therefore, management practices can increase or decrease the formation of different types of aggregates, affecting the structural stability of the soil, either positively or negatively (Fig. 2), and in turn the general sustainability of the soil system (Carter et al. 2003).

The increase of the microaggregate fractions is related to the loss of structural stability and quality of agricultural soils (Carter 2004). Degraded soils have low water infiltration rate, poor aeration, low biological activity and a reduced movement

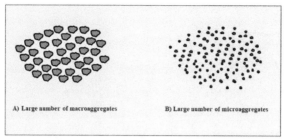

A) Large number of macroaggregates   B) Large number of microaggregates

Fig. 2. State of aggregation of soil under two different agricultural management practices over time: sustainable agriculture (A), and intensive agriculture (B).

of nutrients, which result in poor root development (Traore et al. 2000). Additionally, other studies have indicated that a reduction in macroaggregate formation contributes to an increase in greenhouse gases due to the mineralization of organic carbon. As a result, this process tends to facilitate the erosive processes of soils (Powlson et al. 1987, Gupta and Germida 1988).

From a biological point of view, soil microbial activity is greater when soil OM content is higher. Therefore, the addition of organic materials to the soil increases the quantity of carbon, leading to a greater microbial activity. In turn, this biological response favors the effective productivity of soils due to the flow of nutrients that originated from mineralized organic wastes. The incorporation of organic wastes into the soil (a practice known as amendment) under good moisture and aeration conditions favors the growth and development of microorganisms, which decompose organic materials. This is the reason why some aggressive management practices, such as stubble burning and pesticide application, negatively affect the soil function by reducing microbial biomass and any biological activity.

Organic Amendments and Decomposition of Organic Matter into Soils

The use of organic amendments in agricultural soils is an ancient practice that has evolved along with technological advances in agriculture. Ancient people used animal wastes and vegetable residues as the only nutrient source for their crops. When the use of inorganic fertilizers became a regular farming practice, these amendments became the main supply source of plant nutrients, particularly in intensive agriculture.

In general, organic waste is characterized by a high content of N, P and K. Accordingly, these organic amendments have been used as a complement to the contribution of inorganic sources, and furthermore, constitute a bioremediation practice aimed at recovering the physical, chemical and biological properties of the soil. Organic amendment applications require high rates of OM decomposition to be effective. This task is performed by a wide variety of microorganisms, such as bacteria, fungi, protozoa, actinomycetes, and yeast (Pinochet et al. 2001). Biological stabilization implies an inhibition of the nitrification process. Therefore, the reduction in the biological activity occurring at a certain period of time following the application of organic amendments is a clear indicator of the OM stabilization.

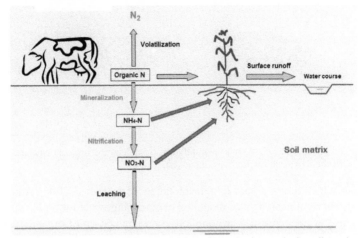

Fig. 3. Chemical transformation and environmental fate of organic nitrogen in the soil matrix.

In fact, the application of organic wastes to the soil first produces an increase in the microbial activity because new biomass and OM are incorporated into the soil, stimulating native microbiota due to the entry of fresh carbon (Ros et al. 2006). Similarly, the N found in organic amendments must be first mineralized before it can be used by plants. This process is performed by bacteria and fungi from the soil, resulting in NH_4-N, which is absorbed by the roots or adsorbed on clay particles. Afterwards, *nitrosomonas* bacteria convert NH_4-N into NO_3-N, which is the chemical form of N most efficiently uptaken by plants (Fig. 3).

There is an interaction between P and N when using organic amendments in soils. Phosphorous is used in the energetic process of biota at cell level, favoring the microbial activity and accelerating the N mineralization (Abdolzadeh et al. 2010). Also, this potentiation effect can be explained by the favorable conditions that roots provide for the development of bacteria and fungi (Ibrikci et al. 1994).

NO_3-N is very mobile in the soil and can become an environmental problem when it exceeds a certain amount, contaminating groundwater and surface water courses or contributing to greenhouse gases. Therefore, organic residues should be applied in a sustainable manner (Sawyer et al. 2003). The use of high amounts of organic wastes can increase the risk of N losses to the environment through leaching, volatilization, denitrification and surface runoff (Paul et al. 1993). Therefore, a soil fertilization program should be based on nitrogen balance.

Sludge from Wastewater Plants, Salmon Aquaculture, and Paper-Pulp Industries

Municipal sludge (also known as biosolids), aquaculture sludge such as those coming from farmed salmon, and cellulose sludge are three types of organic wastes with the potential for promoting soil fertility. The physical and chemical characteristics of these organic residues are summarized in Table 1.

The sludges from municipal wastewater treatment plants correspond to sub-products (liquids, solids, and semisolids) generated during the treatment of wastewater. The P content of these biosolids may vary considerably between 1.2 and 3%, whereas 50 to 90% of the total N may be in an organic form, depending largely on the type of wastewater purification (Sommers 1977).

The lumber industry also produces large amounts of sludge in the form of final solid waste, which is recovered from the wastewater treatment process in cellulose and paper mills (Geng et al. 2007). Sludge from the paper-pulp industry consists mainly of a mixture of cellulose fibers and inorganic materials, such as ashes (San Martín et al. 2016). Depending on the type of process, each ton of product generates between 58 and 234 kg of sludge (Scott and Smith 1995). For example, in the United States, the production of one ton of paper generates about 40–50 kg of sludge (Joyce et al. 1979).

Regarding the aquaculture industry, salmon farming has experienced a significant growth in the last few decades, which has resulted in the production of considerable volumes of organic wastes in the intensive salmon farming systems (Teuber et al. 2005). It is estimated that the salmon industry generates 1.4 tons of organic wastes for every ton of salmon produced, which corresponds mainly to non-consumed food and feces of the fish. This waste falls directly to the tank bottom (land-salmon sludge), lake bottom (lake-salmon sludge) or sea bottom (sea salmon sludge), ultimately causing contamination at a local scale. The chemical composition of salmon sludge varies considerably because the salmon diet changes during fish development. Due to the high concentration of Na, application of sea salmon sludge in the soils may significantly increase the Na content of the soil. Therefore, a prior assessment of the chemical properties of the target soil should be taken into account when this type of organic waste is used as a soil amendment. Unlike wastewater treatment plant and cellulose sludge, farmed salmon sludge has a lower content of

Table 1. Physical and chemical characteristics of sludge from municipal wastewater plants, salmon aquaculture and paper-pulp industry (dry matter basis).

Parameter	Wastewater treatment plants	Salmon aquaculture			Paper pulp industry
		Land farming	Lake farming	Sea farming	
pH (water)	5.9–7.0	6.0–7.0	6.5–6.7	7.2–7.8	5.65–7.3
OM (%)	33.9–61.2	15.1–18.3	18.6–20.7	11.1–14.7	25.3–76.1
Density (g/cm³)	0.73–1.2	–	–	1.05–1.07	0.84
Total N (%)	2.3–5.89	0.30–0.57	0.25–1.35	0.13–0.8	0.59–1.68
P (%)	1.2–3.0	0.78	2.28–3.3	0.74–1.6	0.29–0.31
K (%)	0.1–0.3	0.12	0.10	0.40–0.63	0.0014–0.09
Ca (%)	1.34–11.6	1.55	5.07	2.51–2.62	0.01–0.25
Mg (%)	0.31–0.45	0.38	0.41	1.49–1.65	0.003–0.14
Na (%)	0.05–4.06	0.09	0.28	5.0–11.8	0.01–0.95
EC (dS/m)	3.6	0.2	0.35	27.03	1.93–16.7
Ratio C/N	4.9–18	10.5–15.3	7.5	9.8–17.9	6.0–115.0

References: Sommers (1977), Scott and Smith (1995), Metcalf and Eddy (1995), Mazzarino et al. (1997), Pinochet et al. (2001), Tamoutsidis et al. (2002), Smernik et al. (2003), Geng et al. (2007), Aravena et al. (2007), Celis et al. (2008), Ríos et al. (2012), Celis et al. (2013), Bouabid et al. (2014), San Martín et al. (2016).

OM (approximately 20 percentage dry mass, Table 1), which is probably due to the natural contribution of sand at the bottom of the sea or lakes.

Use of Municipal Sewage Sludge as Soil Amendment

Previous research has reported that biosolids can be used as fertilizers (Shober et al. 2003) because of their high OM content that improves the physical and biological properties of soils (Sims and Pierzynski 2000, Barbarick et al. 2004). The application rate is usually calculated on the basis of the N and P requirements of crops. As a result, there is an increase in soil productivity due to the addition of OM, which occurs after the application of the sludge (Epstein 2002).

The use of biosolids in degraded soils has proven effective in agricultural production as they act as an organic fertilizer. For example, the study by Celis et al. (2008) showed that the production of annual ryegrass in Entisol and Alfisol soils with OM contents of 2.9 and 2.5 percentage, respectively, was similar to the soils amended with 50 tons/ha of biosolids and the soils treated with inorganic fertilization (IF = 160 kg N/ha + 200 kg P/ha + 130 kg K/ha) (Fig. 4).

Soil biota and organic products also contribute to the development of soil structure (Chan et al. 2003). The soil structure has a fundamental function in every soil system and in the supporting process of the soil biota, which can be quantified through the stability of the aggregates of the soil (Bronick and Lal 2005). Thus, aggregate stability is a good indicator of soil structure, which results from the arrangement of particles, flocculation, and cementation (Six et al. 2000). A study by Sandoval-Estrada et al. (2010) found that the application of biosolids, at rates between 100 and 150 tons/ha, to an Entisol soil cultivated with annual ryegrass significantly increased the macroaggregate fraction (Table 2).

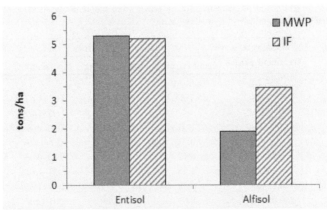

Fig. 4. Dry matter yield (tons/ha) of annual ryegrass (*Lolium multiflorum* L.) expressed as aboveground biomass from two degraded soils amended with municipal water plant sludge (MWP) at 50 tons/ha and inorganic fertilization (IF).

Table 2. Effect of municipal sewage sludge applied at different rates on the physical properties of an Entisol cropped with *Lolium multiflorum*. Data taken from Sandoval-Estrada et al. (2010).

Amendments	Sieve diameter (mm)					Ratio of Ma	AWD (mm)
	4–2	2–1	1–0.5	0.5–0.25	0.25–0.05		
MWP 25	0.6	5.1	23.8	24.1	25.7	53.6 c	0.25 c
MWP 50	0.5	6.2	25.8	25.8	25.0	58.2 bc	0.28 bc
MWP 75	1.6	7.3	26.2	22.8	26.3	57.8 bc	0.31 b
MWP 100	9.7	9.6	23.9	20.9	16.0	64.1 a	0.47 a
MWP 150	6.6	9.7	30.2	17.7	16.4	64.2 a	0.43 a
TC	0.2	5.7	24.2	24.3	25.4	54.4 c	0.25 c

MWP: municipal wastewater plant sludge (25, 50, 75, 100 and 150 tons/ha); TC: no sludge; Ma: macroaggregates ($\geq$ 0.25 mm); AWD: average weight diameter. Means with different letters in a single column are statistically different ($p < 0.05$, post hoc Tukey test).

The higher amount of macroaggregates found in treatments amended with biosolids at higher rates is directly related to the large contribution of OM (48.3%) they provide to the soil. OM has a fundamental role in the formation of aggregates in most types of soils (Tisdall and Oades 1982). Conversely, data in Table 2 indicates that a soil without organic amendments tends to show a greater amount of microaggregates (< 0.25 mm). It is important to note that the application of municipal sludge not only favored the formation of macroaggregates, but also helped to increase the stability of the particles. This stability is given by the average weight diameter (AWD) of the aggregates. According to Le Bissonnais (1996), AWD values $\geq$ 0.4 mm indicate a stable soil structure, which can be observed at biosolid rates of 100 and 150 tons/ha (Table 2).

Additionally, there is evidence that the use of municipal wastewater sludge has a positive effect on the biological properties of degraded granitic soils. A study conducted in an Alfisol (Celis et al. 2011) found that CO_2 due to microbial respiration increases significantly when biosolid amendments are applied at higher rates in a soil cultivated with a fodder species, such as yellow serradella. The results revealed that the roots of this plant has a key role in microbial activity (Fig. 5).

It has been long recognized that addition of organic wastes to the soil stimulates microbial proliferation, which is easily assessed by an increase of soil respiration rate and enzyme activities. For example, a study conducted in Alfisol soils reported high activity of the enzymes β-glucosidase, urease, and acid phosphatase in the soils amended with 60 tons/ha of biosolids compared with unamended soils (Celis et al. 2011). Likewise, other studies performed in Inceptisol and Andisol soils have shown that the application of municipal sludge increases enzyme activity at 140 tons/ha (Pascual et al. 2007) and at 6 g/100 g soil (Arriagada et al. 2009), respectively.

Enzyme activity is strongly correlated with the presence of soil OM because it promotes enzymatic synthesis and secretion by soil microbes, and its role in the stabilization of extracellular enzymes (Tabatabai 1994). These enzymes are fundamental in the C, N, S and P cycles. For example, β-glucosidase activity is essential in the decomposition of OM (Bandick and Dick 1999). Furthermore, urease

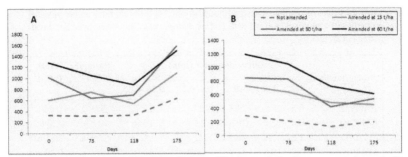

Fig. 5. Microbial respiration (µg CO_2-C/g soil) in an Alfisol amended with sewage sludge and effect of legume yellow serradela (*Ornithopus compressus* L.). A = soil cropped with legume; B = soil not cropped.

catalyzes the hydrolysis of urea into CO_2 and NH_3, which is a very important route of N cycle in soils (Tabatabai 1994). Regarding phosphatases, it is well known that this enzyme activity participates in the mineralization of organic P to generate inorganic soluble forms of P, which are later easily absorbed by plant roots (Richardson 2001).

As an example of why the measurement of soil respiration rate and enzyme activities to assess the impact of organic amendments are important, Fig. 5 illustrates the impact of sewage sludge on biological activity of Alfisol soil cultivated with some leguminous forage species. The role of rhizosphere is a key in the cycling of C, N, and P, especially if soils are physically and biologically degraded (Celis et al. 2011). Similarly, some studies have demonstrated that Alfisol soils amended with urban sludge and cultivated with yellow serradella produced a progressive increase in the aboveground and root biomass of this leguminous forage species (Table 3).

The rhizosphere creates a favorable environment for the development and proliferation of a large amount of soil bacteria and fungi, which generate an increase of soil enzyme activity (Joergensen and Emmerling 2006) and water-stable macroaggregates that improve soil structure (Borie et al. 2000). In the rhizosphere, a great variety of compounds are released by root exudates, which act as cementing agents in the formation of stable aggregates (Schreiner and Bethlenfalvay 1995). Together with this chemical cocktail, soil organisms such as earthworms, filamentous fungi and actinomycetes, among others, can help in this physical process.

Soil microorganisms have an important role in the recovery of degraded soils because of their stimulatory impact on plant growth through substances such as phytohormones and vitamins. These substances stimulate growth by improving soil structure, reducing erosion and improving the availability of nutrients (Waldrop et al. 2003). Other studies have shown similar responses in other types of soils and different crops, such as calcareous soils with a low OM content (1.3%) and cultivated with barley (Hernández et al. 1991), acid soils cultivated with corn, and fodder crops and prairies (Mosquera-Losada et al. 2017). In this context, fungi have an important role in the carbon storage and soil aggregation. The hyphae of the fungi produce a glycoprotein called glomalin, which has a strong cementing capacity and, consequently, high soil stability (Wright and Upadhyaya 1998, Rillig et al. 2002).

A study on a salinic-sodic soil (Aridisol) showed that there is a positive effect of sewage sludge on soil C dynamics as indicated by the microbial activity (Celis et al. 2013). The results showed that the highest respiration rate occurred when soil

Table 3. Aboveground and root biomass after treatment of Alfisol soil cropped with yellow serradella (*Ornithopus compressus* L.) with sewage sludge.

Treatments	Aboveground biomass (g)	Root biomass (g)
Soil not amended	5.2 b	4.3 b
Sewage sludge at 15 tons/ha	7.8 ab	6.5 a
Sewage sludge at 30 tons/ha	8.8 a	7.6 a
Sewage sludge at 60 tons/ha	8.7 a	7.9 a

Different letters in the same column indicate significant differences ($p \leq 0.05$).
Data take from Celis et al. (2011).

was amended with biosolids at the highest rate (270 tons/ha). This is attributed to additional substrate becoming available for the microbial population in the soil with increasing salt concentration, with the effects due to salinity more evident than those due to sodicity. Other studies performed in a degraded arid soil amended with urban wastes have documented similar responses (Ros et al. 2003). Additionally, K-rich wastes such as the sewage sludge (Table 1) amended to saline-sodic soils, may improve the soil chemical properties by reducing Na^+ through electrostatic repulsion, in which K^+ replaces Na^+ to form potash humate (Tchouaffe 2007).

Biosolids improve physical, chemical and biological properties of the soil. In fact, applications in degraded agricultural soils, such as Regosols, Alfisols, Entisols, can improve soil productivity by increasing pH, OM content and levels of N, P and K (Bonmanti et al. 1985, García-Gil 2000, Celis et al. 2008). The increase in OM content facilitates aggregation of soil particles and porosity, which provides a good aeration around the roots of the plants (Darwish et al. 1995). It has also been reported that the use of biosolids combined with other organic waste improves the physical parameters of the soil, particularly in the presence of a growing crop. A study by Sandoval et al. (2012) in an Alfisol showed that biosolids applications at a rate between 25 and 50 tons/ha and mixed with sawdust at a rate of 10, 25 and 50 tons/ha, resulted in an increase of macroaggregate fraction, a greater macroaggregate stability and a higher content of available water in the soil matrix. These parameters became stronger at higher rates and when the soil was cultivated with hybrid ryegrass. Other studies conducted in sandy soils have demonstrated that an increase of soil OM results in an enhanced water available in the soil matrix, as occurred in Inceptisols (Bauer and Black 1994).

Use of Salmon Wastes as Soil Amendment

The use of these organic wastes as amendments requires high rates of OM decomposition until a biologically stabilized by-product is obtained. This biological process is generally performed by a large variety of microorganisms, such as protozoa, bacteria, fungi, actinomycetes, and yeasts (Pinochet et al. 2001). Biological stabilization involves an inhibition of the nitrification process. Nitrates, which are a result of nitrification, are considered as one of the most important pollutants of groundwater and surface water (Sawyer et al. 2003), and their reduction is executed

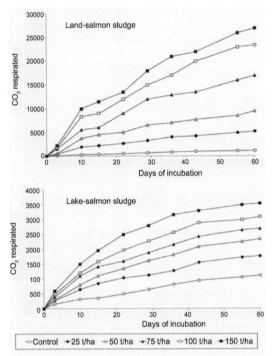

Fig. 6. Accumulated CO_2 evolution (μg CO_2/g dry soil) during *in vitro* incubation of an Entisol (0–20 cm) under different amendments with salmon sludge. Modified from Celis et al. (2009).

by microorganisms, whose activity can be measured through their respiration rate, being a good indicator of sludge stabilization.

Studies dealing with the use of salmon farm sludge in soils are scarce. Nevertheless, Celis et al. (2009) conducted a study in a degraded Entisol with a low OM content (2.9%) and found a progressive increase in the biological activity of the amended soils as the application rates increased (Fig. 6.) One of the biological parameters measured in that study was the soil respiration rate (Alef 1995), which is measured by the production of CO_2 associated with the decomposition of organic waste by multiple metabolic pathways (Sylvia et al. 1998).

The use of farmed salmon sludge in degraded soils may have a certain potential as an organic fertilizer in agricultural production. For example, the study by Celis et al. (2008) demonstrated that the production of annual ryegrass in Entisol and Alfisol soils amended with salmon sludge at different rates was similar to the production obtained with inorganic fertilization application, particularly in the Entisol soil (Table 4). Data in Table 4 show that the impact of salmon sludge on crop yield varied depending on both the type of soil and salmon sludge. Thus, the lake-salmon sludge was a better amendment in Alfisols, whereas land-salmon sludge was more efficient in Entisols.

Another study found that the application of sea salmon wastes to volcanic soils (e.g., Andisols) at a rate of 90 tons/ha caused an increase of total P concentrations up to 229 kg/ha (Teuber et al. 2005). This implies a great benefit for those soils which

Table 4. Dry matter yield of annual ryegrass expressed as aboveground biomass of two degraded soils amended with salmon sludge and inorganic fertilizer. Data taken from Celis et al. (2008).

Amendments	Sludge rate (tons/ha)					Inorganic fertilizer*
	25	**50**	**75**	**100**	**150**	
Entisol						
Land-salmon	5.37a	4.76ab	4.60ab	5.10a	5.10a	5.20a
Lake-salmon	4.80a	5.33a	3.77a	4.40a	5.37a	5.20a
Alfisol						
Land-salmon	1.98b	1.47b	0.22c	0.11c	0.07c	3.45a
Lake-salmon	3.37a	3.59a	3.48a	3.57a	3.04a	3.45a

* (140 kg N/ha + 200 kg P/ha + 130 kg kg/ha); Different letters in same row indicate significant differences ($p < 0.05$).

tend to fix P, as occurs in Andisols. The same study reported that salmon sludge applied at rates of 30, 60 and 90 tons/ha increased dry matter production of ryegrass, reaching values of 5.5, 5.9 and 6.8 tons/ha, respectively. Crop yield was similar to that obtained with inorganic fertilization (5.4 tons/ha), although the high sodium content of sea salmon sludge (Table 1) requires limiting the number of successive applications over time.

Use of Wastes from Paper-Pulp Industry as Soil Amendment

Sludge produced by the pulp and paper industry can improve the physical properties of the soil when applied as a biofertilizer due to its high content of OM, N, P and K (Camberato et al. 2006). In general, an increase in soil OM content is often observed after application of cellulose sludge to Andisol, Alfisol, Entisol and Ultisol soils (O'Brien et al. 2002, Zhang et al. 2004, Gallardo et al. 2007). Amendments with this type of sludge can improve the nutritional level and some physical characteristics of the soil, and even reduce the impact of certain contaminants such as pesticides. Nevertheless, the ammoniacal N could become immobilized in the soil due to the high C/N that this sludge contains (Table 1). In that case, an additional provision of N will be required, but this can overload the soil with this element. In order to prevent this side-effect, it is important to select the appropriate crop, and provide a high rate of OM decomposition with optimum humidity and temperature, so that OM does not accumulate in the soil (Scott and Smith 1995). During the degradation process of NO, N is released into the soil through mineralization and nitrification, which depends on temperature, humidity, pH, and bacterial activity. The release of N is optimum under controlled soil pH conditions, thus there is no overload of this element in the soil. In this way, organic amendments in acid soils require an increase of soil pH, which can be achieved with the application of calcium carbonate. The neutralizing action of this salt is due to the following reactions that take place in the soil, where the presence of water is crucial:

$$CaCO_3 + H_2O \longrightarrow Ca^{+2} + CO_3^{-2} \longrightarrow HCO_3^- + OH^-$$

$$HCO_3^- + H_2O \longrightarrow H_2CO_3 + OH^- \longrightarrow H_2O + CO_2$$

Amendments with cellulose sludge mixed with sawdust help to obtain lower pH values in the soil, which is important in basic soils. In sandy soils, it has been observed that the addition of sawdust increases the retention of water in the soil, therefore favoring the nutritional status of these soils with a low OM content (Grez and Gerding 1995).

Sludge from wastewater treatment plants in the paper-pulp industry is rich in organic matter (Table 1), so it can improve the chemical and physical properties of the soil. This beneficial impact has been described in Andisol and Ultisol soils. For example, an *in vitro* study with these soil types evaluated the effect of different application rates of cellulose sludge (0, 10, 20, 30 and 50 tons/ha) on their physicochemical properties (Aravena et al. 2007). The use of this type of sludge significantly increased the levels of N and OM in both soils, as well as the pH value, with the consequent decrease in the level of Al. It was also observed that the level of Al dropped from 1.02 to 0.23 cmol/kg in the Andisol at a rate of 50 tons/ha of cellulose sludge. On the other hand, sludge application at different rates in the Alfisol significantly improved aggregate stability, an effect that was not clearly observed in the Ultisol. The physical properties of the Ultisol can probably account for this situation since the organic materials were not completely decomposed into the soil during the incubation time (75 d). Furthermore, it has been found that the use of cellulose sludge in Alfisols at application rates of 10, 25, 75, 100 and 150 tons/ha may be advantageous due to the nutrients incorporated into the soil, and because it does not produce negative effects in the germination of seeds (Ríos et al. 2012). The concentrations of Ca and Olsen-P in the soil increased proportionally to the amended rate, while concentrations of Fe and Mn decreased as the application rate increased. It has been observed that elevated levels of Fe and Mn in saturated soils could cause toxicity in plants, as described by Torkashvand et al. (2010). They found that the use of increasing application rates of cellulose sludge decreased the concentration of Fe and Mn in the soil. The application rate > 50 tons/ha may increase the OM up to 1.6 times. This can be considered beneficial because an increase of organic matter in the soil increases water retention and improves stability of water-stable aggregates (Sandoval-Estrada et al. 2010).

Some authors have also reported that an improved water retention capacity in the soil is obtained with the use of cellulose sludge at application rates higher than 20 tons/ha (Aravena et al. 2007). This increase would be due to the significant contribution of OM, which tends to increase macroporosity, and in turn water retention into the soil matrix. The use of pellets of cellulose sludge combined with seaweed has resulted in positive effects on Entisol soils. A study carried out by San Martín et al. (2016) showed that this type of amendment increased the mineralization of N in the soil, as well as the growth rate of lettuce plants and the levels of chlorophyll in the leaves of this species. This finding highlights the potential that these amendments have as biofertilizers, which could be of concern in phytoremediation (see Chapter 6).

Concluding Remarks

The application of sludge as a soil amendment is a suitable strategy for recovering degraded soils because of the high amounts of organic matter and nutrients of sludge. Experimental evidence shows that amendments with biosolids and farmed salmon sludge improve the physicochemical and biological properties of degraded soils with low OM content. According to Jünemann (1969), this kind of soils shows a quick and significant response to organic amendments. In this order of ideas, microorganisms are key actors in the bioremediation of contaminated soils, participating as main actors (e.g., natural attenuation, biostimulation and bioaugmentation) or favoring plant growth and development (phytoremediation). Therefore, microbiota may improve the soil structure and facilitate the availability of nutrients to plants.

Biosolids, salmon sludge, and wastes from the paper industry are more efficient as soil amendments when applied to soils with a well-developed plant cover, so roots create a favorable microenvironment for microbial proliferation. The use of sludge from the pulp-paper industry, which is mainly composed of cellulose fiber with a high content of organic residues, has a higher impact on soil properties as long as it is applied in combination with seaweed. The studies discussed in this chapter demonstrated that soil amendment with wastewater plants, salmon aquaculture or paper-pulp industry may be a valuable tool to recover degraded soils within the context of bioeconomy.

Acknowledgments

Many thanks are due to the personnel of the Chemical Laboratory at Chillán (Universidad de Concepción) for their analytical support, Dr. Mario Briones for all the statistical analyses, and also Peter Lewis for the English revision. We also thank the Research Division of the Universidad de Concepción for its invaluable financial support during all those years we performed research on soil remediation.

References

Abdolzadeh, A., X. Wang, E.J. Veneklaas and H. Lambers. 2010. Effects of phosphorus supply on growth, phosphate concentration and cluster-root formation in three *Lupinus* species. Ann. Bot. 105: 365–374.

Alef, K. 1995. Soil respiration. pp. 214–218. *In*: Alef, K. and P. Nannipieri (eds.). Methods in Applied Soil Microbiology and Biochemistry. Academic Press Limited, San Diego, USA.

Aravena, C., C. Valentin, M. Diez, M. Mora and F. Gallardo. 2007. Application of sludge from cellulose treatment plant: Effect in some physical and chemical properties of volcanic soils. R.C. Suelo Nutr. Veg. 7: 1–14.

Arriagada, C., P. Pacheco, G. Pereira, A. Machuca, M. Alvear and J. Ocampo. 2009. Effect of arbuscular mycorrhizal fungal inoculation on *Eucalyptus globulus* seedlings and some soil enzyme activities under application of sewage sludge amendment. R.C. Suelo Nutr. Veg. 9: 89–101.

Bandick, A.K. and R.P. Dick. 1999. Field management effects on soil enzyme activities. Soil Biol. Biochem. 31: 1471–1479.

Barbarick, K., K. Doxtader, E. Rédente and R. Brobst. 2004. Biosolids effects on microbial activity in shrubland and grassland soils. Soil Sci. 169: 176–187.

Bauer, A. and A.L. Black. 1994. Quantification of the effect of soil organic matter content on soil productivity. Soil Sci. Soc. America J. 58: 185–193.

Borie, F., R. Rubio, A. Morales and C. Castillo. 2000. Relationships between arbuscular mycorrhizal hyphal density and glomalin production with physical and chemical characteristics of soils under no-tillage. Rev. Chil. Hist. Nat. 73: 749–756.

Bouabid, G., B. Wassate, K. Touaj, D. Nahya, K. El Falaki and M. Azzi. 2014. Effluents treatment plants sludge characterization in order to be used as solid fuels. J. Mater. Environ. Sci. 5: 1583–1590.

Bonmanti, M., J. Pujola, F. Sana, M. Soliva, M. Felipo, B. Garau et al. 1985. Chemical properties populations of nitrite oxidizers, urease and phosphatase activities in sewage sludge-amended soil. Plant Soil 84: 79–91.

Bronick, C. and R. Lal. 2005. Soil structure and management: A review. Geoderma 124: 3–22.

Camberato, J., B. Gagnon, D. Angers, M. Chantigny and L. Pan. 2006. Pulp and paper mill by-products as soil amendments and plant nutrient sources. Canadian J. Soil Sci. 784: 641–653.

Carter, M., H. Kunelius, J. Sanderson, J. Kimpinski, H. Platt and M. Bolinder. 2003. Productivity parameters and soil health dynamics under long-term 2-years potato rotations in Atlantic Canada. Soil Tillage Res. 72: 153–168.

Carter, M.R. 2004. Researching structural complexity in agricultural soils. Soil Tillage Res. 79: 1–6.

Celis, J., M. Sandoval and R. Barra. 2008. Plant response to salmon wastes and sewage sludge used as organic fertilizer on two Chilean degraded soils under greenhouse conditions. Chilean J. Agric. Res. 68: 274–283.

Celis, J., M. Sandoval and E. Zagal. 2009. Evolution of microbial respiratory activity in a Patagonian soil amended with salmon-farming sludge. Arch. Med. Vet. 41: 275–279.

Celis, J., M. Sandoval, A. Machuca and P. Morales. 2011. Biological activity in a degraded Alfisol amended with sewage sludge and cropped with yellow serradela (*Ornithopus compressus* L.). Chilean J. Agric. Res. 71: 164–172.

Celis, J., M. Sandoval, B. Martínez and C. Quezada. 2013. Effect of organic and mineral amendments upon soil respiration and microbial biomass in a saline-sodic soil. Ciencia e Investigación Agraria 49: 571–580.

Chan, K.Y., D.P. Heenan and H.B. So. 2003. Sequestration of carbon and changes in soil quality under conservation tillage on light-textured soils in Australia: A review. Aust. J. Exp. Agric. 43: 325–334.

Darwish, O.H., N. Persaud and D.C. Martens. 1995. Effect of long-term application of animal manure on physical properties of three soils. Plant Soil 176: 289–295.

Elliott, E. 1986. Aggregate structure and carbon, nitrogen and phosphorus in native and cultivated soils. Soil Sci. Soc. Am. J. 50: 627–633.

Epstein, E. 2002. Land application of sewage sludge and biosolids. Lewis Publishers, CRC Press, Boca Raton, Florida, USA.

FAO. 2011. The state of the world's land and water resources for food and agriculture (SOLAW)—managing systems at risk. Food and Agriculture Organization of the United Nations, Rome and Earthscan, London.

Gallardo, F., M. Mora and M. Diez. 2007. Kraft mill sludge to improve vegetal production in Chilean Andisol. Water Sci. Technol. 55: 31–37.

García-Gil, J.C. 2000. Long-term effects of municipal solid waste compost application on soil enzyme activities and microbial biomass. Soil Biol. Biochem. 32: 1907–1913.

Geng, X., S.Y. Zhang and J. Deng. 2007. Characteristics of paper mill sludge and its utilization for the manufacture of medium density fiberboard. Wood Fiber Sci. 39: 345–351.

Grez, R. and V. Gerding. 1995. Application of sawdust from the forest industry in order to improve soils. Bosque 16: 115–119.

Gupta, U. and J. Germida. 1988. Distribution of microbial biomass and its activity in different soil aggregate size classes as affected by cultivation. Soil Biol. Biochem. 20: 777–786.

Hernández, T., J.L. Moreno and F. Costa. 1991. Influence of sewage sludge application on crop yields and heavy metal availability. Soil Sci. Plant Nutr. 37: 201–210.

Ibrikci, H., N. Comerford, E. Hanlon and J. Rechcigl. 1994. Phosphorus uptake by bahiagrass from Spodosols: Modeling of uptake from different horizons. Soil Sci. Soc. Am. J. 58: 139–143.

Jastrow, J.D., R.M. Miller and T.W. Boutton. 1996. Carbon dynamics of aggregate-associated organic matter estimated by carbon-13 natural abundance. Soil Sci. Soc. Am. J. 60: 801–807.

Joergensen, R. and C. Emmerling. 2006. Methods for evaluating human impact on soil microorganisms based on their activity, biomass, and diversity in agricultural soils. J. Plant Nutr. Soil Sci. 169: 295–309.

Joyce, T.W., A.A. Webb and H.S. Dugal. 1979. Quality and composition of pulp and paper mill primary sludge. Resour. Recovery Conserv. 4: 99–103.

Jünemann, O. 1969. Dinámica de la mineralización e inmovilización potenciales del nitrógeno y del carbono en dos suelos aluviales del valle Central (Chile). Tesis Ing. Agrónomo, Universidad Católica de Chile, Santiago, Chile.

Le Bissonnais, Y. 1996. Aggregate stability and assessment of soil crustability and erodibility: I. theory and methodology. Eur. J. Soil Sci. 47: 425–437.

Mazzarino, M., I. Walter, G. Costa, F. Laos, L. Roselli and P. Satti. 1997. Plant response to fish farming wastes in volcanic soils. J. Environ. Qual. 26: 522–528.

Metcalf and Eddy. 1995. Wastewater Engineering: Treatment, Disposal and Reuse. McGraw-Hill, New York.

Mosquera-Losada, R., A. Amador-García, N. Muñóz-Ferreiro, J. Santiago-Freijanes, N. Ferreiro-Domínguez, R. Romero-Franco et al. 2017. Sustainable use of sewage sludge in acid soils within a circular economy perspective. Catena 149: 341–348.

Oades, J.M. and G. Waters. 1991. Aggregate hierarchy in soils. Aust. J. Soil Res. 29: 815–828.

O'Brien, T.A., S.J. Herbert and A.V. Barker. 2002. Growth of corn in varying mixtures of paper mill sludge and soil. Commun. Soil Sci. Plan. 33: 635–646.

Pascual, I., C. Antolin, C. García, A. Polo and M. Sánchez-Díaz. 2007. Effect of water deficit on microbial characteristics in soil amended with sewage sludge or inorganic fertilizer under laboratory conditions. Bioresour. Technol. 98: 29–37.

Paul, J.W., E.G. Beauchamp and X. Zhang. 1993. Nitrous and nitric oxide emissions during nitrification and denitrification from manure-amended soil in laboratory. Canadian J. Soil Sci. 73: 539–553.

Pinochet, D., P. Artacho and P. Azúa. 2001. Potencialidad como abono orgánico de los desechos sólidos subproductos del cultivo de especies salmonídeas. Agro. Sur. 29: 78–82.

Powlson, D., P. Brookes and B. Christensen. 1987. Measurement of soil biomass provides an early indication of changes in total soil organic matter due to straw incorporation. Soil Biol. Biochem. 19: 159–164.

Rillig, M.C., S.F. Wright and V.T. Eviner. 2002. The role of arbuscular mycorrhizal fungi and glomalin in soil aggregation: comparing effects of five plant species. Plant Soil 238: 325–333.

Ríos, D., C. Pérez and M. Sandoval. 2012. Phytotoxic effect of paper pulp sludge on Alfisol soil. J. Soil Sci. Plant Nutr. 12: 315–327.

Ros, M., M. Hernández and C. García. 2003. Soil microbial activity after restoration of a semiarid soil by organic amendments. Soil Biol. Biochem. 35: 463–469.

Ros, M., J. Pascual, C. García, M. Hernández and M. Insam. 2006. Hydrolases activities, microbial biomass and bacterial community in a soil after long-term amendment with different compost. Soil Biol. Biochem. 38: 3443–3452.

Sandoval, M., J. Celis and C. Bahamondes. 2012. Effect of sewage sludge and sawdust in association with hybrid ryegrass (*Lolium* × *hybrydum* HAUSSKN.) on soil macroaggregates and water content. Chilean J. Agric. Res. 72: 568–573.

Sandoval-Estrada, M., N. Stolpe, E. Zagal, M. Mardones and J. Celis. 2008. No-tillage organic carbon contribution and effects on an Andisol structure from the Chilean Andean foothills. Agrociencia 42: 139–149.

Sandoval-Estrada, M., J.E. Celis-Hidalgo, N. Stolpe-Lau and J. Capulín-Grande. 2010. Effect of sewage sludge and salmon wastes amendments on the structure of an Entisol and Alfisol in Chile. Agrociencia 44: 503–515.

San Martín, V., P. Undurraga, C. Quezada, J. Celis and M. Sandoval. 2016. Effect of pellets made of waste materials from the paper industry enhanced with seaweed (*Ulva lactuca* L.) on N mineralization and lettuce production. Chilean J. Agric. Res. 76: 363–370.

Sawyer, C.N., P.L. McCarty and G.F. Parkin. 2003. Chemistry for Environmental and Engineering Science. McGraw Hill Inc., New York.

Schreiner, P.R. and G.J. Bethlenfalvay. 1995. Mycorrhizal interactions in sustainable agriculture. Crit. Rev. Biotechnol. 15: 271–285.

Scott, G.M. and A. Smith. 1995. Sludge characteristics and disposal alternatives for the pulp and paper industry. Proc. Intern. Environ. Conf. Atlanta, USA 269–279.

Shetty, K., B. Hetrick, D. Figge and A. Schwab. 1994. Effects of mycorrhizae and other soil microbes on revegetation of heavy-metal contaminated mine spoil. Environ. Pollut. 86: 181–188.

Shober, A., R. Stehouwer and K. MacNeal. 2003. On-farm assessment of biosolid effects on soil and crop quality. J. Environ. Qual. 32: 1873–1880.

Sims, J.T. and G.M. Pierzynski. 2000. Assessing the impact of agricultural, municipal, and industrial by-products on soil quality. pp. 237–261. *In*: Power, J.F. (ed.). Beneficial Uses of Land Applied Agricultural, Municipal and Industrial By-Products. SSSA Spec. Publ. 6. Madison, Wisconsinn, USA.

Smernik, R.J., I.W. Olivera and G. Merringtona. 2003. Characterization of sewage sludge organic matter using solid-state carbon-13 nuclear magnetic resonance spectroscopy. J. Environ. Qual. 32: 1516–1522.

Sommers, LE. 1977. Chemical composition of sewage sludge and analysis of their potential use as fertilizers. J. Environ. Qual. 6: 225–232.

Sylvia, D., J. Fuhrmann, P. Hartel and D. Zuberer. 1998. Principles and applications of soil microbiology. Prentice Hall, Inc. New Jersey, USA.

Six, J., K. Paustian, E. Elliott and C. Combrink. 2000. Soil Structure and organic matter: I. distribution of aggregate-size classes and aggregate-associated carbon. Soil Sci. Soc. Am. J. 64: 681–689.

Tabatabai, M.A. 1994. Soil enzymes. pp. 775–778. *In*: Weaver, R.W., Angle, J.S. and Bottomley, P.S. (eds.). Methods of Soil Analysis. Microbiological and Biochemical Properties. Soil Science Society of America, Madison, Wisconsin, USA.

Tamoutsidis, E., I. Papadopoulos, I. Tokatlidis, S. Zotis and T. Mavropoulos. 2002. Wet sewage sludge application effect on soil properties and element content of leaf and root vegetables. J. Plant Nutr. 25: 1941–1955.

Tang, C. and Q. Yu. 1999. Impact of chemical composition of legume residues and initial soil pH of a soil after residue incorporation. Plant Soil 215: 29–38.

Teuber, N., M. Alfaro, F. Salazar and C. Bustos. 2005. Sea salmon sludge as fertilizer: effects on a volcanic soil and annual ryegrass yield and quality. Soil Use Manage. 21: 32–434.

Tisdall, J. and J. Oades. 1982. Organic matter and water-stable aggregates in soils. J. Soil Sci. 33: 141–163.

Tchouaffe, N.F. 2007. Strategies to reduce the impact of salt on crops (rice, cotton and chilli) production: a case study of the tsunami-affected area of India. Desalination 206: 524–530.

Torkashvand, A.M., N. Haghighat and V. Shadparvar. 2010. Effect of paper mill lime sludge as an acid soil amendment. Sci. Res. Essays 5: 130–1306.

Traore, O., V. Groleau-Renaud, S. Plantureux, A. Tubeileh and V. Boeuf-Tremblay. 2000. Effect of root mucilage and modeled root exudates on soil structure. Eur. J. Soil Sci. 51: 575–581.

Waldrop, M.P., J.G. McColl and R.F. Powers. 2003. Effects of forest postharvest management practices on enzyme activities in decomposing litter. Soil Sci. America J. 67: 1250–1256.

Wright, S.F. and A. Upadhyaya. 1998. A survey of soils for aggregate stability and glomalin, a glycoprotein produced by hyphae of arbuscular mycorrhizal fungi. Plant Soil 198: 97–107.

Zhang, S., S. Wang, X. Shan and H. Mu. 2004. Influences of lignin from paper mill sludge on soil properties and metal accumulation in wheat. Biol. Fert. Soils 40: 237–242.

Chapter 9

Bioremediation of Pesticide-Contaminated Soils by using Earthworms

Juan C. Sanchez-Hernandez

INTRODUCTION

Soil degradation has been described as "adverse changes in soil properties and processes leading to a reduction in ecosystem services" (Palm et al. 2007). Such changes are caused by multiple physical, chemical and biological stressors derived from both natural and anthropogenic sources. Among these, pesticides are major chemical stressors that threaten the quality and fertility of agricultural soils. Many studies have demonstrated that soil is an environmental sink for pesticides and acts as a secondary pollution source for other environmental compartments such as groundwater and surface water (e.g., rivers, lakes and coastal areas). However, the environmental risk associated with contaminated soils depends on many physicochemical and biological processes that govern the transport and transformation of pesticides (Gavrilescu 2005, Mirsal 2008, Arias-Estévez et al. 2008, Köhne et al. 2009, Odukkathil and Vasudevan 2013). Nevertheless, the persistence of pesticides in soil is a serious threat to organisms living in both belowground and aboveground systems and which are important for soil functioning.

Bioremediation has become a common means of removing pesticide residues from soil and thus reducing their toxicity. Bioremediation consists of the use of

Laboratory of Ecotoxicology, Faculty of Environmental Sciences and Biochemistry, University of Castilla-La Mancha, 45071 Toledo, Spain.
Email: juancarlos.sanchez@uclm.es

organisms, mainly microorganisms and plants, to facilitate the transformation and degradation (or immobilization) of environmental contaminants (Masciandaro et al. 2013). Compared to sophisticated engineering-based methods, bioremediation is an inexpensive, eco-friendly approach (Cummings 2010). However, the use of organisms such as bacteria and fungi in bioremediation has several challenges such as those associated with the metabolic adaptation of microorganisms to the target contaminants (Springael and Top 2004, Singh and Walker 2006). This is true of the organophosphorus pesticide chlorpyrifos. Degradation of this pesticide by chemical hydrolysis and microbial activity generates the metabolites chlorpyrifos-oxon and 3,5,6-trichloro-2-pyridinol (Racke 1993). However, these metabolites may be toxic to microorganisms (Singh and Walker 2006, Fang et al. 2009, John and Shaike 2015), thus reducing their mineralization capacity and the biodegradation of the parent compound (chlorpyrifos). Likewise, soil properties such as pH may prevent metabolic adaptation of organophosphorus microbial degraders to repeated application of organophosphorus pesticides (Singh et al. 2003). In addition to metabolic adaptation, bioremediation with microorganisms generally requires some type of biostimulation (i.e., addition of nutrients) to maintain the microorganisms in the long-term or even to accelerate degradation of the contaminants (Tyagi et al. 2011). However, biostimulation must be carried out carefully because excess organic matter in the soil may reduce the accessibility of pesticides to microbes, thus hampering the biodegradation process. This limited bioavailability of pesticides can be resolved by the addition of surfactants (Cheng et al. 2018) or suitable surfactant-producing microbes (Odukkathil and Vasudevan 2016, Morillo and Villaverde 2017). On the other hand, the addition of organic amendments to contaminated soils may lead to the pesticides being adsorbed onto exogenous dissolved organic matter, thereby increasing the pesticide mobility (Song et al. 2008, Bolan et al. 2011).

Ecological interactions between microbial populations, such as competition and predation, may negatively affect the success of bioremediation. One innovative strategy to prevent the side-effects of ecological interactions on bioremediation is the use of purified preparations of detoxifying enzymes extracted from microorganisms (Sutherland et al. 2004, Nair and Jayachandran 2017, Sharma et al. 2018). However, this approach also has some limitations such as the high cost of cell-free enzyme production, degradation of enzymes, and the accessibility of pesticides to the active site of enzymes (Sutherland et al. 2004, Scott et al. 2011). Together, these disadvantages and potential bioremediation solutions move the future of this technology to an integrated approach, whereby multiple biological vectors of contaminant degradation are applied jointly to the contaminated site (Masciandaro et al. 2013, Megharaj and Naidu 2017). In this context, if the degradation of pesticides requires the intervention of extracellular enzymes, microorganisms and plants, as well as other potential soil decomposers (e.g., earthworms and springtails), use of a combination of these biological entities would be the ideal way to facilitate the bioaccessibility, dispersion, uptake and metabolism of pesticides (Megharaj and Naidu 2017). However, this "artificial" ecosystem created *in situ* with the aim of facilitating pesticide degradation would require continual monitoring of the potentially adverse effects on other soil components.

The scientific literature provides numerous examples of bioremediation strategies for removing soil contaminants. Many reviews (Megharaj et al. 2011, Tyagi et al. 2011, Rayu et al. 2012), books (Margesin and Schinner 2005, Mirsal 2008, Bharagava 2017) and special issues on this topic (Singh and Naidu 2012) have systematically tackled the bioremediation of contaminated soils from different perspectives, ranging from technological and innovative viewpoints to more ecological visions. Compiling a list of these studies or discussing the latest advances in the knowledge of bioremediation is outside the scope of this chapter. The goal of this chapter is to highlight the importance of a group of soil organisms that have been shown in the last decade to have great potential in the bioremediation of contaminated soils, i.e., earthworms. The chapter is divided into four sections. The first two sections provide an overview of pesticide consumption worldwide and the main impacts of pesticides on the agroecosystem. These two sections will outline why pesticide monitoring and remediation are still of current concern within the context of sustainable agriculture and environmental protection. The third section will examine how earthworms can decrease the concentration and toxicity of pesticide residues in soil. This chemical-biological interaction in two parts of the drilosphere (the soil environment under the influence of earthworm activity) will be considered, i.e., in internal (gastrointestinal tract) and external (casts, middens and burrow walls) microenvironments. The final section will summarize the key findings of bioremediation of pesticide-contaminated soils by using earthworms and will describe new lines of exciting research for the coming years.

Global Pesticide Consumption: An Overview

Pesticides remain key players to combat agricultural pests. A rapid inspection of the statistical data provided by the Organisation for Economic Co-operation and Development (OECD 2013) shows that pesticide consumption has increased in many countries during the past decade (Fig. 1). Within the European Union (EU), Spain, France, Italy, Germany and Poland are the Member States with the highest consumption of pesticides (Fig. 2). Although some of these countries have reduced substantially their pesticide consumption since 2011, others have not changed or even have experienced a slight increase in the consumption of plant protection products.

Organophosphates (OPs) are among the most common pesticides used in agriculture. One of the most common OP insecticides is chlorpyrifos. For example, in the USA the consumption of chlorpyrifos in 2012 ranged between 2.2 and 3.6 million of kg of active ingredient (Atwood and Paisley-Jones 2017). In Chile, the estimated amounts of chlorpyrifos and diazinon sold in 2012 was around 8.5×10^5 and 12×10^5 kg of active ingredient, respectively (SAG 2012). Residues of both of these OP pesticides are frequently detected in agricultural products from Chile. For example, analysis of 71 olive oil samples collected during 2007–09 revealed chlorpyrifos residues in 48 samples (concentrations ranging between 0.005 and 0.563 µg/g), and diazinon in 15 samples (0.004 to 0.143 µg/g) (Fuentes et al. 2010).

Consumption of OPs is also high in India and China. The former has a long history of pesticide use, and production began in the 1960s (Abhilash and Singh 2009). Again, chlorpyrifos is one of the most frequently used OPs in Indian agriculture,

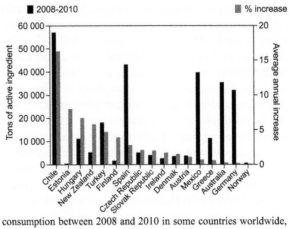

Fig. 1. Pesticide consumption between 2008 and 2010 in some countries worldwide, and percentage of increase (data taken from OECD 2013).

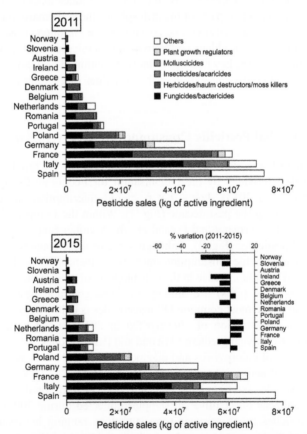

Fig. 2. Consumption of plant protection products in some European Union Member States in 2011 and 2015. Inset graph shows the percentage of variation during that period of time (data taken from Eurostat in 16/03/2018, http://ec.europa.eu/eurostat, data corresponding to molluscicide consumption in Greece for 2011 and in Poland for 2015 are not available).

with an estimated 7,354 metric tons used during the period 2005–2010 (Kumar et al. 2016). As a result, this OP is frequently detected in environmental samples as diverse as soil, surface water, groundwater and vegetables (Kumar et al. 2016, and references within). Chen et al. (2016) provide an illustrative example of the high consumption of OPs in China, reporting that 71% of 810 samples of tea were contaminated by OPs, with an average concentration of chlorpyrifos of 82 µg/kg, and a maximum concentration of 378 µg/kg. Similarly, Yu et al. (2016) detected up to eleven OP pesticides in 214 samples of vegetables collected in crops in Changchun (Jilin province, NE China) in 2014. The most frequently detected OPs in these samples were dichlorvos (84.3 µg/kg), methamidophos (82.2 µg/kg), diazinon (50.4 µg/kg), parathion (15.6 µg/kg), omethoate (8.8 µg/kg), phorate (7.8 µg/kg), parathion-methyl (7.0 µg/kg), dimethoate (6.7 µg/kg), fenthion (6.5 µg/kg), fenitrothion (4.1 µg/kg), and malathion (1.7 µg/kg). Three of these pesticides (parathion, methamidophos, and parathion-methyl) are currently banned in China, and 23.4% of vegetable samples had concentrations higher than the maximum residue limits (MRLs) established by the Standardized Administration of China (Yu et al. 2016). Similar findings were also reported by Liu et al. (2016), who detected concentrations of OP residues (acephate, dimethoate, chlorpyrifos, and parathion-methyl) ranging between 19.0 and 74.0 µg/kg in 44 samples of nuts collected from local farmers between 2013 and 2014. In the study, 4.5% of the nut samples exceeded the MRLs for OPs established by China, and 6.8% of the samples exceeded the MRLs for OPs established by the EU (Liu et al. 2016).

Within the EU, Spain is an illustrative example of how the agri-environmental schemes introduced to protect biodiversity (Council Regulation No. EEC 2078/92) have not had a significant impact on some of the measures, such as the reduction of pesticide and fertilizer inputs. Among the EU Member States, Spain has the highest consumption of plant protection products (Fig. 2). These statistical data are corroborated by the presence of pesticide residues in soil (Sánchez-González et al. 2013, Padilla-Sánchez et al. 2015), sediments (Masiá et al. 2013, Ccanccapa et al. 2016a, 2016b), groundwater and surface waters (Sánchez-González et al. 2013, Masiá et al. 2013, Herrero-Hernández et al. 2017, Menchen et al. 2017) and non-target organisms (Ccanccapa et al. 2016b).

In summary, these studies and statistical data suggest that pesticide consumption continues to increase in some parts of the world and that OP pesticides still form an important part of the chemical cocktail used to combat agricultural pests. Furthermore, some studies predict that this scenario will tend to worsen in the coming decades because of global climate change (Noyes et al. 2009, Delcour et al. 2015). It has been predicted that the increased temperature and the increasing frequency of strong rainfall events will lead to the enhancement of pesticide consumption due to several factors. First, higher temperatures will facilitate the degradation of pesticides such as OPs, probably because of enhanced microbial proliferation, a higher rate of pesticide volatilization, and increased run-off in those regions where the frequency of precipitation is expected to increase (Delcour et al. 2015). Second, the toxicity of pesticides may also increase (Noyes et al. 2009). For example, OP pesticides become more toxic after an oxidative desulfuration reaction whereby the sulfur bound to the phosphorus atom of the OP molecule is replaced by an oxygen atom. This transformation may occur in the environment by hydrolysis (Racke 1993, Racke et al. 1996) and by the action of ultraviolet light (Chambers et al. 2010a). Therefore, it would be not surprising if the rate of formation of highly toxic OP metabolites in the field increase under a global warming scenario. Third, climate change will probably affect the geographical distribution and abundance of agricultural pests, thereby threatening crop yields (Noyes et al. 2009). It can be assumed that the amount and variety of pesticides will increase in order to counteract these warming-induced environmental threats.

Impact of Pesticides in the Agroecosystem

Pesticides may reach the soil through two main routes: intentionally through direct application to soil, whereby pesticides are generally dissolved in the irrigation water or applied as a granular formulation; and indirectly via the washout of pesticide-treated foliage or via the pesticide fraction that reaches the soil accidentally during terrestrial and aerial crop spraying (Gavrilescu 2005, Arias-Estévez et al. 2008). Many physicochemical and biological processes in the soil contribute to dissipating pesticides. However, the intensive use of pesticides, marked by continual application at high doses, may prolong the presence of residues in soil and trigger multiple environmental impacts. Adverse effects on non-target organisms, changes in soil microbiological and biochemical processes, and contamination of groundwater and surface waters are the most well-documented impacts. Below, these three environmental impacts are illustrated with examples that justify both the need to monitor the side-effects of pesticides and the use of bioremediation measures to reduce or prevent the accumulation of pesticides in soil.

It is now widely recognized that pesticides represent a serious threat to non-target organisms in the agroecosystem, such as earthworms, pollinators (e.g., bees) and natural enemies of pests. Because of they are in continuous and direct contact with soil, earthworms are recommended for use in standardized toxicity tests to assess the environmental risk associated with pesticides (Spurgeon et al. 2003, Stanley and Preetha 2016), and for monitoring pesticide toxicity in a post-authorization phase (Newman et al. 2006). Earthworms alter physicochemical and biological properties of soil by continuous feeding and burrowing activities (Jouquet et al. 2006). Therefore, knowledge of how pesticides affect these activities is essential for understanding the ecological consequences of these compounds, at least, on a local scale. Accordingly, many standardized toxicity tests have been developed to examine the impact of pesticides on earthworm behavior, earthworm life cycle traits and biomarkers (Sanchez-Hernandez 2006, Pelosi et al. 2014, Stanley and Preetha 2016, Velki and Ečimović 2017). Furthermore, soil-dwelling earthworm species (e.g., *Lumbricus terrestris* and *Aporrectodea caliginosa*) seem to be more sensitive to pesticides than epigeic earthworms (e.g., *Eisenia fetida*) that live on the soil surface (Pelosi et al. 2013a, 2014).

In the case of organophosphorus pesticides, earthworm burrowing and casting activities are both sensitive exposure biomarkers (Jouni et al. 2018). Similarly, the capacity of earthworms to avoid contaminated soils (a toxicity assay known as "the avoidance behavior response test", ISO 2008) is another sublethal and sensitive indicator of OP exposure (Martínez Morcillo et al. 2013, Rico et al. 2016, Berenstein et al. 2017, Vasantha-Srinivasan et al. 2018). Many other examples of the adverse effects of pesticides, and other contaminants, on earthworms have been described in review papers by Stanley and Preetha (2016), Sanchez-Hernandez (2006), Velki and Ečimović (2017) and Pelosi et al. (2014). Despite the effort made to develop new toxicity tests and biomarkers with earthworms, the consequences of pesticide-exposed earthworms on soil properties have not been well established (Pelosi et al. 2014, Velki and Ečimović 2017).

Many pesticides induce changes in soil microbial activity and biomass (Gianfreda and Rao 2008), although such changes largely depend on many environmental and biological factors such as pesticide type and dose, soil properties, the sensitivity of microorganisms to pesticides and their metabolites, and time of exposure. Soil enzymes are also molecular targets of pesticides (Riah et al. 2014). Most soil enzymes are produced in direct response to microbial foraging, and they are involved in the biogeochemical cycling of nutrients (Shaw and Burns 2006, Wallenstein and Burns 2011). Accordingly, soil enzyme activities are measured as an indicator of microbial functional diversity and soil quality (Burns et al. 2013). Because pesticides can interact directly with enzyme molecules or indirectly by altering microbial biomass and activity (Gianfreda and Rao 2008), it is difficult to predict the behavior of enzymes in pesticide-contaminated soils (Riah et al. 2014). Chapter 12 provides additional information about the use of soil enzymes as indicators of soil pollution.

Finally, the accumulation of pesticides in soil may also lead to a serious risk of contamination of groundwater and surface waters. For example, a two-decade survey in the USA revealed that the herbicides atrazine, simazine, prometon and metolachlor were the most frequently detected pesticides (53% of a total of 2,542 samples), although the concentrations have tended to decrease over the years, probably because of limited use of these agrochemicals (Toccalino et al. 2014). A similar survey was performed in Catalonia (Spain), where 29 aquifers were monitored over a period of 4 years (Köck-Schulmeyer et al. 2014); the herbicides simazine, atrazine and diuron were the most frequently detected pesticides (> 50% of samples) from a total of 233 samples. Considering insecticides, the organophosphorus diazinon (0.32–30.8 ng/l, min–max concentrations), dimethoate (0.24–2,277 ng/l), fenitrothion (8.15–19.5 ng/l) and malathion (2.57–86.6 ng/l) were detected in respectively 51%, 10%, 3% and 7% of groundwater samples. Triazine and terbuthylazine were also frequently (24%) detected in 314 samples of groundwater collected from Spain between 2010 and 2013 (Menchen et al. 2017). These examples show that the intensive use of agrochemicals contaminates groundwater, as indicated by the frequency of positive samples (with at least one pesticide detected), by the concentrations, which occasionally exceed regulatory limits (Köck-Schulmeyer et al. 2014), and by persistence of the compounds after their use on agricultural lands was banned or restriction (Toccalino et al. 2014).

Awareness of the aforementioned environmental risks has led to the development of methods aimed at reducing the persistence of pesticide residues in soil. In the next two sections, we discuss how earthworms can assist in the bioremediation of pesticides in agricultural soil contaminated by agrochemicals, particularly organophosphorus pesticides.

Use of Earthworms in the Bioremediation of Pesticide-Contaminated Soils

In general, bioremediation can be understood as the use of living organisms to remove contaminants from contaminated sites (Megharaj et al. 2011). Nowadays, bioremediation is the preferred approach to recovering degraded land because it has a minimal impact on soil physicochemical and biological properties relative to remediation strategies that involve chemical or physical treatment of the contaminated soil (Morillo and Villaverde 2017). Furthermore, bioremediation is a cost-effective and environmentally-friendly option, although it also has some disadvantages. For example, bioremediation generally takes a long time to cause a significant decrease in contaminant concentrations in polluted sites, particularly when plants are used for this purpose (phytoremediation). On some occasions, the toxicity of pesticides (or their metabolites) to microorganisms used in bioremediation reduce the probability of success. The particular requirements (e.g., soil moisture, aeration and nutrients) that microorganisms may need to effectively degrade pesticides is another important factor limiting bioremediation. The main advantages and limitations of bioremediation technology in pesticide-polluted soils are summarized in Table 1.

Microorganisms are the key biological vectors in the bioremediation of soil contaminated by organic pollutants, although other soil organisms such as earthworms play a significant role in pesticide degradation. A brief description of the impact of these organisms on soil structure and function is provided below to explain how earthworms can contribute to eliminating pesticide residues in soil.

Earthworms are considered soil engineers because of their impact on soil physicochemical and biological properties (Jouquet et al. 2006). Their continuous burrowing and feeding activities represent a major driving force for enhancing and creating soil microhabitats, with significant consequences in the below- and aboveground systems. Increased soil porosity and aggregate formation are perhaps the most immediate physical changes induced by earthworms (Shipitalo and Le Bayon 2004). These physical changes lead to increased water infiltration, soil aeration and development of plant roots (Capowiez et al. 2014). Among the biological effects, earthworms are able to improve plant productivity, stimulating the development of both the plant roots and shoots, although the effect depends on the plant species considered (Lavelle et al. 1999, Scheu 2003). For example, a meta-analysis conducted by Xiao et al. (2018) revealed that plant growth increased by 20% in soils under the influence of earthworms, as a result of changes in soil texture, nutrient mineralization and microbial communities (Lavelle et al. 1999, Scheu 2003, Brown et al. 2004, van Groenigen et al. 2014). However, other such studies have questioned the beneficial impact of earthworms on plants or, at least, on the biodiversity of native plant species in forest ecosystems (Craven et al. 2017). Moreover, changes in plant performance have direct consequences on herbivores, so that earthworms may indirectly alter the aboveground herbivore populations (Wurst 2010). More interestingly, these soil

Table 1. Main advantages and limitations of *in situ* bioremediation of pesticide-contaminated soils.

Strategy	Advantages	Limitations
Microorganisms[1-4]	- Possibility of using three methodological approaches: natural attenuation, biostimulation and bioaugmentation.	- Need of increasing basic knowledge on microbial metabolic processes.
	- Cost-effective (natural attenuation and biostimulation).	- High concentrations of pesticides, and their metabolites, may be toxic to microorganisms.
	- High capability to degrade a wide variety of pesticides.	- Microorganisms require optimal soil conditions for growth (e.g., nutrients, soil moisture, temperature and pH), and often bioremediation results a slow or incomplete process.
	- Possibility of increasing the degradation of highly hydrophobic pesticides by adding synthetic or biosurfactants.	- Occasionally, microbial metabolism of pesticides may produce highly toxic metabolites to non-target organisms.
	- Enhanced biodegradation of native microorganisms may occur after multiple episodes of organophosphorus pesticide contamination.[5]	- Sorption of pesticides to organic matter reduces biodegradation.
	- Microorganisms (especially fungi) have a vast variety of both intracellular and extracellular (exoenzymes) enzymes able to degrade many types of pesticides.[6]	- Ecological interactions (predation and competition) between native and inoculated microorganisms (bioaugmentation) may reduce bioremediation efficiency.
Plants[1,2]	- Cost-effective and eco-friendly methodology.	- Significant removal of contaminants often achieved at long-term scale.
	- Improve physicochemical and biological properties of soil.	- Sensitivity of plants to pesticides (and their metabolites).
	- Increase local above- and below-ground biodiversity.	- Plant survival highly dependent on soil features and climatic conditions.
	- Favor the development of complementary bioremediation actors such as microorganisms (rhizosphere).	- Effective impact of phytoremediation limited only to the root zone.
		- Disposal of plant wastes after harvesting.
Cell-free enzymes[2,7,8]	- Enzymatic bioremediation may be a recycling process as long as enzymes are immobilized in solid supports.	- High costs of production.
	- Enhancement of catalytic properties of enzymes by engineering enzyme molecules.	- Need of continual monitoring of enzyme activity because of protein degradation in soil.
	- Possibility of applying a cocktail of pesticide-detoxifying enzymes.	- Genetically engineered enzymes often provide a limited substrate range.
	- Solid supports (e.g., nanomaterials, biochar) may be used to stabilize enzymes in soil.	- Enzyme activity may change drastically with soil properties (e.g., pH, temperature, metal ions, etc.).
		- The enzyme may be bound to organomineral complexes of soil, thus reducing substrate (pesticide) availability.

Table 1 contd. ...

...Table 1 contd.

Strategy	Advantages	Limitations
Earthworms[1,2]	- They stimulate microbial proliferation, therefore increasing the biodegradation capacity of contaminated soils. - They may contribute to pesticide degradation through gastrointestinal secretion of detoxifying enzymes. - Earthworm activity can increase pesticide bioaccessibility to microbial degradation.	- Sensitivity of earthworms to toxic effects from pesticides in soil (earthworms may avoid soils with high pesticide concentrations). - Earthworm survival highly dependent on soil features (e.g., moisture and temperature) and food availability. - Earthworm activity may increase persistence of pesticides (and their metabolites) in soil by changes in the dynamic of soil organic matter.

References: [1]Morillo and Villaverde (2017), [2]Marican and Durán-Lara (2017), [3]Cycon et al. (2017), [4]Boopathy (2000), [5]Singh (2009), [6]Harms et al. (2011), [7]Sutherland et al. (2004) and [8]Sanchez-Hernandez (2018).

organisms may favor plant resistance to pests by increasing plant chemical defenses (Wurst 2010, Xiao et al. 2018).

The most frequent impact of earthworm activity on the belowground organisms is that exerted on soil microorganisms. Earthworms play an important role in the dispersion and proliferation of microbes. It is generally observed that the soil under the influence of earthworms is characterized by high levels of microbial activity and biomass (Hoang et al. 2016a, 2016b). Therefore, higher soil enzyme activities are found in these bioturbed soils than in earthworm-free soils (Tao et al. 2009, Sanchez-Hernandez et al. 2014c, Athmann et al. 2017).

There is currently some debate as to whether earthworm activity contributes to carbon sequestration in soil or whether earthworms act as biological catalysts of reactions leading to greenhouse gas emissions. In a meta-analysis of relevant studies, Lubbers et al. (2013) concluded that earthworm activity may increase soil N_2O and CO_2 emissions by respectively 42% and 33%. However, although this environmental effect was significant in the short-term, a reduction in CO_2 emissions was observed in the longer term (> 200 days) (Lubbers et al. 2013). This observation has been corroborated by other researchers, who showed that earthworms had a higher impact on carbon stabilization than on carbon mineralization (Zhang et al. 2013).

However, not all earthworm species are able to perform these ecological services to the same extent. Earthworms are traditionally divided into three functional groups (Lavelle et al. 1998, Orgiazzi et al. 2016): epigeic, endogeic and anecic (Fig. 3). Epigeic earthworms are surface-dwellers that feed on the organic matter accumulated on the soil surface (leaf litter, decaying plant roots). As they rarely burrow into the soil, they ingest few soil particles. Species in this functional group are frequently found in forest soils, living in the litter layer. Endogeic species are soil-dwellers and obtain their nutrients by ingesting large amounts of soil, and they are referred to as geophagous. These earthworms burrow intensively in the uppermost 10–15 cm of soil, elaborating temporary horizontal burrows (Capowiez et al. 2014). Anecic species are large earthworms that construct long, generally permanent vertical burrows. The species belonging to this ecological group feed on decaying organic residues that drag into their burrows, although they also ingest some mineral soil

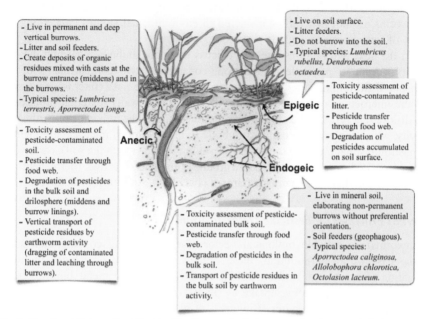

- Live in permanent and deep vertical burrows.
- Litter and soil feeders.
- Create deposits of organic residues mixed with casts at the burrow entrance (middens) and in the burrows.
- Typical species: *Lumbricus terrestris, Aporrectodea longa.*

- Live on soil surface.
- Litter feeders.
- Do not burrow into the soil.
- Typical species: *Lumbricus rubellus, Dendrobaena octaedra.*

- Toxicity assessment of pesticide-contaminated litter.
- Pesticide transfer through food web.
- Degradation of pesticides accumulated on soil surface.

Epigeic

- Toxicity assessment of pesticide-contaminated soil.
- Pesticide transfer through food web.
- Degradation of pesticides in the bulk soil and drilosphere (middens and burrow linings).
- Vertical transport of pesticide residues by earthworm activity (dragging of contaminated litter and leaching through burrows).

Anecic

Endogeic

- Toxicity assessment of pesticide-contaminated bulk soil.
- Pesticide transfer through food web.
- Degradation of pesticides in the bulk soil.
- Transport of pesticide residues in the bulk soil by earthworm activity.

- Live in mineral soil, elaborating non-permanent burrows without preferential orientation.
- Soil feeders (geophagous).
- Typical species: *Aporrectodea caliginosa, Allolobophora chlorotica, Octolasion lacteum.*

Fig. 3. Functional features (in callouts) of the three ecological groups of earthworms (epigeic, endogeic and anecic), and their potential use in pesticide ecotoxicology (in post-it labels). Pictorial representation elaborated from Lavelle et al. (1998).

and even collect green leaves that incorporate into the burrows (Brown 1995, Brown et al. 2000, Römbke et al. 2005, Griffith et al. 2013). Anecic earthworms usually form a deposit (a midden) of organic residues mixed with casts (i.e., faeces) at the entrance of their burrows (Brown et al. 2004). The middens are considered hotspots of organic matter decomposition and faunal diversity (Butt and Lowe 2007, Stroud et al. 2016, Nuutinen et al. 2017).

Earthworms can participate in the degradation of pesticides, either directly through the release of pesticide-detoxifying enzymes in their gastrointestinal tract (Sanchez-Hernandez et al. 2009), or indirectly through stimulation and dispersion of microorganisms that are capable of degrading pesticides (Hickman and Reid 2008, Rodriguez-Campos et al. 2014). However, earthworm-mediated degradation of pesticides is limited to the luminal microenvironment of the gastrointestinal tract and to the soil under their influence (burrow walls, casts and middens). Therefore, not all earthworm species will have a similar impact on pesticide degradation, which will depend on the feeding habits and microhabitats of the three ecological groups of earthworms (Fig. 3). Theoretically, endogeic and anecic earthworms will have a greater impact on pesticide degradation than epigeic earthworms, because the former consume greater amounts of soil and display more intense burrowing activity than the latter. Similarly, epigeic and anecic earthworms should be more capable of degrading pesticides that are accumulated in decaying plant debris, because the life style of the former is closely linked to organic residues accumulated on the soil surface, and because the feeding habit of the latter involves dragging plant debris from the soil surface into the burrows. However, these ecological characteristics of earthworms are not generally considered in environmental studies concerning the ability of earthworms to degrade pesticides.

In the next two sections, we will discuss the main findings regarding the role of earthworms in pesticide degradation, considering two compartments of the drilosphere: the luminal microenvironment of the digestive canal (internal processes), and the microenvironment associated with the burrow walls (external processes). The drilosphere is defined as the soil environment under the influence of earthworm activity, which includes the earthworm body (i.e., the luminal microenvironment of the gastrointestinal tract and the earthworm skin in contact with the soil), earthworm casts (deposited on surface and belowground), middens, burrows (permanent and temporary) and diapause chambers (Brown et al. 2000, Andriuzzi et al. 2013).

Internal Processes of Pesticide Detoxification

Earthworms are suitable biological vectors for remediating contaminated soils. In the case of metal pollution, these organisms can induce changes in metal speciation (Sizmur and Hodson 2009), thereby increasing metal bioavailability and consequently facilitating uptake by plants—a functional feature that has been exploited in phytoremediation (Bityutskii et al. 2016). Such earthworm-induced alteration of metal speciation is mainly attributed to changes in soil properties such as pH, dissolved organic carbon, and microbial proliferation, which, in turn, mobilize the fraction of metals bound to organic matter. Nevertheless, the precise mechanisms whereby earthworms contribute to the remediation of metal-polluted soils are not fully understood (Sizmur and Hodson 2009). In the case of organic contaminants, cooperation between earthworms and microorganisms is essential to promote degradation of these chemicals (Hickman and Reid 2008, Rodriguez-Campos et al. 2014). Some studies have demonstrated that symbiont microorganisms inhabiting the earthworm gastrointestinal tract, may degrade organic pollutants such as crude oils, polycyclic aromatic hydrocarbons, polychlorinated biphenyls, and pesticides (Hickman and Reid 2008). For example, a *Rhodococcus* MTCC bacterial strain isolated from the gastrointestinal tract of *Metaphire posthuma* was able to degrade the organochlorine pesticide endosulfan (Verma et al. 2006). Using the same earthworm species, Ramteke and Hans (1992) isolated some bacterial strains (*Flavobacterium* spp., *Pseudomonas* spp. and *Acromobacter* spp.) that degraded the organochlorine lindane and its isomers. However, the pesticide-degrading capacity was only demonstrated in culture, and so it was not clear whether these bacteria strains were able to degrade these organochlorine pesticides *in vivo*, i.e., in the luminal microenvironment of earthworms. On the other hand, some studies have reported that earthworm activity does not alter the persistence of pesticides in the soil (Sanchez-Hernandez et al. 2018). Changes in soil aggregates and the higher organic carbon content in casts and burrow walls probably contribute to the adsorption of pesticides, thus hampering their biodegradation (Hickman and Reid 2008).

Past studies have demonstrated that the earthworm gastrointestinal tract can secrete glycolytic digestive enzymes, thus contributing, together with enzymes produced by gut symbionts, to the decomposition of ingested organic matter (Lattaud et al. 1998, Garvín et al. 2000, Nozaki et al. 2009, 2013). Interestingly, gastrointestinal secretion of an important group of pesticide-detoxifying enzymes called carboxylesterases (EC 3.1.1.1) was demonstrated by Sanchez-Hernandez et al. (2009), corroborating past findings in *L. terrestris* (Prentø 1987) and other invertebrate species (Mommsen 1978, Geering and Freyvogel 1975, Turunen and Chippendale 1977, Turunen 1978). Carboxylesterases activity play an important role in the metabolism of xenobiotics in mammals (Hatfield et al. 2016). These esterases interact with organophosphorus, carbamates and synthetic pyrethroid pesticides, thus contributing to their detoxification (Jackson et al. 2011). The interaction between carboxylesterases and these three groups of pesticides is illustrated in Fig. 4. In the case of synthetic pyrethroids, carboxylesterases catalyze the hydrolysis of these pesticides to yield the corresponding alcohol and carboxylic acid (Sogorb and Vilanova 2002, Ross et al. 2006). In the case of organophosphorus compounds, carboxylesterases inactivate the oxidized metabolite of this class of pesticides, named "oxon" (Chambers et al. 2010a), by irreversible binding between the active site of the enzyme and the oxon compound (Maxwell 1992). Although the toxic metabolite is inactivated, the enzyme will also be non-functional. Therefore, this mechanism of organophosphorus detoxification is considered non-catalytic, and it is highly dependent on the affinity of the pesticide for the active site of the enzyme, and the number of carboxylesterase molecules able to interact with the pesticide (Chanda et al. 1997). In the case of carbamate compounds, carboxylesterases interact with this class of pesticides in a similar way as with organophosphorus compounds, although the inhibition mainly depends largely on the species and the carbamate, although it is weaker than inhibition by oxon metabolites (Jackson et al. 2011).

The luminal carboxylesterase activity described in earthworms is sensitive to inhibition by oxon metabolites of organophosphorus pesticides (Sanchez-Hernandez et al. 2009). Extracellular detoxification should have a direct consequence, i.e., reduced intestinal uptake of these toxic metabolites. Indeed, Sanchez-Hernandez et al. (2014b) demonstrated that luminal carboxylesterases are efficient molecular scavengers of chlorpyrifos as long as this organophosphorus pesticide is previously transformed into the toxic metabolite chlorpyrifos-oxon. A conceptual model that suggests both internal and external processes of pesticide detoxification in earthworms is shown in Fig. 5. Although the mutualistic relationship between earthworms and symbiont microorganisms is key in the degradation of pesticide residues, earthworm activity also creates external hot-spots of microbial activity in the casts, middens and burrow walls. For example, Liu et al. (2011) found that the endogeic species *Aporrectodea caliginosa* stimulated the proliferation of *Alphaprotobacteria*, which actively degraded the herbicide 2-methyl-4-chlorophenoxyacetic acid in casts and burrow walls, but apparently not in the gastrointestinal content. The authors argued that the anoxic conditions in the luminal microenvironment of earthworm digestive canal was the main reason for the lower rate of degradation of this herbicide (Liu et al. 2011).

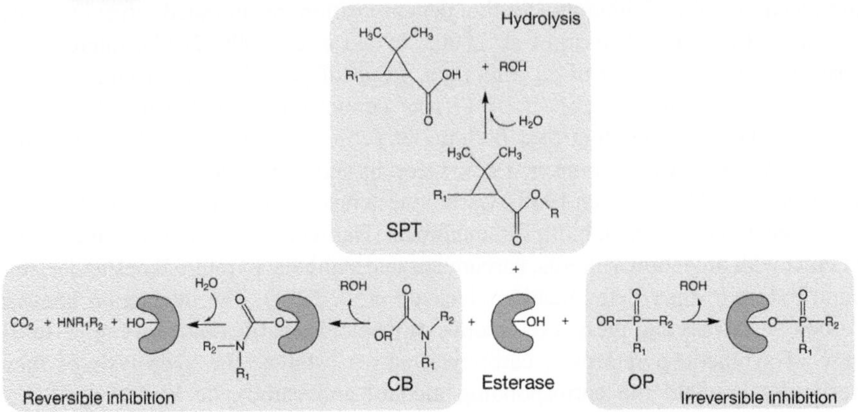

Fig. 4. Interactions of carboxylesterases with carbamate (CB), organophosphorus (OP) and synthetic pyrethroid (SPT) pesticides. Inhibition of esterases by CB compounds yields a carbamylated complex which is unstable, and the enzyme recovers its activity rapidly in the presence of water. Organophosphorus pesticides inhibit irreversibly the esterase activity by the formation of a stable phosphorylated complex. In both cases the interaction between the enzyme and the pesticide releases an alcohol (ROH). In the case of synthetic pyrethroids, the enzyme catalyzes the hydrolysis of these pesticides to yield the corresponding alcohol and carboxylic acid. Scheme elaborated from Sogorb and Vilanova (2002), and Thompson and Richardson (2004).

In summary, the available data suggest that the earthworm gastrointestinal tract acts as a biochemical reactor than can directly (via secretion of detoxifying enzymes) or indirectly (via symbiont microbiota) degrade pesticides present in ingested material (soil and plant debris). Although many studies have attributed a significant role of earthworms in the degradation of compounds as diverse as the herbicide atrazine, the organochlorine pesticide lindane and the organophosphorus pesticide chlorpyrifos, the real capacity of earthworms to significantly degrade pesticides under field conditions has not yet been established.

External Processes of Pesticide Detoxification

Earthworm casts, middens and burrows are the three components of the drilosphere characterized by a high microbial and enzymatic activities. For example, protease, β-glucosidase, alkaline phosphatase and dehydrogenase activities are generally higher in the earthworm casts than in the bulk soil (Lipiec et al. 2016). The physicochemical properties of casts (high organic matter content and abundance of fine particles) explain the greater enrichment of extracellular enzymes in this compartment than in the bulk soil. Extracellular enzymes are stabilized and remain active when bond to the soil organomineral complexes (Nannipieri et al. 1996). This is probably also true of earthworm casts. Nevertheless, extracellular enzymes seem to be less active in aged casts, probably because of factors such as antagonistic interactions between microbes (Tiunov and Scheu 2000) and nutrient depletion (Aira et al. 2010). However, physicochemical properties of casts are highly dependent on the soil type and earthworm species. Clause et al. (2014) found that pH, K and Mg concentrations and total N content were significantly affected by the soil type and, to a lesser extent,

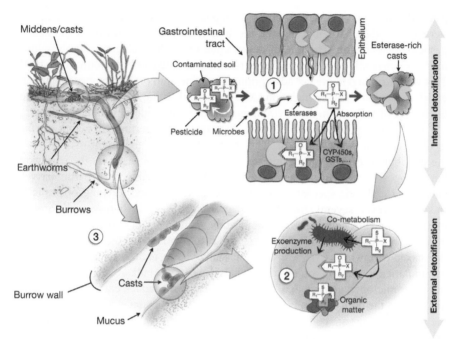

Fig. 5. Pictorial representation of the proposed conceptual model on the role of earthworms in the biodegradation of organophosphorus (OP) pesticides. This earthworm-assisted biodegradation would imply internal (gastrointestinal tract, [1]), and external processes performed in the casts (and middens) deployed on the soil surface [2] or inside the burrows [3]. The internal processes would involve an enzymatic detoxification by the luminal carboxylesterase activity, which has a double origin from symbiont microbes and gut epithelium. In addition, the fraction of OP that may be absorbed by the gut epithelium would be bound to intracellular carboxylesterases. This detoxification capability of OP pesticide by carboxylesterases is an efficient mechanism as long as the parental compound (e.g., chlorpyrifos, P=S) is metabolized to the oxon metabolite (e.g., chlorpyrifos-oxon, P=O), which displays a higher affinity by the active site of the enzyme. Microorganisms present in the gastrointestinal tract, casts, middens and burrow walls are responsible for this OP bioactivation. The external processes are performed in casts, middens and burrow walls that receive an important contribution of carboxylesterases via gastrointestinal secretion and microbial production (exoenzymes). In this enzymatic detoxification, we must not forget that high content of organic matter and fine particles (clays) in casts also contribute to bind OP pesticide (and metabolites). In summary, carboxylesterases would act as external bioscavengers for OP pesticides that, together to organic matter, would reduce their bioavailability and toxicity in soil.

the earthworm species, with *L. terrestris* having a greater impact than *Allolobophora chlorotica* and *Aporrectodea rosea*. Although these researchers did not measure any microbial properties, it can be speculated that casts with different physicochemical properties will boost different types of microbial activity and community structure.

Earthworm middens are small mounds built by anecic earthworms accumulating plant litter mixed with casts around the entrance of their burrows (Brown et al. 2004). These structures are microhabitats in which organic matter decomposition is facilitated by decomposers belonging to meso- and microfauna (Orazova et al. 2003, Stroud et al. 2016, Nuutinen et al. 2017). Middens should be local hotspots of contaminant degradation because of the intense microbial activity that occurs in these sites. Together with casts deposited on the soil surface, middens should receive pesticides that are sprayed on crops or applied directly to the soil by irrigation. Very few studies have investigated the functional role of earthworm middens in pesticide degradation. The rate of degradation of the herbicide atrazine was lower in *L. terrestris* middens than in the surrounding untilled soil (Akhouri et al. 1997). This was attributed to greater adsorption of atrazine onto organic matter in middens, which prevented its biodegradation (Alekseeva et al. 2006). Despite these examples, the impact of middens in the environmental fate of pesticides remain unknown.

One of the most evident and immediate impacts of earthworm activity on soil is the increased water filtration. This physical effect is the result of changes in the distribution of soil aggregates and in macroporosity caused by the earthworm burrowing activity. The tunnel network created by earthworms, particularly by anecic and endogeic species, may favor the movement and leaching of pesticides in soil. However, several studies have demonstrated that the burrow walls are, like casts and middens, hotspots of microbial activity and nutrient mineralization (Görres et al. 2001, Jégou et al. 2001, Furlong et al. 2002). In fact, the burrowing activity of the anecic earthworm *L. terrestris*, for instance, has a greater impact on microbial communities in the soil around the tunnels than in the bulk soil, thereby resulting in higher microbial activity and biomass as well as associated extracellular enzyme activities in the burrow linings (Hoang et al. 2016a, Athmann et al. 2017). Likewise, the burrow linings are composed of fragmented litter dragged down by earthworms and mixed with mucous and casts excreted by earthworms as they pass through the burrows (Binet et al. 2006). The accumulation of organic matter in the burrow linings thus facilitates the adsorption of pesticides (Worrall et al. 1997, Alekseeva et al. 2006). Together the findings of these studies suggest that the chemical composition of earthworm burrows reduces pesticide mobility as a result of two complementary processes: sorption on to organic ligands and biodegradation. These processes are illustrated in the studies by Farenhorst and colleagues (Farenhorst et al. 2000a, 2000b, Farenhorst and Bowman 2000), who demonstrated that *L. terrestris* reduced the potential leaching of atrazine and metabolites in the soil. The authors concluded that the feeding activity of this earthworm species, and the higher proportion of organic carbon in the burrow walls than in the bulk soil are determining factors in reducing the vertical transport of atrazine and increasing the non-extractable fraction of this herbicide.

In summary, earthworm casts, middens and the burrow walls are metabolically active microenvironments that may promote pesticide degradation and immobilization. Three major processes may have a significant impact on the environmental fate of pesticides in these drilosphere compartments. First, the accumulation of organic matter in these microsites will increase the binding of hydrophobic pesticides to organic ligands, thereby reducing their bioavailability to microorganisms (biodegradation) and increasing their persistence. This effect would explain why inoculating pesticide-contaminated soils with earthworm did not have a greater impact on the persistence of atrazine (Farenhorst et al. 2000b) or chlorpyrifos (Sanchez-Hernandez et al. 2018) than in contaminated, earthworm-free soils. Second, earthworm activity may induce proliferation of indigenous soil microorganisms able to degrade pesticides. For example, microorganisms and genes coding pesticide-degrading enzymes that can metabolize the herbicides atrazine (Monard et al. 2010) and chlorophenoxyacetic acid (Liu et al. 2013) have been isolated from soils influenced by earthworm activity. Third, earthworms can accumulate pesticide residues by absorption via their skin and gastrointestinal epithelium (Katagi and Ose 2015). In this context, a recent study by Briones and Álvarez-Otero (2018) showed that the thickness of both the cuticle and epidermis layers are very different in the three ecological groups of earthworms (epigeic, anecic and endogeic). This finding leads to the possibility of testing the hypothesis that the tegument thickness contributes to interspecific differences in pesticide accumulation by earthworms. In addition, detoxification and the sensitivity of earthworms to pesticides may also contribute to the persistence of pesticides in soil. Together with microbial degradation of pesticides, earthworms could contribute to the bioremediation process via beneficial impacts on microbial dispersion and proliferation and also via their metabolic capacity to degrade pesticides (Rodríguez-Castellanos and Sanchez-Hernandez 2007, Sanchez-Hernandez et al. 2014a, Katagi and Ose 2015).

Soil extracellular enzymes represent a significant fraction of the multiple forms in which enzymes may occur in the soil (Nannipieri et al. 2002, Dick et al. 2011). These extracellular enzymes play an important role in nutrient cycling, thus promoting soil fertility (Wallenstein and Burns 2011, Burns et al. 2013). Likewise, some of these extracellular enzymes such as laccases, peroxidases and carboxylesterases can metabolize organic pollutants as diverse in chemical structure as phenolic compounds, polycyclic aromatic hydrocarbons, polychlorinated biphenyls, azo dyes and pesticides (Rao et al. 2014, Gianfreda et al. 2016). In addition, carboxylesterases can hydrolyze some polyester polymers (Zumstein et al. 2017, Wei and Zimmermann 2017). In the particular case of organophosphate pesticides, past studies have reported that soil carboxylesterase activity can hydrolyze malathion and bind several organophosphate pesticides (Satyanarayana and Getzin 1973). More recently, the stability and reactivity of carboxylesterases have been investigated in soils inoculated with earthworms. For example, *L. terrestris* caused two- to four-fold increases in soil carboxylesterase activity relative to that in earthworm-free soils. Moreover, the esterase activity remained stable in the long-term after removal of the earthworms, probably because the extracellular enzyme was bound to organo-mineral complexes of soil, thus guaranteeing its stability (Sanchez-Hernandez et al. 2014c). More interestingly, the earthworm-induced carboxylesterase activity acted as an efficient molecular scavenger of organophosphorus pesticides, thus inactivating the highly toxic oxon metabolites (Sanchez-Hernandez et al. 2015). This is one of the few examples that illustrate a direct interaction between soil enzymes and pesticides. In fact, *in vitro* trials showed that soil carboxylesterase activity was irreversibly inhibited by the oxon metabolite of organophosphorus pesticides, thus reducing their availability and toxic effects on other biological targets in soil. A microcosm study with chlorpyrifos-contaminated soils showed that inoculation of these soils with *L. terrestris* caused a gradual increase in the activity of several soil enzymes (acid phosphatase, alkaline phosphatase and β-glucosidase) (Sanchez-Hernandez et al. 2018), which suggested that this earthworm species can be used in the bioremediation of organophosphorus-contaminated soils.

Together these findings on the detoxifying role of soil carboxylesterase activity allow us to propose the conceptual model illustrated in Fig. 5: earthworms may contribute to organophosphorus detoxification in soil through internal (gastrointestinal environment) and external (burrow walls, casts and middens) processes. These processes involve the direct interaction between the active site of carboxylesterases and the organophosphorus molecule. However, this chemical interaction is more efficient when the pesticides are transformed into their oxon metabolites. Most commercial organophosphorus pesticides are formulated as phosphorothioates, phosphorodithioates, phosphonothioates, phosphonodithioates and phosphoramidothioates, all of which have a sulfur atom bound to the phosphorus by a coordinate covalent bond (Chambers et al. 2010a). In this molecular configuration (P=S), the insecticide displays a weak affinity for the active site of soil carboxylesterases (Sanchez-Hernandez et al. 2017). However, the affinity increases strongly when the sulfur atom is substituted by oxygen (P=O, oxon metabolite). Although the desulfuration reaction is catalyzed by either cytochrome P450-dependent monooxygenases or flavin-dependent monooxygenases (Hodgson 2010, Chambers et al. 2010b), it also takes place in the environment via the action of ultraviolet light (Chambers et al. 2010a) or via hydrolysis (Racke 1993). Therefore, luminal carboxylesterase activity in the earthworm gastrointestinal tract will be able to bind the oxon metabolites of organophosphorus that are present in the ingested soil (and plant debris). The gastrointestinal epithelium and gut symbionts are the major sources of this luminal carboxylesterase activity, which reduces the uptake of pesticide in the digestive tract of earthworms (Sanchez-Hernandez et al. 2014b). On the other hand, enzyme activity, including carboxylesterase activity, is high in microsites formed by earthworms (casts, burrow walls and middens). It therefore seems reasonable to hypothesize that these structures will contribute to immobilizing oxon metabolites of organophosphorus pesticides by binding them the active site of extracellular carboxylesterases. Although past studies have demonstrated that the walls of earthworm burrows may act as a chemical barrier to herbicide leaching (Farenhorst et al. 2000a, 2000b, Farenhorst and Bowman 2000), this effect has not been confirmed for other classes of pesticides. Pending future studies, it is expected that carboxylesterases in casts and in the burrow walls may reduce the bioavailability and mobility of organophosphorus pesticides in soil.

Concluding Remarks

This chapter has considered the importance of earthworms, particularly anecic and endogeic species, in the bioremediation of pesticide-contaminated soils. Some studies have served as examples to illustrate how these soil organisms may act as biological vectors of microbial proliferation and extracellular enzyme production for detoxification purposes. However, future research should focus on validating the capacity of earthworms as bioremediators in the field. In this respect, the answers to the following two questions await empirical data:

(1) Which earthworm species are most suitable for the bioremediation of pesticide-contaminated soil? Earthworm sensitivity to pesticide residues is a key point in the efficacy of bioremediation. Indeed, some studies have documented species-specific differences in pesticide sensitivity, which also depend on the type of

pesticide and soil physicochemical properties (Edwards and Bohlen 1992, Pelosi et al. 2013a, Pelosi et al. 2014). As a starting point, the use of standardized sublethal bioassays with soil samples from pesticide-contaminated lands should be implemented to save time, cost and effort before any bioremediation action is undertaken. For example, the standardized behavior response test (ISO 2008) could be used to assess whether particular earthworm species is a suitable candidate in a specific bioremediation scenario. The cast production test (Capowiez et al. 2010) and the burrowing behavior test (Dittbrenner et al. 2011) could also be used for prior assessment of the compatibility between earthworms and the contaminated soil. Likewise, potential candidate species should be selected from the most abundant earthworms in the agroecosystem, such as *L. terrestris*, *A. caliginosa* and *Allolobophora chlorotica* (Pelosi et al. 2013b). Although the main role of earthworms in bioremediation is to promote microbial degraders and facilitate biodegradation of contaminants by increasing their availability to microorganisms (Hickman and Reid 2008), the earthworms themselves may also act as contaminant degraders. Indeed, earthworms may accumulate soil contaminants via their tegument and gastrointestinal tract, and they may actively metabolize organic contaminants such as pesticides (Stenersen 1984, Katagi and Ose 2015).

(2) What is the most appropriate strategy for introducing earthworms in contaminated land? Although inoculation of contaminated or degraded soils with earthworms is the most obvious approach (Baker et al. 2006, Butt 2008), the inoculation of uncontaminated agricultural soil with earthworms, or even the stimulation of native earthworm populations are attractive preventive strategies. For example, organic mulching and green manure management commonly used in organic agriculture to promote soil fertility and increase crop yields (Eilittä et al. 2007, Sharma et al. 2017, Kader et al. 2017), may also increase earthworm activity and biomass. Field studies by Fonte et al. (2009) and Frøseth et al. (2014) encourage this type of functional approach. In the first study, mulching with tomato residues increased the biomass of *A. caliginosa*, *A. chlorotica* and *Megascolecid* sp. (Fonte et al. 2009). In the second study, a higher density and biomass of earthworms (*A. caliginosa, Lumbricus rubellus, Aporrectodea rosea* and *L. terrestris*) was recorded in soils in which green manure was left on the soil surface (Frøseth et al. 2014). The management of organic waste to increase earthworm activity and abundance in agricultural land may be an eco-friendly strategy for mitigating the impact of pesticides from early stages.

Related to the latter question, some strategies emerge to improve earthworm-assisted bioremediation. In the last two decades, biochar has been revealed to be a promising technology for increasing soil fertility and facilitating degradation, or retention, of environmental contaminants. Some studies have demonstrated that earthworms and biochar can co-exist in soil (Elmer et al. 2015). Furthermore, earthworm activity in biochar-amended soil favors the retention of extracellular detoxifying enzymes on the biochar surface (Sanchez-Hernandez et al. 2018). This functional tandem may also be an interesting and innovative approach to explore in the coming years. We thank also Yvan Capowiez for his comments and suggestions on an earlier version of this manuscript.

Acknowledgements

We thank the Ministerio de Economía y Competitividad, Spanish Government (grants no. CTM2011-25788/TECNO and CTM2014-53915-R) for the financial support of this research.

References

Abhilash, P.C. and N. Singh. 2009. Pesticide use and application: An Indian scenario. J. Hazard. Mater. 165: 1–12.

Aira, M., C. Lazcano, M. Gómez-Brandón and J. Domínguez. 2010. Ageing effects of casts of *Aporrectodea caliginosa* on soil microbial community structure and activity. Appl. Soil Ecol. 46: 143–146.

Akhouri, N.M., E.J. Kladivko and R.F. Turco. 1997. Sorption and degradation of atrazine in middens formed by *Lumbricus terrestris*. Soil Biol. Biochem. 29: 663–666.

Alekseeva, T., P. Besse, F. Binet, A.M. Delort, C. Forano, N. Josselin et al. 2006. Effect of earthworm activity (*Aporrectodea giardi*) on atrazine adsorption and biodegradation. Eur. J. Soil Sci. 57: 295–307.

Andriuzzi, W.S., T. Bolger and O. Schmidt. 2013. The drilosphere concept: Fine-scale incorporation of surface residue-derived N and C around natural *Lumbricus terrestris* burrows. Soil Biol. Biochem. 64: 136–138.

Arias-Estévez, M., E. López-Periago, E. Martínez-Carballo, J. Simal-Gándara, J.-C. Mejuto and L. García-Río. 2008. The mobility and degradation of pesticides in soils and the pollution of groundwater resources. Agric. Ecosyst. Environ. 123: 247–260.

Athmann, M., T. Kautz, C. Banfield, S. Bauke, D.T.T. Hoang, M. Lüsebrink et al. 2017. Six months of *L. terrestris* L. activity in root-formed biopores increases nutrient availability, microbial biomass and enzyme activity. Appl. Soil Ecol. 120: 135–142.

Atwood, D. and C. Paisley-Jones. 2017. Pesticide industry sales and usage: 2008–2012 market estimates. U.S. Environmental Protection Agency, Washington, DC.

Baker, G.H., G. Brown, K. Butt, J.P. Curry and J. Scullion. 2006. Introduced earthworms in agricultural and reclaimed land: their ecology and influences on soil properties, plant production and other soil biota. Biol. Invasions 8: 1301–1316.

Berenstein, G., S. Nasello, É. Beiguel, P. Flores, J. Di Schiena, S. Basack et al. 2017. Human and soil exposure during mechanical chlorpyrifos, myclobutanil and copper oxychloride application in a peach orchard in Argentina. Sci. Total Environ. 586: 1254–1262.

Bharagava, R.N. 2017. Environmental Pollutants and their Bioremediation Approaches. CRC Press Francis & Taylor Group, Boca Raton, USA.

Binet, F., A. Kersanté, C. Munier-Lamy, R.-C. Le Bayon, M.-J. Belgy and M.J. Shipitalo. 2006. Lumbricid macrofauna alter atrazine mineralization and sorption in a silt loam soil. Soil Biol. Biochem. 38: 1255–1263.

Bityutskii, N., P. Kaidun and K. Yakkonen. 2016. Earthworms can increase mobility and bioavailability of silicon in soil. Soil Biol. Biochem. 99: 47–53.

Bolan, N.S., D.C. Adriano, A. Kunhikrishnan, T. James, R. McDowell and N. Senesi. 2011. Dissolved organic matter: biogeochemistry, dynamics, and environmental significance in soils. Adv. Agron. 110: 1–75.

Briones, M.J.I. and R. Álvarez-Otero. 2018. Body wall thickness as a potential functional trait for assigning earthworm species to ecological categories. Pedobiologia 67: 26–34.

Brown, G.G. 1995. How do earthworms affect microfloral and faunal community diversity. Plant Soil 170: 209–231.

Brown, G.G., I. Barois and P. Lavelle. 2000. Regulation of soil organic matter dynamics and microbial activityin the drilosphere and the role of interactions with other edaphic functional domains. Eur. J. Soil Biol. 36: 177–198.

Brown, G.G., B.M. Doube and C.A. Edwards. 2004. Functional interactions between earthworms, microorganisms, organic matter, and plants. pp. 213–239. *In*: Edwards, C.A. (ed.). Earthworm Ecology. CRC Press, Boca Raton, USA.

Burns, R.G., J.L. DeForest, J. Marxsen, R.L. Sinsabaugh, M.E. Stromberger, M.D. Wallenstein et al. 2013. Soil enzymes in a changing environment: Current knowledge and future directions. Soil Biol. Biochem. 58: 216–234.

Butt, K.R. and C.N. Lowe. 2007. Presence of earthworm species within and beneath *Lumbricus terrestris* (L.) middens. Eur. J. Soil Biol. 43: S57–S60.

Butt, K.R. 2008. Earthworms in soil restoration: Lessons learned from United Kingdom case studies of land reclamation. Restor. Ecol. 16: 637–641.

Capowiez, Y., N. Dittbrenner, M. Rault, R. Triebskorn, M. Hedde and C. Mazzia. 2010. Earthworm cast production as a new behavioural biomarker for toxicity testing. Environ. Pollut. 158: 388–393.

Capowiez, Y., S. Sammartino and E. Michel. 2014. Burrow systems of endogeic earthworms: Effects of earthworm abundance and consequences for soil water infiltration. Pedobiologia 57: 303–309.

Ccanccapa, A., A. Masiá, V. Andreu and Y. Picó. 2016a. Spatio-temporal patterns of pesticide residues in the Turia and Júcar Rivers (Spain). Sci. Total Environ. 540: 200–210.

Ccanccapa, A., A. Masiá, A. Navarro-Ortega, Y. Picó and D. Barceló. 2016b. Pesticides in the Ebro River basin: Occurrence and risk assessment. Environ. Pollut. 211: 414–424.

Chambers, H.W., E.C. Meek and J.E. Chambers. 2010a. Chemistry of organophosphorus insecticides. pp. 1395–1398. *In*: Krieger, R. (ed.). Hayes' Handbook of Pesticide Toxicology. Elsevier, London, U.K.

Chambers, J.E., E.C. Meek and H.W. Chambers. 2010b. The metabolism of organophosphorus insecticides. pp. 1399–1407. *In*: Krieger, R. (ed.). Hayes' Handbook of Pesticide Toxicology. Elsevier, London, U.K.

Chanda, S.M., S.R. Mortensen, V.C. Moser and S. Padilla. 1997. Tissue-specific effects of chlorpyrifos on carboxylesterase and cholinesterase activity in adult rats: An *in vitro* and *in vivo* comparison. Fund. Appl. Toxicol. 38: 148–157.

Chen, H., Z. Hao, Q. Wang, Y. Jiang, R. Pan, C. Wang et al. 2016. Occurrence and risk assessment of organophosphorus pesticide residues in Chinese tea. Human Ecol. Risk Assess. 22: 28–38.

Cheng, M., G. Zeng, D. Huang, C. Yang, C. Lai, C. Zhang et al. 2018. Tween 80 surfactant-enhanced bioremediation: toward a solution to the soil contamination by hydrophobic organic compounds. Crit. Rev. Biotechnol. 38: 17–30.

Clause, J., S. Barot, B. Richard, T. Decaëns and E. Forey. 2014. The interactions between soil type and earthworm species determine the properties of earthworm casts. Appl. Soil Ecol. 83: 149–158.

Craven, D., M.P. Thakur, E.K. Cameron, L.E. Frelich, R. Beauséjour, R.B. Blair et al. 2017. The unseen invaders: introduced earthworms as drivers of change in plant communities in North American forests (a meta-analysis). Glob. Chang. Biol. 23: 1065–1074.

Cummings, S.P. 2010. Bioremediation: Methods and Protocols. Humana Press, New York, USA.

Delcour, I., P. Spanoghe and M. Uyttendaele. 2015. Literature review: Impact of climate change on pesticide use. Food Res. Inter. 68: 7–15.

Dick, R.P., R.G. Burns and R.P. Dick. 2011. A brief history of soil enzymology research. pp. 1–34. *In*: Dick, R.P. (ed.). Methods of Soil Enzymology. Soil Science Society of America, SSSA Book Series, Madison, USA.

Dittbrenner, N., I. Moser, R. Triebskorn and Y. Capowiez. 2011. Assessment of short and long-term effects of imidacloprid on the burrowing behaviour of two earthworm species (*Aporrectodea caliginosa* and *Lumbricus terrestris*) by using 2D and 3D post-exposure techniques. Chemosphere 84: 1349–1355.

Edwards, C.A. and P.J. Bohlen. 1992. The effects of toxic chemicals on earthworms. Rev. Environ. Contam. Toxicol. 125: 23–99.

Eilittä, M., J. Mureithi and R. Derpsch. 2004. Green Manure/Cover Crop Systems of Smallholder Farmers: Experiences from Tropical and Subtropical Regions. Kluwer Academic Publishers, Dordrecht, Netherlands.

Elmer, W.H., C.V. Lattao and J.J. Pignatello. 2015. Active removal of biochar by earthworms (*Lumbricus terrestris*). Pedobiologia 58: 1–6.

Fang, H., Y. Yu, X. Chu, X. Wang, X. Yang and J. Yu. 2009. Degradation of chlorpyrifos in laboratory soil and its impact on soil microbial functional diversity. J. Environ. Sci. 21: 380–386.

Farenhorst, A. and B.T. Bowman. 2000. Sorption of atrazine and metolachlor by earthworm surface castings and soil. J. Environ. Sci. Health 35B: 157–173.

Farenhorst, A., E. Topp, B.T. Bowman and A.D. Tomlin. 2000a. Earthworm burrowing and feeding activity and the potential for atrazine transport by preferential flow. Soil Biol. Biochem. 32: 479–488.

Farenhorst, A., E. Topp, B.T. Bowman and A.D. Tomlin. 2000b. Earthworms and the dissipation and distribution of atrazine in the soil profile. Soil Biol. Biochem. 32: 23–33.

Frøseth, R.B., A.K. Bakken, M.A. Bleken, H. Riley, R. Pommeresche, K. Thorup-Kristensen et al. 2014. Effects of green manure herbage management and its digestate from biogas production on barley yield, N recovery, soil structure and earthworm populations. Eur. J. Agron. 52: 90–102.

Fuentes, E., M.E. Báez and J. Díaz. 2010. Survey of organophosphorus pesticide residues in virgin olive oils produced in Chile. Food Add. Contam. 3: 101–107.

Furlong, M.A., D.R. Singleton, D.C. Coleman and W.B. Whitman. 2002. Molecular and culture-based analyses of prokaryotic communities from an agricultural soil and the burrows and casts of the earthworm *Lumbricus rubellus*. Appl. Environ. Microbiol. 68: 1265–1279.

Garvín, M.H., C. Lattaud, D. Trigo and P. Lavelle. 2000. Activity of glycolytic enzymes in the gut of *Hormogaster elisae* (Oligochaeta, Hormogastridae). Soil Biol. Biochem. 32: 929–934.

Gavrilescu, M. 2005. Fate of pesticides in the environment and its bioremediation. Eng. Life Sci. 5: 497–526.

Geering, K. and T.A. Freyvogel. 1975. Lipase activity and stimulation mechanism of esterases in the midgut of female *Aedes aegypti*. J. Insect Physiol. 21: 1251–1256.

Gianfreda, L., M.A. Rao, R. Scelza and M. De la Luz Mora. 2016. Role of enzymes in environment cleanup/remediation. pp. 133–155. *In*: Dhillon, G.S. and S. Kaur (eds.). Agro-Industrial Wastes as Feedstock for Enzyme Production. Elsevier, London, U.K.

Gianfreda, L. and M.A. Rao. 2008. Interactions between xenobiotics and microbial and enzymatic soil activity. Crit. Rev. Environ. Sci. Technol. 38: 269–310.

Görres, J.H., M.C. Savin and J.A. Amador. 2001. Soil micropore structure and carbon mineralization in burrows and casts of an anecic earthworm (*Lumbricus terrestris*). Soil Biol. Biochem. 33: 1881–1887.

Griffith, B., M. Türke, W.W. Weisser and N. Eisenhauer. 2013. Herbivore behavior in the anecic earthworm species *Lumbricus terrestris* L. Eur. J. Soil Biol. 55: 62–65.

Hatfield, M.J., R.A. Umans, J.L. Hyatt, C.C. Edwards, M. Wierdl, L. Tsurkan et al. 2016. Carboxylesterases: General detoxifying enzymes. Chem.-Biol. Interact. 259: 327–331.

Herrero-Hernández, E., M.S. Rodríguez-Cruz, E. Pose-Juan, S. Sánchez-González, M.S. Andrades and M.J. Sánchez-Martín. 2017. Seasonal distribution of herbicide and insecticide residues in the water resources of the vineyard region of La Rioja (Spain). Sci. Total Environ. 609: 161–171.

Hickman, Z.A. and B.J. Reid. 2008. Earthworm assisted bioremediation of organic contaminants. Environ. Int. 34: 1072–1081.

Hoang, D.T.T., J. Pausch, B.S. Razavi, I. Kuzyakova, C.C. Banfield and Y. Kuzyakov. 2016a. Hotspots of microbial activity induced by earthworm burrows, old root channels, and their combination in subsoil. Biol. Fert. Soils 52: 1105–1119.

Hoang, D.T.T., B.S. Razavi, Y. Kuzyakov and E. Blagodatskaya. 2016b. Earthworm burrows: Kinetics and spatial distribution of enzymes of C-, N- and P-cycles. Soil Biol. Biochem. 99: 94–103.

Hodgson, E. 2010. Metabolism of pesticides. pp. 893–921. *In*: Krieger, R. (ed.). Hayes' Handbook of Pesticide Toxicology. Elsevier, London, U.K.

ISO. 2008. Soil quality—avoidance test for determining the quality of soils and effects of chemicals on behaviour–Part 1: test with earthworms (*Eisenia fetida* and *Eisenia andrei*). ISO/DIS 17512-1. International Organization for Standardization, Geneva, Switzerland.

Jackson, C.J., J.G. Oakeshott, J.C. Sanchez-Hernandez and C.E. Wheelock. 2011. Carboxylesterases in the metabolism and toxicity of pesticides. pp. 57–75. *In*: Satoh, T. and R.C. Gupta (eds.). Anticholinesterase Pesticides: Metabolism, Neurotoxicity, and Epidemiology. Wiley, New Jersey, USA.

Jégou, D., S. Schrader, H. Diestel and D. Cluzeau. 2001. Morphological, physical and biochemical characteristics of burrow walls formed by earthworms. Appl. Soil Ecol. 17: 165–174.

John, E.M. and J.M. Shaike. 2015. Chlorpyrifos: pollution and remediation. Environ. Chem. Lett. 13: 269–291.

Jouni, F., J.C. Sanchez-Hernandez, C. Mazzia, M. Jobin, Y. Capowiez and M. Rault. 2018. Interspecific differences in biochemical and behavioral biomarkers in endogeic earthworms exposed to ethyl-parathion. Chemosphere 202: 85–93.

Jouquet, P., J. Dauber, J. Lagerlöf, P. Lavelle and M. Lepage. 2006. Soil invertebrates as ecosystem engineers: Intended and accidental effects on soil and feedback loops. Appl. Soil Ecol. 32: 153–164.

Kader, M.A., M. Senge, M.A. Mojid and K. Ito. 2017. Recent advances in mulching materials and methods for modifying soil environment. Soil Till. Res. 168: 155–166.

Katagi, T. and K. Ose. 2015. Toxicity, bioaccumulation and metabolism of pesticides in the earthworm. J. Pest. Sci. 40: 69–81.

Köck-Schulmeyer, M., A. Ginebreda, C. Postigo, T. Garrido, J. Fraile, M. López de Alda et al. 2014. Four-years advanced monitoring program of polar pesticides in groundwater of Catalonia (NE-Spain). Sci. Total Environ. 470-471: 1087–1098.

Köhne, J.M., S. Köhne and J. Simůnek. 2009. A review of model applications for structured soils: b) Pesticide transport. J. Contam. Hydrol. 104: 36–60.

Kumar, S., G. Kaushik and J.F. Villarreal-Chiu. 2016. Scenario of organophosphate pollution and toxicity in India: A review. Environ. Sci. Pollut. Res. 23: 9480–91.

Lattaud, C., S. Locati, P. Mora, C. Rouland and P. Lavelle. 1998. The diversity of digestive systems in tropical geophagous earthworms. Appl. Soil Ecol. 9: 189–195.

Lavelle, P., I. Barois, E. Blanchart, G. Brown, L. Brussaard, T. Decaëns et al. 1998. Earthworms as a resource in tropical agroecosystems. Nat. Res. 34: 26–41.

Lavelle, P., L. Brussaard and P.F. Hendrix. 1999. Earthworm Management in Tropical Agroecosystems. CABI Publishing. Wallingford, U.K.

Lipiec, J., M. Frąc, M. Brzezińska, M. Turski and K. Oszust. 2016. Linking microbial enzymatic activities and functional diversity of soil around earthworm burrows and casts. Front. Microbiol. 7: 1361.

Liu, Y., D. Shen, S. Li, Z. Ni, M. Ding, C. Ye et al. 2016. Residue levels and risk assessment of pesticides in nuts of China. Chemosphere 144: 645–651.

Liu, Y.J., S.J. Liu, H.L. Drake and M.A. Horn. 2011. Alphaproteobacteria dominate active 2-methyl-4-chlorophenoxyacetic acid herbicide degraders in agricultural soil and drilosphere. Environ. Microbiol. 13: 991–1009.

Liu, Y.J., S.J. Liu, H.L. Drake and M.A. Horn. 2013. Consumers of 4-chloro-2-methylphenoxyacetic acid from agricultural soil and drilosphere harbor cadA, r/sdpA, and tfdA-like gene encoding oxygenases. FEMS Microbiol. Ecol. 86: 114–129.

Lubbers, I.M., K.J. van Groenigen, S.J. Fonte, J. Six, L. Brussaard and J.W. van Groenigen. 2013. Greenhouse-gas emissions from soils increased by earthworms. Nat. Clim. Chang. 3: 187–194.

Margesin, R. 2005. Determination of enzyme activities in contaminated soil. pp. 309–320. *In*: Margesin, R. and F. Schinner (eds.). Manual of Soil Analysis: Monitoring and Assessing Soil Bioremediation. Springer-Verlag, Berlin, Germany.

Margesin, R. and F. Schinner. 2005. Manual for Soil Analysis—Monitoring and Assessing Soil Bioremediation. Springer-Verlag, Berlin, Germany.

Martínez Morcillo, S., J.L. Yela, Y. Capowiez, C. Mazzia, M. Rault and J.C. Sanchez-Hernandez. 2013. Avoidance behaviour response and esterase inhibition in the earthworm, *Lumbricus terrestris*, after exposure to chlorpyrifos. Ecotoxicology 22: 597–607.

Masciandaro, G., C. Macci, E. Peruzzi, B. Ceccanti and S. Doni. 2013. Organic matter–microorganism–plant in soil bioremediation: a synergic approach. Rev. Environ. Sci. Biotechnol. 12: 399–419.

Masiá, A., J. Campo, P. Vázquez-Roig, C. Blasco and Y. Picó. 2013. Screening of currently used pesticides in water, sediments and biota of the Guadalquivir River Basin (Spain). J. Hazard Mater. 263: 95–104.

Maxwell, D.M. 1992. The specificity of carboxylesterase protection against the toxicity of organophosphorus compounds. Toxicol. Appl. Pharmacol. 114: 306–312.

Megharaj, M. and R. Naidu. 2017. Soil and brownfield bioremediation. Microb. Biotechnol. 10: 1244–1249.

Megharaj, M., B. Ramakrishnan, K. Venkateswarlu, N. Sethunathan and R. Naidu. 2011. Bioremediation approaches for organic pollutants: A critical perspective. Environ. Int. 37: 1362–1375.

Menchen, A., J.L. Heras and J.J. Alday. 2017. Pesticide contamination in groundwater bodies in the Júcar River European Union Pilot Basin (SE Spain). Environ. Monit. Assess. 189: 146.

Mirsal, I. 2008. Soil Pollution: Origin, Monitoring and Remediation. Springer-Verlag, Berlin, Germany.

Mommsen, T.P. 1978. Digestive enzymes of a spider (*Tegenaria atrica* Koch)—III. Esterases, phosphatases, nucleases. Comp. Biochem. Physiol. 60A: 377–382.

Monard, C., F. Martin-Laurent, M. Devers-Lamrani, O. Lima, P. Vandenkoornhuyse and F. Binet. 2010. atz gene expressions during atrazine degradation in the soil drilosphere. Mol. Ecol. 19: 749–759.

Morillo, E. and J. Villaverde. 2017. Advanced technologies for the remediation of pesticide-contaminated soils. Sci. Total Environ. 586: 576–597.

Nair, I.C. and K. Jayachandran. 2017. Enzymes for bioremediation and biocontrol. pp. 75–97. *In*: Sugathan, S., N.S. Pradeep and S. Abdulhameed (eds.). Bioresources and Bioprocess in Biotechnology. Springer Singapore, Singapore, India.

Nannipieri, P., P. Sequi and P. Fusi. 1996. Humus and enzyme activity. pp. 293–328. *In*: Piccolo, A. (ed.). Humic Substances in Terrestrial Ecosystems. Elsevier, Amsterdam, Netherlands.

Nannipieri, P., E. Kandeler and P. Ruggiero 2002. Enzyme activities and microbiological and biochemical processes in soil. pp. 1–33. *In*: Burns, R.G. and R.P. Dick (eds.). Enzymes in the Environment: Activity, Ecology, and Applications. Marcel Dekker, New York, USA.

Newman, M.C., M. Crane and G. Holloway. 2006. Does pesticide risk assessment in the European Union assess long-term effects. Rev. Environ. Contam. Toxicol. 187: 1–65.

Noyes, P.D., M.K. McElwee, H.D. Miller, B.W. Clark, L.A. Van Tiem, K.C. Walcott et al. 2009. The toxicology of climate change: environmental contaminants in a warming world. Environ. Int. 35: 971–986.

Nozaki, M., C. Miura, Y. Tozawa and T. Miura. 2009. The contribution of endogenous cellulase to the cellulose digestion in the gut of earthworm (*Pheretima hilgendorfi*: Megascolecidae). Soil Biol. Biochem. 41: 762–769.

Nozaki, M., K. Ito, C. Miura and T. Miura. 2013. Examination of digestive enzyme distribution in gut tract and functions of intestinal caecum, in megascolecid earthworms (Oligochaeta: Megascolecidae) in Japan. Zoolog. Sci. 30: 710–715.

Nuutinen, V., K.R. Butt, J. Hyväluoma, E. Ketoja and J. Mikola. 2017. Soil faunal and structural responses to the settlement of a semi-sedentary earthworm *Lumbricus terrestris* in an arable clay field. Soil Biol. Biochem. 115: 285–296.

Odukkathil, G. and N. Vasudevan. 2013. Toxicity and bioremediation of pesticides in agricultural soil. Rev. Environ. Sci. Biotechnol. 12: 421–444.

Odukkathil, G. and N. Vasudevan. 2016. Residues of endosulfan in surface and subsurface agricultural soil and its bioremediation. J. Environ. Manag. 165: 72–80.

OECD. 2013. OECD Compendium of Agri-environmental Indicators, OECD Publishing, Paris.

Orazova, M.K., T.A. Semenova and A.V. Tiunov. 2003. The microfungal community of *Lumbricus terrestris* middens in a linden (*Tilia cordata*) forest. Pedobiologia 47: 27–32.

Orgiazzi, A., R.D. Bardgett, E. Barrios, V. Behan-Pelletier, M.J.I. Briones, J.-L. Chotte et al. 2016. Global Soil Biodiversity Atlas. Luxembourg: European Commission, Publications Office of the European Union.

Padilla-Sánchez, J.A., R. Romero-González, P. Plaza-Bolaños, A. Garrido Frenich and J.L. Martínez Vidal. 2015. Residues and organic contaminants in agricultural soils in intensive agricultural areas of Spain: A three years survey. Clean–Soil Air Water 43: 746–753.

Palm, C., P. Sanchez, S. Ahamed and A. Awiti. 2007. Soils: A contemporary perspective. Ann. Rev. Environ. Resour. 32: 99–129.

Pelosi, C., S. Joimel and D. Makowski. 2013a. Searching for a more sensitive earthworm species to be used in pesticide homologation tests—A meta-analysis. Chemosphere 90: 895–900.

Pelosi, C., L. Toutous, F. Chiron, F. Dubs, M. Hedde, A. Muratet et al. 2013b. Reduction of pesticide use can increase earthworm populations in wheat crops in a European temperate region. Agric. Ecosyst. Environ. 181: 223–230.

Pelosi, C., S. Barot, Y. Capowiez, M. Hedde and F. Vandenbulcke. 2014. Pesticides and earthworms. A review. Agron. Sust. Develop. 34: 199–228.

Prentø, P. 1987. Distribution of 20 enzymes in the midgut region of the earthworm, *Lumbricus terrestris* L., with particular emphasis on the physiological role of the chloragog tissue. Comp. Biochem. Physiol. 87A: 135–142.

Racke, K.D. 1993. Environmental fate of chlorpyrifos. Rev. Environ. Contam. Toxicol. 131: 1–150.

Racke, K.D., K.P. Steele, R.N. Yoder, W.A. Dick and E. Avidov. 1996. Factors affecting the hydrolytic degradation of chlorpyrifos in soil. J. Agric. Food Chem. 44: 1582–1592.

Ramteke, P.W. and R.K. Hans. 1992. Isolation of hexachlorocyclohexane (HCH) degrading microorganisms from earthworm gut. J. Environ. Sci. Health 27A: 2113–2122.

Rao, M.A., R. Scelza, F. Acevedo, M.C. Diez and L. Gianfreda. 2014. Enzymes as useful tools for environmental purposes. Chemosphere 107: 145–162.

Rayu, S., D.G. Karpouzas and B.K. Singh. 2012. Emerging technologies in bioremediation: constraints and opportunities. Biodegradation 23: 917–926.

Riah, W., K. Laval, E. Laroche-Ajzenberg, C. Mougin, X. Latour and I. Trinsoutrot-Gattin. 2014. Effects of pesticides on soil enzymes: a review. Environ. Chem. Lett. 12: 257–273.

Rico, A., C. Sabater and M.-Á. Castillo. 2016. Lethal and sub-lethal effects of five pesticides used in rice farming on the earthworm *Eisenia fetida*. Ecotoxicol. Environ. Saf. 127: 222–229.

Rodriguez-Campos, J., L. Dendooven, D. Alvarez-Bernal and S.M. Contreras-Ramos. 2014. Potential of earthworms to accelerate removal of organic contaminants from soil: A review. Appl. Soil Ecol. 79: 10–25.

Rodríguez-Castellanos, L. and J.C. Sanchez-Hernandez. 2007. Earthworm biomarkers of pesticide contamination: Current status and perspectives. J. Pest. Sci. 32: 360–371.

Römbke, J., S. Jänsch and W. Didden. 2005. The use of earthworms in ecological soil classification and assessment concepts. Ecotoxicol. Environ. Saf. 62: 249–265.

Ross, M.K., A. Borazjani, C.C. Edwards and P.M. Potter. 2006. Hydrolytic metabolism of pyrethroids by human and other mammalian carboxylesterases. Biochem. Pharmacol. 71: 657–669.

Sánchez-González, S., E. Pose-Juan, E. Herrero-Hernández, A. Álvarez-Martín, M.J. Sánchez-Martín and S. Rodríguez-Cruz. 2013. Pesticide residues in groundwaters and soils of agricultural areas in the Águeda River Basin from Spain and Portugal. Int. J. Environ. Anal. Chem. 93: 1585–1601.

Sanchez-Hernandez, J.C. 2006. Earthworm biomarkers in ecological risk assessment. Rev. Environ. Contam. Toxicol. 188: 85–126.

Sanchez-Hernandez, J.C., C. Mazzia, Y. Capowiez and M. Rault. 2009. Carboxylesterase activity in earthworm gut contents: Potential (eco)toxicological implications. Comp. Biochem. Physiol. 150C: 503–511.

Sanchez-Hernandez, J.C., C. Narvaez, P. Sabat and S. Martínez Mocillo. 2014a. Integrated biomarker analysis of chlorpyrifos metabolism and toxicity in the earthworm *Aporrectodea caliginosa*. Sci. Total Environ. 490: 445–455.

Sanchez-Hernandez, J.C., M. Aira and J. Domínguez. 2014b. Extracellular pesticide detoxification in the gastrointestinal tract of the earthworm *Aporrectodea caliginosa*. Soil Biol. Biochem. 79: 1–4.

Sanchez-Hernandez, J.C., S. Martínez Morcillo, J. Notario del Pino and P. Ruiz. 2014c. Earthworm activity increases pesticide-sensitive esterases in soil. Soil Biol. Biochem. 75: 186–196.

Sanchez-Hernandez, J.C., J. Notario del Pino and J. Domínguez. 2015. Earthworm-induced carboxylesterase activity in soil: Assessing the potential for detoxification and monitoring organophosphorus pesticides. Ecotoxicol. Environ. Saf. 122: 303–312.

Sanchez-Hernandez, J.C. and M. Sandoval. 2017. Effects of chlorpyrifos on soil carboxylesterase activity at an aggregate-size scale. Ecotoxicol. Environ. Saf. 142: 303–311.

Sanchez-Hernandez, J.C., M. Sandoval and A. Pierart. 2017. Short-term response of soil enzyme activities in a chlorpyrifos-treated mesocosm: Use of enzyme-based indexes. Ecol. Indic. 73: 525–535.

Sanchez-Hernandez, J.C., J. Notario del Pino, Y. Capowiez, C. Mazzia and M. Rault. 2018. Soil enzyme dynamics in chlorpyrifos-treated soils under the influence of earthworms. Sci. Total Environ. 612: 1407–1416.

Satyanarayana, T. and L.W. Getzin. 1973. Properties of a stable cell-free esterase from soil. Biochemistry 12: 1566–1572.

Scheu, S. 2003. Effects of earthworms on plant growth: patterns and perspectives: the 7th international symposium on earthworm ecology Cardiff Wales 2002. Pedobiologia 47: 846–856.

Scott, C., C. Begley, M.J. Taylor, G. Pandey, V. Momiroski, N. French et al. 2011. Free-enzyme bioremediation of pesticides: A case study for the enzymatic remediation of organophosphorous insecticide residues. pp. 155–174. *In*: Kean, S.G., B.L. Brian, P.L. Thomas and J. Gan (eds.). Pesticide Mitigation Strategies for Surface Water Quality. ACS Publications, Washington, DC, USA.

Sharma, B., A.K. Dangi and P. Shukla. 2018. Contemporary enzyme based technologies for bioremediation: A review. J. Environ. Manag. 210: 10–22.

Sharma, P., Y. Laor, M. Raviv, S. Medina, I. Saadi, A. Krasnovsky et al. 2017. Green manure as part of organic management cycle: Effects on changes in organic matter characteristics across the soil profile. Geoderma 305: 197–207.

Shaw, L.J. and R.G. Burns. 2006. Enzyme activity profiles and soil quality. pp. 158–182. *In*: Bloem, J., D.W. Hopkins and A. Benedetti (eds.), Microbiological Methods for Assessing Soil Quality. CABI. Publishing, Oxfordshire, UK.

Shipitalo, M. and R.C. Le Bayon. 2004. Quantifying the effects of earthworms on soil aggregation and porosity. pp. 183–200. *In*: Edwards, C.A. (ed.). Earthworm Ecology. CRC Press, Boca Raton, USA.

Singh, B.K., A. Walker, J.A.W. Morgan and D.J. Wright. 2003. Effects of soil pH on the biodegradation of chlorpyrifos and isolation of a chlorpyrifos-degrading bacterium. Appl. Environ. Microbiol. 69: 5198–5206.

Singh, B.K. and A. Walker. 2006. Microbial degradation of organophosphorus compounds. FEMS Microbiol. Rev. 30: 428–471.

Singh, B.K. and R. Naidu. 2012. Cleaning contaminated environment: A growing challenge. Biodegradation 23: 785–786.

Sizmur, T. and M.E. Hodson. 2009. Do earthworms impact metal mobility and availability in soil?—A review. Environ. Pollut. 157: 1981–1989.

Sogorb, M.A. and E. Vilanova. 2002. Enzymes involved in the detoxification of organophosphorus, carbamate and pyrethroid insecticides through hydrolysis. Toxicol. Lett. 128: 215–228.

Song, N.H., L. Chen and H. Yang. 2008. Effect of dissolved organic matter on mobility and activation of chlorotoluron in soil and wheat. Geoderma 146: 344–352.

Springael, D. and E.M. Top. 2004. Horizontal gene transfer and microbial adaptation to xenobiotics: new types of mobile genetic elements and lessons from ecological studies. Trends Microbiol. 12: 53–58.

Spurgeon, D.J., J.M. Weeks and C.A.M. van Gestel. 2003. A summary of eleven years progress in earthworm ecotoxicology: The 7th international symposium on earthworm ecology Cardiff Wales 2002. Pedobiologia 47: 588–606.

Stanley, J. and G. Preetha. 2016. Pesticide toxicity to earthworms: exposure, toxicity and risk assessment methodologies. pp. 277–350. *In*: Stanley, J. and G. Preetha (eds.). Pesticide Toxicity to Non-target Organisms. Springer, Dordrecht, Netherlands.

Stenersen, J. 1984. Detoxification of xenobiotics by earthworms. Comp. Biochem. Physiol. 78C: 249–252.

Stroud, J.L., D.E. Irons, J.E. Carter, C.W. Watts, P.J. Murray, S.L. Norris et al. 2016. *Lumbricus terrestris* middens are biological and chemical hotspots in a minimum tillage arable ecosystem. Appl. Soil Ecol. 105: 31–35.

Sutherland, T.D., I. Horne, K.M. Weir, C.W. Coppin, M.R. Williams, M. Selleck et al. 2004. Enzymatic bioremediation: From enzyme discovery to applications. Clinical Exper. Pharmacol. Physiol. 31: 817–821.

Tao, J., B. Griffiths, S. Zhang, X. Chen, M. Liu, F. Hu et al. 2009. Effects of earthworms on soil enzyme activity in an organic residue amended rice–wheat rotation agro-ecosystem. Appl. Soil Ecol. 42: 221–226.

Tiunov, A.V. and S. Scheu. 2000. Microbial biomass, biovolume and respiration in *Lumbricus terrestris* L. cast material of different age. Soil Biol. Biochem. 32: 265–275.

Toccalino, P.L., R.J. Gilliom, B.D. Lindsey and M.G. Rupert. 2014. Pesticides in groundwater of the United States: Decadal-scale changes, 1993–2011. Ground Water 52: 112–125.

Thompson, C.M. and R.J. Richardson. 2004. Anticholinesterase insecticides. pp. 89–127. *In*: Marrs, T.C. and B. Ballantyne (eds.). Pesticide Toxicology and International Regulation. John Wiley & Sons, West Sussex, U.K.

Turunen, S. and G.M. Chippendale. 1977. Esterase and lipase activity in the midgut of *Diatraea grandiosella*: digestive functions and distribution. Insect Biochem. 7: 67–71.

Turunen, S. 1978. Multiplicity of tissue esterases in *Pieris brassicae* (L.) (Lepidoptera, Pieridae). Ann. Zool. Fenn. 15: 89–93.

Tyagi, M., M.M. da Fonseca and C.C. de Carvalho. 2011. Bioaugmentation and biostimulation strategies to improve the effectiveness of bioremediation processes. Biodegradation 22: 231–241.

van Groenigen, J.W., I.M. Lubbers, H.M. Vos, G.G Brown, G.B. De Deyn and K.J. van Groenigen. 2014. Earthworms increase plant production: A meta-analysis. Sci. Rep. 4: 6365.

Vasantha-Srinivasan, P., S. Senthil-Nathan, A. Ponsankar, A. Thanigaivel, M. Chellappandian, E.S. Edwin et al. 2018. Acute toxicity of chemical pesticides and plant-derived essential oil on the behavior and development of earthworms, *Eudrilus eugeniae* (Kinberg) and *Eisenia fetida* (Savigny). Environ. Sci. Pollut. Res. Int. 25: 10371–10382.

Velki, M. and S. Ečimović. 2017. Important issues in ecotoxicological investigations using earthworms. Rev. Environ. Contam. Toxicol. 239: 157–184.

Verma, K., N. Agrawal, M. Farooq, R.B. Misra and R.K. Hans. 2006. Endosulfan degradation by a Rhodococcus strain isolated from earthworm gut. Ecotoxicol. Environ. Saf. 64: 377–381.

Wallenstein, M.D. and R.G. Burns. 2011. Ecology of extracellular enzyme activities and organic matter degradation in soil: A complex community-driven process. pp. 35–55. *In*: Dick, R.P. (ed.). Methods of Soil Enzymology. Soil Science Society of America, Madison, Wisconsin, USA.

Wei, R. and W. Zimmermann. 2017. Microbial enzymes for the recycling of recalcitrant petroleum-based plastics: how far are we. Microb. Biotechnol. 10: 1308–1322.

Worrall, F., A. Parker, J.E. Rae and A.C. Johnson. 1997. The role of earthworm burrows in pesticide transport from ploughlands. Toxicol. Environ. Chem. 61: 211–222.

Wurst, S. 2010. Effects of earthworms on above- and belowground herbivores. Appl. Soil Ecol. 45: 123–130.

Xiao, Z., X. Wang, J. Koricheva, A. Kergunteuil, R.-C. Le Bayon, M. Liu et al. 2018. Earthworms affect plant growth and resistance against herbivores: A meta-analysis. Funct. Ecol. 32: 150–160.

Yu, R., Q. Liu, J. Liu, Q. Wang and Y. Wang. 2016. Concentrations of organophosphorus pesticides in fresh vegetables and related human health risk assessment in Changchun, Northeast China. Food Cont. 60: 353–360.

Zhang, W., P.F. Hendrix, L.E. Dame, R.A. Burke, J. Wu, D.A. Neher et al. 2013. Earthworms facilitate carbon sequestration through unequal amplification of carbon stabilization compared with mineralization. Nat. Commun. 4: 2576.

Zumstein, M.T., D. Rechsteiner, N. Roduner, V. Perz, D. Ribitsch, G.M. Guebitz et al. 2017. Enzymatic hydrolysis of polyester thin films at the nanoscale: effects of polyester structure and enzyme active-site accessibility. Environ. Sci. Technol. 51: 7476–7485.

Chapter 10

Biochar Mitigates the Impact of Pesticides on Soil Enzyme Activities

Juan C. Sanchez-Hernandez

INTRODUCTION

Biochar, a carbonaceous material produced by pyrolysis of organic biomass, is currently used in various environmental applications. It is used as an organic amendment for promoting soil fertility and quality (Ding et al. 2016, Yang et al. 2017, Tan et al. 2017); it also produced in order to enhance carbon storage, and applied to soil biochar also reduces greenhouse gas emissions (Qambrani et al. 2017, Kammann et al. 2017); it has physicochemical characteristics that make it suitable as a sorbent material for the remediation of contaminated soils (Oliveira et al. 2017) and waters (Sizmur et al. 2017); and finally it is used as a catalyst in bioenergy refinery systems (Qian et al. 2015, Lee et al. 2017). During the last decade, an intensive effort has been made by scientific community to develop this technology in terms of standardizing production, methods and defining the physicochemical characteristics of the final material (guideline documents can be downloaded free, from the International Biochar Initiative website, www.biochar-international.org). However, the impact of biochar on soil organisms (micro-, meso- and macrofauna) demands further research and methods for its ecotoxicological assessment need standardization (Lehmann et al. 2011, Domene 2016). The available data are not conclusive regarding the short-

Laboratory of Ecotoxicology, Institute of Environmental Sciences, University of Castilla-La Mancha, 45071 Toledo, Spain.
Email: juancarlos.sanchez@uclm.es

and long-term side-effects of biochar application. In addition, although there are some guidelines for the sustainable and safe production of biochar before its use as a soil amendment (EBC 2012, Initiative 2015), no toxicity testing procedures are available for predicting or regulating the potentially adverse effects on soil fauna such as earthworms and springtails. The few studies that have examined the potential toxicity of biochar to soil organisms highlight the urgent need for advances in this area to enable the safe use of biochar to soil (Kuppusamy et al. 2016). For example, Marks et al. (2014) showed that the adverse effects of biochar on soil fauna were highly dependent on the type of biochar, the animal species and the toxicity endpoint. Some studies have used the standardized avoidance behavior response test (ISO 2008) to characterize the impact of biochar on soil in terms of habitat function (Li et al. 2011, Weyers and Spokas 2011, Busch et al. 2012). Altogether, these findings encourage the establishment of a framework of exposure/effect risk assessment to enable evaluation of the short- and long-term impacts of biochar on soil function.

Soil enzymes, which are essential biomolecules in soil nutrient cycling (Burns et al. 2013), have received little attention as molecular targets of biochar toxicity. Indeed, soil enzymes catalyze most chemical reactions involved in the transformation and decomposition of organic matter, and "without their presence, the soil would no longer function properly" (Dick 2011). Accordingly, the impact of biochar on soil enzyme activities is ecologically relevant because soil enzymes play a significant role in organic matter cycling (Gianfreda and Ruggiero 2006, Wallenstein and Burns 2011) and bioremediation (Gianfreda et al. 2016). It is tempting to suggest a twofold approach to using soil enzymes as a toxicity endpoint. On the one hand, soil enzymes can be used as biomarkers of the microbiological and biochemical recovery of polluted soils under a biochar-assisted remediation action and, on the other hand, the predictive assessment of the ecological consequences of biochar may include the measurement of soil enzyme activities.

The purpose of this chapter is, therefore, to examine the current knowledge on the impact of biochar on soil enzyme activities, which are considered good indicators of soil microbial diversity (Caldwell 2005), soil pollution (Riah et al. 2014, Rao et al. 2014) and changes in land use (Dick 1997, Bastida et al. 2008, Bowles et al. 2014). One of the most attractive environmental applications of biochar is its potential capacity for pollution remediation. In the last decade, numerous studies have demonstrated that biochar acts as a sorbent for metals (Hmid et al. 2015, Qian et al. 2016, Ahmad et al. 2016, Abdelhadi et al. 2017, Sizmur et al. 2017), pharmaceuticals (Williams et al. 2015, Peiris et al. 2017) and other organic pollutants (Xie et al. 2015, Xiao et al. 2016) and also stimulates biodegradation of organic pollutants (Zhu et al. 2017). Perhaps surprisingly, the impact of biochar on pesticide-contaminated soils has received less attention, even though pesticides represent an important threat to soil function (see Chapters 1 and 9).

This chapter will tackle some important issues concerning the association between soil enzymes and biochar, although considering pesticides as environmental stressors. Studies conducted in several laboratories, including ours, will be discussed in an attempt to answer the following questions: (1) Which soil enzymes are most sensitive to biochar applications, and what type of response occurs? (2) Does biochar facilitate the stabilization of soil extracellular enzymes? (3) Is there a synergistic relationship between biochar and soil enzymes, that enhances enzymatic bioremediation? (4) What are the ecological consequences of biochar-induced changes in soil enzyme activities?

The first section of this chapter will summarize the main findings on the impact of pesticides on soil enzyme activities. In particularly, the response of soil enzyme activities will be examined in soil treated with organophosphorus pesticides. Although most studies have explored the effects of biochar on soil physicochemical properties, the effect on enzyme activities has been less frequently studied. Therefore, the second section will provide an overview of soil enzyme responses to biochar application. The third section, presented as a case study, will illustrate how certain soil enzyme activities can be used as indicators of the impact of biochar on the availability and toxicity of pesticides. This case study will focus on the organophosphorus pesticide chlorpyrifos. Because this pesticide displays a high affinity for the active site of esterase enzymes, carboxylesterase activity will be proposed as a biomarker of organophosphorus contamination and also as an enzymatic mechanism of organophosphorus detoxification. In the final section we will summarize the main conclusions of this case study and suggest future lines of research in relation to the use of biochar to mitigate the impact of pesticides on soil biochemistry.

The Impact of Pesticides on Soil Enzyme Activities: Organophosphorus Insecticides and Carboxylesterases

Organophosphorus compounds are an important group of plant protection products used in agriculture. Current statistical data place these compounds among the agrochemicals most commonly used worldwide to combat agricultural pests (see Chapter 9). Development of environmentally-friendly technologies to reduce the accumulation of these toxic chemicals in soil is still needed for their rational use and to reduce their toxicity to non-target organisms.

Among the multiple terrestrial ecotoxicity tests aimed at authorizing the use of plant protection products in the European Union, few assess the impact of active substances on soil microbial and biochemical functions. For example, in relation to terrestrial ecotoxicology, the guidance document on the authorization of plant protection products only includes soil nitrification and carbon mineralization as target soil functions in the battery of ecotoxicity tests recommended (European Commision 2002). However, pesticides and other environmental pollutants induce changes in a wide range of soil enzyme activities other than those involved in C- and N-cycling. Although soil enzyme activities are considered suitable indicators of soil deterioration (Margesin 2005, Bastida et al. 2008, Rao et al. 2014, Paz-Ferreiro and Fu 2016), these biochemical endpoints are rarely used as microbial biomarkers of pesticide contamination (Sanchez-Hernandez et al. 2017, Guasch et al. 2017, Cravo-Laureau et al. 2017). Nevertheless, some comprehensive reviews have dealt with the toxic effects of pesticides on soil enzyme activities (Gianfreda and Rao 2008, Hussain et al. 2009, Riah et al. 2014), suggesting that enzymes may be sensitive indicators of short-term changes induced by pesticides. On the other hand, these studies show that the response of soil enzymes to pesticides is unpredictable, thus challenging the development of toxicity criteria based on changes in the soil enzymatic profile (Riah et al. 2014). Such uncertainty in the enzyme response is derived from a number of factors related to the location of the enzyme in soil and the chemical nature of the enzyme-pesticide interaction.

With regard to the first source of uncertainty, enzymes are located in the soil in two main forms (Nannipieri et al. 2002, Wallenstein and Burns 2011): (1) intracellularly in living, resting and dead cells and in cell debris; and (2) extracellularly, associated with organic colloids and inorganic complexes, or free in the soil solution. Extracellular enzymes are also known as "abiontic enzymes" because their activity is no longer under biological control (Nannipieri et al. 2018). In the extracellular fraction, enzymes in the soil solution display low persistence because they are highly exposed to microbial degradation and proteolysis, whereas the most stable forms are those enzymes adsorbed onto organomineral complexes (Nannipieri et al. 2002). This may explain why the latter fraction is generally the most abundant in soil, thus significantly contributing to the total enzyme activity (Nannipieri et al. 2018). The location of the enzyme in the soil matrix clearly affects the response to pesticides due to differences in availability. This is true of the enzymes that directly interact with pesticides through their active site. In the case of intracellular enzymes, their response to pesticides takes place as long as pesticides are absorbed by cells and are not metabolized by microbial detoxification systems. Likewise, the physicochemical properties of pesticides and the soil properties are key factors defining the availability of pesticides to interact with extra- and intracellular enzymes.

Regarding the second source of uncertainty, the interaction between soil enzymes and pesticides may be direct or indirect. Pesticides, and their metabolites, can interact directly with intra- and extracellular enzymes by binding to their active sites (Sanchez-Hernandez and Sandoval 2017) or by inducing changes in their protein conformation (Gianfreda and Rao 2008). The most common response of this direct interaction is inhibition of the enzyme activity. On the other hand, pesticides can alter the activity of soil enzymes via their effect on the biomass, biodiversity and activity of soil microorganisms. Because microorganisms represent the main source of soil enzymes, any pesticide-induced impact on their communities will imply short-term changes in the activity of soil enzymes. Therefore, the sensitivity of microorganisms to pesticides determine whether the enzyme activity increases, decreases or remains unchanged. For example, many organophosphorus pesticides can induce enhanced biodegradation (Singh and Walker 2006, Singh 2009), a term that defines the increasing capacity of soil microorganisms to degrade these pesticides after extensive and repeated application. Enhanced biodegradation would therefore be expected to induce an increase in the activity of enzymes involved in pesticide detoxification (Singh 2009, Harms et al. 2011) and microbial foraging (Wallenstein and Burns 2011, Burns et al. 2013). However, the organophosphorus chlorpyrifos is not able to induce the phenomenon of enhanced biodegradation (Racke et al. 1990, Singh et al. 2002). Although the mechanism is still not fully understood, the metabolites generated during chlorpyrifos degradation (e.g., 3,5,6,-trichloro-2-pyridinol) may inhibit microbial proliferation (Singh and Walker 2006). In the particular case of chlorpyrifos, therefore, a reduced or unchanged response in soil enzyme activity may occur, relative to pesticide-free soils or soils treated with pesticides that induce enhanced biodegradation.

In addition to the environmental and biological factors that modulate the response of soil enzymes to pesticides, we must bear in mind that other potential variables such as pesticide concentration, duration of pesticide exposure, soil physicochemical properties, and the type of enzyme also contribute to the enzyme response. In relation to the type of enzyme, Riah et al. (2014) found that dehydrogenase, phosphatases, cellulase, urease, arylsulfatase and β-glucosidase are the enzymes most commonly used to assess the impact of pesticides on soil biochemical function. However, considering the studies reviewed by Riah et al. (2014), it is not possible to predict how these enzymes will respond to pesticide exposure. Some authors have therefore used simple enzymatic indexes such as the geometric index (Lessard et al. 2014, Paz-Ferreiro and Fu 2016), the treated-soil quality index (Mijangos et al. 2010) and the integrated biological response index (Sanchez-Hernandez et al. 2017) to facilitate interpretation of enzyme responses. Nevertheless, use of these enzyme-based indexes for pollution monitoring should meet some recommended criteria. First, more than one enzyme activity representing the main sol biogeochemical cycles (C, N, S and P) should be included in the enzymatic index as suggested by Lessard et al. (2014). This approach would thus satisfy the need to combine information derived from different soil microbial and biochemical processes. Second, the enzymatic indexes should integrate other biological and physicochemical properties of soil (Bastida et al. 2008) or supplementary information about biological or biochemical properties other than the enzyme activities (Trasar-Cepeda et al. 2000). Third, the enzymatic index outcomes will reliably fit our hypothesis only when measurement of the enzyme activity is accurate. Accordingly, the recommendations by Dick (2011), Tabatabai and Dick (2002) and Nannipieri et al. (2018) regarding the need to optimize soil enzyme assays should be heeded, especially when traditional procedures are implemented under a high-throughput approach such as microplate-scale assays.

In a review paper, Riah et al. (2014) reported that organophosphorus pesticides inhibit the activity of most soil enzymes, probably because of the toxic action of this class of pesticides on microorganisms. Indeed, Sanchez-Hernandez et al. (2017) demonstrated that the inhibition of acid phosphatase and β-glucosidase activities in soils treated with chlorpyrifos was not due to a direct interaction between these hydrolases and the pesticide, but to changes in microbial activity and biomass. By contrast, the carboxylesterase enzyme was directly inhibited by chlorpyrifos, as reported for other organophosphorus pesticides (Satyanarayana and Getzin 1973, Sanchez-Hernandez et al. 2014). Therefore, the activity of this esterase was identified as a specific biomarker for monitoring organophosphorus availability and toxicity in soil, because its response to these pesticides is direct and does not require microbial intervention (Sanchez-Hernandez et al. 2017).

Carboxylesterases (EC 3.1.1.1) are serine hydrolases that act on ester, amide and carbamate bonds (Wheelock and Nakagawa 2010). They have been described in many organisms, ranging from microorganisms (Singh 2014) and plants (Gershater and Edwards 2007) to vertebrates (Wheelock et al. 2008), including humans (Satoh and Hosokawa 1998, Taketani et al. 2007). Besides being ubiquitous, carboxylesterases are involved in many physiological processes. For example, human carboxylesterases participate in lipid metabolism (Ross et al. 2010, Lian et al. 2018). These enzymes are also able to inactivate certain hormones such as ghrelin (De Vriese et al. 2007) and insect juvenile hormone (Kamita and Hammock 2010). Furthermore, past studies have suggested an important role of carboxylesterases in spermatogenesis of the male reproductive system (Mikhailov and Torrado 1999). From a toxicological viewpoint, carboxylesterases play an important role in drug metabolism (Casey Laizure et al. 2013, Thomsen et al. 2014), prodrug activation (Hosokawa 2008), and pesticide detoxification (Ross et al. 2006, Ross and Crow 2007, Gershater et al. 2007, Jackson et al. 2011, Hatfield et al. 2016). More recently, some studies have shown that carboxylesterases may degrade some types of polyester polymers (Zumstein et al. 2017), representing an exciting area of future research in the field of microplastic pollution.

In pesticide ecotoxicology, carboxylesterases are common biomarkers of exposure in non-target organisms (Wheelock et al. 2008). Their activity is generally more sensitive to inhibition by organophosphorus pesticides than that of other esterases such as brain acetylcholinesterase (i.e., the neural target of organophosphorus toxicity, Casida and Quistad 2004). The high affinity of organophosphorus for the active site of carboxylesterases has led to these enzymes being suggested for use as non-catalytic detoxifiers or bioscavengers (Maxwell 1992). Nowadays, this scavenging ability continues to be observed in many studies on organophosphorus toxicology (e.g., Otero and Kristoff 2016). Therefore, species-specific variations in the sensitivity to organophosphorus may be partly attributed to the levels of activity of this constitutive enzyme, as well as to the affinity of pesticide molecules for the active site of the enzyme (Chanda et al. 1997). Under this assumption, Wheelock et al. (2008) proposed the development of an index of pesticide susceptibility based on the levels of background activity of carboxylesterases. However, and as suggested by these authors, the primary targets of pesticide toxicity (e.g., inhibition of brain acetylcholinesterase in the case of organophosphorus insecticides or blockage of sodium channels in the case of pyrethroid insecticides) should not be excluded from the assessment of interspecific sensitivity to pesticides. Together these findings lead us to envisage that carboxylesterases may behave in this way in organophosphorus-contaminated soils. This poses the question as to whether background levels of soil carboxylesterase modulate the bioavailability and toxicity of organophosphorus pesticides in soil, or whether soil carboxylesterase activity could act as a biomarker of pesticide susceptibility in soil. However, the most interesting point regarding this association is that soil carboxylesterase activity could be exploited as an enzymatic bioremediation system.

Carboxylesterase activity has been described in soil microorganisms (Brown et al. 1997, Singh et al. 2014, Singh 2014), and its hydrolytic activity should therefore be expected to be detected in soil. In fact, the pioneering studies by Getzin and Rosefield (1971) and Satyanarayana and Getzin (1973) described a soil carboxylesterase that can hydrolyze the organophosphorus malathion. However, since these results, few advances have been made in developing the carboxylesterase-mediated detoxification as an enzymatic bioremediation system (Burns and Edwards 1980, Gianfreda and Rao 2008, Gianfreda et al. 2016). The absence of catalytic

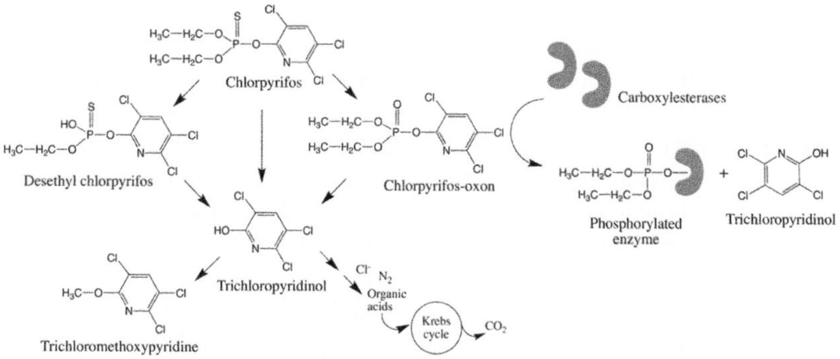

Fig. 1. Degradation of chlorpyrifos in the environment (elaborated from Racke 1993, Singh and Walker 2006, Solomon et al. 2014). Soil carboxylesterases are proposed as molecular scavengers for the metabolite chlorpyrifos-oxon (Sanchez-Hernandez et al. 2017), thus contributing to chlorpyrifos degradation in soil.

decomposition on organophosphorus pesticides other than malathion is probably the main reason for the apparent lack of interest in this topic.

Recent evidence suggests that soil carboxylesterase activity exerts a non-catalytic mechanism of organophosphorus detoxification (Fig. 1), based on irreversible inhibition comparable to that occurring in organisms (Sanchez-Hernandez et al. 2017). However, in addition to the microbial origin of soil carboxylesterases (Singh 2014), some studies have demonstrated that earthworms can also contribute to the pool of soil carboxylesterase activity via the production of gastrointestinal secretions and casts (feces) (Sanchez-Hernandez et al. 2009). Therefore, the burrowing, feeding and casting activities of earthworms can increase the carboxylesterase activity in earthworm-inoculated soils (Sanchez-Hernandez et al. 2014), even in pesticide-contaminated soils (Sanchez-Hernandez et al. 2018). Together, the findings of these studies suggest that earthworms could be used as biological vectors of carboxylesterase dispersion in soil, thus increasing the probability of the organophosphorus molecule being retained by the enzyme. Nevertheless, the precise molecular mechanisms underlying the inhibition of carboxylesterase activity by organophosphorus pesticides remain unknown. Therefore, pending advances in this area, the ecological consequences of inhibiting soil carboxylesterases and which soil processes regulate the production of this esterase enzyme are not known. Such knowledge would help us to promote the role of these enzymes as molecular scavengers of pesticides by acting on the processes that increase their concentration in soil.

Impact of Biochar on Soil Enzyme Activities

Biochar is currently used as an organic soil amendment (Ding et al. 2016). Physical characteristics such as open porosity and large surface area make biochar a suitable support for boosting microbial colonization and proliferation (Lehmann et al. 2011, Thies et al. 2015). Moreover, biochar is a source of nutrients (Ippolito et al. 2015, Tan et al. 2017, Yang et al. 2017), and soil enzyme activities are therefore generally expected to increase in biochar-amended soils as a consequence of microbial foraging. Nevertheless, the data reported in the relevant literature are somewhat contradictory and do not provide robust conclusions regarding the effect of biochar on soil enzyme activities. Indeed, Kuppusamy et al. (2016) suggested that the dynamic of enzyme activity in biochar-amended soils should be considered as a future area of biochar research in the coming years. In the following paragraphs, therefore, we examine the state of knowledge on this emerging issue, identifying some future research lines.

The most recent research shows that the responses of soil enzymes to biochar depend on the following variables: type of biochar, dose of application, duration of incubation of the biochar and soil, type of enzyme and type of soil. The type of biochar is clearly one of the most important variables regarding the modulation of soil enzyme responses. Structural and chemical differences in biochar are due to both the type of feedstock and the pyrolysis conditions (Ippolito et al. 2015). For example, the pyrolysis temperature affects the structure, composition and surface functionality of biochar (Weber and Quicker 2018). In general, pyrolysis temperatures > 450°C produce biochar with a higher surface area and microporosity, higher density of aromatic groups and lower oxygen-containing functional groups on the surface, lower polarity and higher pH than biochar generated at pyrolysis temperatures < 450°C (Sizmur et al. 2017). Moreover, biochar produced at low temperature contains an important amount of unpyrolyzed organic matter susceptible to microbial decomposition. These marked physicochemical differences will determine the capacity of the biochar to alter soil enzyme activity.

The study by Gascó et al. (2016) is an illustrative example of how biochar produced at different temperatures has contrasting effects on soil enzyme activities. These researchers incubated a sandy loam soil with 8% (w/w) biochar elaborated from pig manure, produced at two temperatures (300°C and 500°C), for 219 days. Compared to the control (biochar free) soils, the 300°C-biochar significantly increased the dehydrogenase activity, whereas the 500°C-biochar did not alter the activity of this oxydoreductase enzyme. Interestingly, the β-glucosidase, phosphomonoesterase and phosphodiesterase activities were significantly lower in soils amended with 500°C-biochar than in control soils (Gascó et al. 2016). Similar findings were obtained with soils incubated (for 74 days at 25°C and 60% humidity) with 8% w/w biochar obtained by pyrolysis of urban waste at 300°C and 500°C (Benavente et al. 2018). In this study, the phosphatase and dehydrogenase activities were higher in the soil amended with 300°C-biochar than in control (biochar free) soils and soils amended with 500°C-biochar. However, β-glucosidase activity was lower in all biochar-treated soils than in the control soils, irrespective of the dose and type of biochar. Moreover, the 300°C-biochar displayed phytotoxicity (germination index using *Lepidium sativum*), which suggested the presence of toxic compounds (e.g., metals) in this type of biochar. The study by Benavente et al. (2018) clearly showed that the chemical nature of the feedstock (urban waste) is critical for producing an environmentally compatible biochar for use as a soil fertility. The effect of biochar produced at different pyrolysis temperatures on soil enzyme activities was also explored by Subedi et al. (2016). These researchers incubated acidic silt-loam and alkaline sandy soils for 150 days in the presence of two types of biochar (2%, w/w) produced from animal manure (poultry litter and swine manure); both feedstocks were pyrolyzed at 400°C and 600°C. The researchers found a general increase in dehydrogenase activity in the soils amended with both types of 400°C-biochar, which was lower in the sandy soil amended with 600°C-biochar than in the control (biochar free) soils. However, as in the aforementioned studies, the β-glucosidase activity was lower in soils treated most biochars than in the respective control soils (Subedi et al. 2016). Similarly, Vithanage et al. (2018) compared the impact of a woody biomass pyrolyzed at 300, 500 and 700°C on soil enzyme activity, but introduced a new variable, i.e., the biochar dose. These researchers found that application of 1 and 2.5% biochar (w/w) caused a maximum increase in catalase and dehydrogenase activities relative to those unamended soils. They also confirmed previously reported decreased enzyme activities in soils amended with biochar produced at high pyrolysis temperatures (Vithanage et al. 2018). The inhibitory response of soil enzymes to biochar produced at high temperatures may explain the lack of significant variation in soil treated with biochars produced by pyrolysis of oak wood and bamboo at 600°C for 2 h (Demisie et al. 2014). The activity of β-glucosidase, dehydrogenase and urease did not vary significantly between the different treatments (0.5, 1 and 2% w/w) after incubation for 372 days at 25°C. Only slight decreases in dehydrogenase and β-glucosidase activities were reported for soil amended with 2.0% bamboo-derived biochar and 2.0% oak wood-derived biochar respectively (Demise et al. 2014). Altogether, these findings suggest that biochar produced at high temperatures (> 450°C) is not suitable for inducing soil enzyme production. Nevertheless, a study by Paz-Ferreiro et al. (2014) suggested an elegant means of mitigating the negative effects of high temperature-derived biochar, i.e., by the co-application of biochar with earthworms. Indeed, a recent study in our laboratory demonstrated that incubation of biochars derived from spent coffee ground and pine needle with the earthworms *Lumbricus terrestris* and *Aporrectodea caliginosa* favored the retention of extracellular enzymes on the biochar surface (Sanchez-Hernandez 2018).

Another variable that modulates the response of soil enzymes is the duration of soil-biochar incubation. For example, Wang et al. (2015) found maximum β-glucosidase, β-cellobiosidase, β-xylosidase, aminopeptidase, and α-glucosidase activities in soils incubated for a short time (14 days) with biochar produced at both 450° and 600°C (Wang et al. 2015). However, with the exception of aminopeptidase activity, the enzyme activities decreased after longer incubation times (90 days). In the same study, in soils incubated with biochar produced at low pyrolysis temperatures (300°C), enzyme activities were maximal after incubation for 90 days, confirming the general assumption that biochar produced at low temperature stimulates soil enzyme activities. In a short-term laboratory experiment (45 days) conducted to assess the capacity of biochar to act as a metal sorbent, biochar produced by pyrolysis of crop residues at 500°C, and applied at a dose of 5% (w/w), did not change microbial community and dehydrogenase activity of metal-polluted soils, although it retained Pb and As (Igalavithana et al. 2017).

The response of soil enzymes does not always depend on the biochar dose. For example, dehydrogenase and β-glucosidase activities did not differ in soils incubated with 1% and 2.4% (w/w, dry mass) straw-derived biochar after 100 d of incubation (Wu et al. 2013), although the biochar was generated by fast pyrolysis at 450°C. However, in both biochar treatments, the urease activity was significantly lower than in control (biochar free) soils. According to the authors, the more recalcitrant C in biochar could account for an absence of microbial proliferation and, consequently, the lack of response of C-cycling enzyme activities. The lower urease activity was consistent with lower N_2O emission caused by biochar (Wu et al. 2013). Although still not fully characterized, the generally held view is that differences in labile organic matter and microbial foraging strategies, in relation to the form of C substrates, are the main causes of the inverse responses (inhibition or induction) generally observed in soil enzymes in biochar-amended soils (Subedi et al. 2016). In this context, the concept of "charosphere" has been recently proposed to explain the chemical and biological processes of organic carbon decomposition occurring at the interface between biochar and soil (Luo et al. 2013). This microenvironment was also suggested by Pei et al. (2017) to explain the significant increase in soil enzyme activities and soil respiration rate in biochar-amended soils relative to biochar-free soils after incubation for almost 8 months (at 20°C and 60% of water-holding capacity). The proximity between microorganisms, organic substrates and enzymes at the charosphere may explain the greater efficiency of soil carbon decomposition as well as the long-term persistence of extracellular enzymes in biochar-amended soils relative to unamended soils (Pei et al. 2017).

One limitation in most biochar studies involving soil enzyme responses is the occurrence of confounding variables that complicate the real effect of biochar on soil enzyme activities. For instance, Wang et al. (2015) added a urea solution (200 mg N/kg soil) to biochar-amended soils in order to examine the response of extracellular enzyme activities to a combination of nitrogen fertilization (urea) and biochar amendment. These researchers hypothesized that urea could stimulate microbial proliferation and contribute to the dynamics of enzyme activities during the incubation period. Indeed, the β-glucosidase, β-cellobiosidase, β-xylosidase, α-glucosidase, aminopeptidase and N-acetyl-glucosaminidase activities were very similar in soil amended with urea and soil amended with both 300°C-biochar and urea (Wang et al. 2015). The beneficial impact of biochar on soil fertility and crop production has recently been demonstrated to be enhanced by the co-application of chemical fertilizers (Song et al. 2018). In a pot experiment, these researchers examined the synergistic effect of NPK fertilizer and maize straw-derived biochar produced at different temperatures (300°C, 450°C and 600°C) on wheat yield and several soil microbial parameters (microbial diversity and biomass, and soil enzyme activities). The co-application of NPK fertilizer and the 450°C-biochar was the most efficient treatment due to a greater increase in crop production, microbial biomass and enzyme activities than in the other treatments. Nevertheless, the response of soil enzyme activities depended on the temperature at which the biochar was produced. The β-xylosidase, N-acetyl-glucosaminidase and phosphatase activities decreased with increasing pyrolysis temperatures (Song et al. 2018). In a study by Vithanage et al. (2018), biochar-amended soils were incubated (6 weeks) in the presence of tomato plants (*Lycopersicon esculentum*), as part of a wider study, in which the rhizosphere could may have contributed to changes in extracellular enzyme activities in addition to those derived from biochar. This may also have occurred in the study by Subedi et al. (2016), who examined the impact of two types of biochar produced at different temperatures in soils cultivated with Italian ryegrass (*Lolium multiflorum*). Likewise, Bhattacharjya et al. (2015) incubated biochar-amended soils in the presence of rice (*Oryza sativa*) and wheat (*Triticum aestivum*), so that the effect of the both biochars alone (pine needle- and *Lantana*-derived biochar) on extracellular enzyme activities was difficult to assess. Although most studies indicate that the physicochemical properties of biochar are the major drivers of soil enzyme responses, a better understanding of the impact of biochar itself in soil biochemical processes is required (Gascó et al. 2016, Benavente et al. 2018). This would be important in the restoration of degraded and polluted soils using biochar as a soil amendment, or in combination with other amendments such as lime (Novak et al. 2018).

Field studies have provided controversial results regarding the impact of biochar on soil enzyme activities. For example, Mierzwa-Hersztek et al. (2016) compared the performance of biochar derived from poultry litter and the raw feedstock as soil amendments. They found no differences in the changes in dehydrogenase and urease activities produced in response to biochar doses of 2.5 and 5 t dry mass ha^{-1}, poultry litter (5 t dry mass ha^{-1}) and control (no fertilizer input). The authors argued that the doses of biochar used were probably too low to have a significant effect on soil enzyme activities. However, a field study by Liao et al. (2016) showed that doses (2.25 and 4.5 t ha^{-1}) of cotton straw biochar, similar to those tested by Mierzwa-Hersztek et al. (2016), caused an increase in soil enzyme activities related to C-cycling. Moreover, the highest dose of this biochar caused a significant increase in soil microbial biomass, basal respiration, community composition and C- and N-cycling enzymes relative to those in the control soils. In a long-term field study, Elzobair et al. (2016b) also found no significant differences in six soil enzymes activities related to C-, N- and P-cycling between hardwood biochar-amended (22.4 Mg ha^{-1} dry wt) and dairy manure-amended (42 Mg ha^{-1} dry wt) plots sampled one and four years after application. Although statistically non-significant, the mean activities of β-glucosidase, β-D-cellobiosidase, β-xylosidase, N-acetyl-β-glucosaminidase and phosphatase were higher in biochar-treated plots after 4 years of application than in the respective control plots. Therefore, biochar could have beneficial effects in terms of microbial stimulation and enzyme production as well as enzyme stability on biochar surface (Elzobair et al. 2016a, Sanchez-Hernandez 2018).

Under field conditions, biochar dose of 20–100 Mg ha^{-1} had no significant long-term effect on the arylsulfatase activity of biochar-amended soils relative to control (biochar free) soils (Sun et al. 2014). Nevertheless, the results of this study were considered with caution because the product of arylsulfatase activity, i.e., 4-nitrophenol, was strongly bound to the biochar particles, thus underestimating the real enzyme activity. Indeed, this methodological limitation to the assays in which the enzymatic reaction medium contains biochar particles was already described by Bailey et al. (2011). Using an *in vitro* approach, these researchers observed that biochar particles adsorbed some colorimetric and fluorescent substrates commonly used in the assays, rendering them less accessible to the active site of the enzymes and consequently underestimating the enzyme activity. Nonetheless, previous studies have demonstrated that biochar does not alter arylsulfatase activity, irrespectively of the biochar type and time of soil-biochar incubation (Paz-Ferreiro et al. 2012, Yoo and Kang 2012).

In summary, the type of biochar is a critical variable that should be considered in measuring potentially stimulatory effects on microbial and soil enzyme activities. As discussed above, the type of feedstock and the pyrolysis conditions (temperature) significantly affect the structural and chemical properties of the biochar. Biochar produced at low pyrolysis temperatures (e.g., < 450°C) is recommended for promoting soil enzyme activities. Nevertheless, the co-application of conventional fertilizers (e.g., urea or NPK) or even biological vectors of microbial proliferation and dispersion such as earthworms, is highly recommended in soils amended with biochar elaborated at high pyrolysis temperatures.

Biochar Reduces the Impact of Chlorpyrifos on Soil Carboxylesterase Activity: A Case Study

This case study focuses on a microcosm experiment that examined the impact of biochar on the rate of chlorpyrifos degradation. Moreover, the experimental design allowed us to explore whether the presence of biochar reduced the impact of chlorpyrifos on soil carboxylesterase activity. The biochar was elaborated by pyrolysis (450°C for 2 h) of spent-coffee grounds (SCGs) in a muffle furnace. The physicochemical features and the procedure used to pyrolyze this organic waste have been already described (Sanchez-Hernandez 2018). The experimental design of this case study comprised two steps. In the first step, soil samples were incubated with 1% and 10% w/w (dry mass) of SCG-derived biochar for 45 days at 20°C in darkness. Biochar-free soil was used as control. In the second step, the biochar-amended and control soil samples were spiked with 0, 10, 20, 40 and 80 µg chlorpyrifos g^{-1} wet soil. Four replicate samples were used for each pesticide concentration and biochar dose, and the resultanting 60 test samples were incubated in individual Petri dishes at 20°C and darkness for 80 days. A subsample was removed periodically from each replicate sample (t = 0, 10, 30 and 80 d) to determine the chlorpyrifos concentration (and its metabolites chlorpyrifos-oxon and 3,5,6-trichloro-2-pyridinol) as well as the soil carboxylesterase activity. This enzyme was selected because it interacts directly with chlorpyrifos-oxon as described above. The esterase activity was used as an indicator or biomarker of pesticide availability in the test soils.

The analytical procedures used to determine chlorpyrifos residues and carboxylesterase activity are detailed in Box 1. Chlorpyrifos-oxon was probably not detected in the soil samples because of the alkaline pH of the soil, which facilitated degradation of this metabolite (see the chromatograms in Box 1). Nevertheless, the inhibition of soil carboxylesterase activity corroborated the formation of chlorpyrifos-oxon during the 80 days incubation period. Previous studies by our research group involving the *in vitro* incubation of soil carboxylesterase activity in the presence of chlorpyrifos, chlorpyrifos-oxon and 3,5,6-trichloro-2-pyridinol, have demonstrated a dose-dependent inhibition of this enzyme activity with chlorpyrifos-oxon alone (median inhibitory concentration or IC50 = 5.34 µM).

In this case study, the chlorpyrifos degradation rates were significantly affected by the biochar dose (Fig. 2). Although the chlorpyrifos concentration decreased in all treatments during the 80-d incubation time, the degradation rate was higher in the biochar-free soils than in the soils amended with 1 and 10% biochar. For

Box 1. Chemical analysis of chlorpyrifos in soil and carboxylesterase activity assay.

Chlorpyrifos residues. Extraction of chlorpyrifos and its metabolites followed the QuEChERS (Quick, Easy, Cheap, Effective, Rugged and Safe) method described in the AOAC official method 2007.01 (Lehotay et al. 2007), with some modifications by Asensio-Ramos et al. (2010). Samples (2.5 g wet soil) were shaken for 1 min with 5 ml acetonitrile (HPLC-grade) containing 1% acetic acid, Na-acetate and magnesium sulfate; followed by sonication (5 min, 50 W) and centrifugation (4,500 × g for 5 min). The supernatants (1 ml) were cleaned up by dispersive solid phase extraction with 50 mg primary-secondary amine (40 m, particle size) and were then centrifuged at 10,000 × g for 5 min. Supernatants were filtered (0.45 μm), and aliquots (20 μl) injected into the HPLC system. Chlorpyrifos

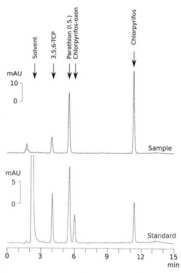

was separated on an Agilent Eclipse Plus LC-18 column (0.46 × 150 mm, 5 μm particle size) at a flow rate of 0.8 ml/min with the following solvent program: 70% acetonitrile (solvent A)/30% H_2O (solvent B) at t = 0 min, increase to 100% A in 8 min and maintain for 2 min, followed by decrease to 70% A in 1 min, and maintain for 4 min for equilibration. The detector was set to 290 nm (bandwidth = 8, reference wavelength = 360 nm). The average rate of recovery of chlorpyrifos from spiked soils was 98%. Parathion was used as the internal standard (I.S.).

Carboxylesterase activity. This enzyme activity was measured in a reaction medium consisting of 140 μl of 0.1 M Tris-HCl (pH = 7.4), 10 μl of 1-naphthyl butyrate (final conc., 2 mM) and the sample (soil:water suspensions), following the method described by Sanchez-Hernandez et al. (2018). The microplates containing the reaction medium and soil samples were agitated for 60 min at 20°C in an orbital thermostatized shaker (Elmi® Skyline DTS-2, 800 rpm), before being centrifuged at 2,500 × g for 10 min and 10°C. A secondary reaction with the diazonium salt Fast Red ITR was performed by transferring supernatants (150 μl) to clean microplates, to which 75 μl of 0.1% (w/v) Fast Red ITR dissolved in 1:1 (v/v) SDS (2.5%)/Triton X-100 (2.5%) was added. The plate was then incubated in the dark for 20 min and the product of the reaction was determined in a 4100 LabTech® microplate reader (Ortenberg, Germany) at 530 nm. The activity was expressed as μmol of 1-naphthol formed per hour and gram of dry mass, estimated from a calibration curve constructed from responses of 1-naphthol standard solutions in the presence of the sample (and biochar), to correct the adsorption of the chromogen to soil colloids and biochar. Controls (substrate-free) and blanks (soil-free) were used to correct the background absorbance and non-enzymatic hydrolysis of the substrates, respectively.

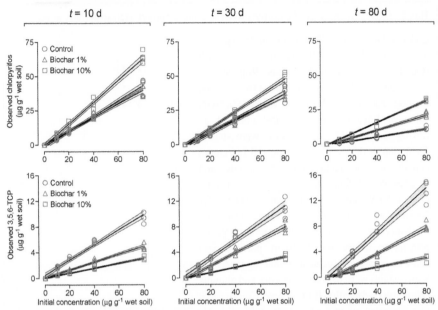

Fig. 2. Observed concentrations of chlorpyrifos and 3,5,6-trichloropyridinol (3,5,6-TCP) in soils amended with biochar produced from spent coffee grounds (n = 4 replicates per treatment), 10, 30 and 80 days after acute treatment with chlorpyrifos (0, 10, 20, 40, 80 μg/g wet soil). Comparison of slopes between each sampling time (10, 30 and 80 d) and chemical (chlorpyrifos and 3,5,6-TCP) were run using GraphPad Prism software (ver. 7.00, GraphPad Software, La Jolla California, USA) to test the null hypothesis that regression lines are parallel. The slopes were statistically different ($p < 0.0001$) in all cases.

example, at the end of the incubation time, the mean (± standard deviation, n = 4) chlorpyrifos concentrations (11.8 ± 1.25, 21.5 ± 1.83 and 31.7 ± 1.00 μg/g wet soil in respectively the control, biochar 1% and biochar 10% treatments) were much lower than the initial concentration (80 μg/g wet soil in all treatments). The marked difference in the chlorpyrifos degradation rate between treatments was confirmed by the formation of the metabolite 3,5,6-trichloro-2-pyridinol (Fig. 2). Considering the same experimental soils (spiked with 80 μg/g wet soil chlorpyrifos), the mean concentrations of 3,5,6-trichloro-2-pyridinol were 13.4 ± 1.70, 7.91 ± 0.71 and 2.95 ± 0.51 μg/g wet soil in respectively the control, biochar 1% and biochar 10% treatments at the end of the incubation time.

The slow degradation of chlorpyrifos in the biochar-amended soils was concomitant with the lower degree of inhibition of soil carboxylesterase activity (Fig. 3). The response of this esterase activity to chlorpyrifos exposure followed a non-linear model (Sanchez-Hernandez and Sandoval 2017):

$$E = \frac{100}{1 + \left(\dfrac{I}{I_{50}}\right)^{b}}$$

where E represents the percentage of enzyme activity in the chlorpyrifos-treated soil relative to the controls (100%), I is the chlorpyrifos concentration (mg/kg), I_{50} is the chlorpyrifos concentration that inhibits the enzyme activity by 50% of its initial

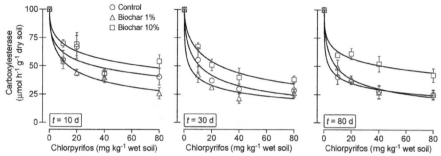

Fig. 3. Variation in response of soil carboxylesterase activity to chlorpyrifos treatment (0–80 μg/g wet soil) over time. Data are mean values and standard deviation of four replicates. Non-linear regressions were statistically significant ($r^2 = 0.69$–0.98, $p < 0.05$).

activity, and b is a coefficient that describes the slope of the dose–response curve (Motulsky and Christopoulos 2003). The I_{50} corresponds to the "ecological dose" (ED_{50}) defined by Babich et al. (1983) as "the concentration of a toxicant that inhibits a microbe-mediated ecological process by 50%". This toxicological parameter has been used by many researchers to estimate the impact of environmental pollutants, particularly metals, on several enzyme activities such as arylsulfatase (Haanstra and Doelman 1991), dehydrogenase and urease (Moreno et al. 2001, Gao et al. 2010), and phosphatase (Renella et al. 2003). In this case study, we used the ED_{50} of carboxylesterase activity as a measurement of chlorpyrifos-oxon availability. This metabolite is the primary cause of the toxicity of chlorpyrifos to soil fauna (Barron and Woodburn 1995) and microorganisms (Wang et al. 2010). Therefore, environmental monitoring of chlorpyrifos-oxon following agricultural application of chlorpyrifos, or during the remediation of chlorpyrifos-contaminated soils, is toxicologically important. However, chlorpyrifos-oxon persists for a short time in the environment (Mackay et al. 2014), and hydrolysis is the main route of dissipation (Racke 1993). This explains why chlorpyrifos-oxon is rarely detected by chemical analysis of environmental samples. Nonetheless, the high affinity of the compound for the active site of carboxylesterases makes measurement of the hydrolytic activity of this enzyme a suitable biomarker of chlorpyrifos-oxon bioavailability (Wheelock et al. 2008, Sanchez-Hernandez et al. 2014).

The higher ED_{50} values in the biochar (10% w/w)-amended soils than in control soils indicate a remediation role for biochar (Fig. 4). This leads to the need to determine whether the lower impact of chlorpyrifos on carboxylesterase activity is due to a lower rate of degradation of the pesticide or to binding of chlorpyrifos-oxon to biochar surface, thereby reducing its accessibility to carboxylesterases. Although the experimental design did not enable solid evidences to be obtained, the results of this microcosm trial suggest the slower degradation of chlorpyrifos by biochar as the most plausible reason. Although abiotic (hydrolysis) and microbial degradation of chlorpyrifos yields chlorpyrifos-oxon and 3,5,6-trichloro-2-pyridinol, the former metabolite is more susceptible to hydrolysis than chlorpyrifos (Racke 1993), which rapidly generates 3,5,6-trichloro-2-pyridinol (Fig. 1). The hydrolytic decomposition of chlorpyrifos-oxon to 3,5,6-trichloro-2-pyridinol is even faster in alkaline environments than caused by biochar treatment. Therefore, the slower formation of

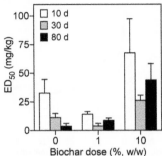

Fig. 4. Values (mean ± SEM) of ecological dose (ED_{50}) defined as the chlorpyrifos concentration that inhibits soil carboxylesterase activity by 50% (see text for further details).

3,5,6-trichloro-2-pyridinol in the biochar-amended soils than in control soils (Fig. 2) suggest slower degradation of chlorpyrifos rather than retention of chlorpyrifos-oxon onto the biochar surface. In fact, soil carboxylesterase activity was inhibited in the biochar-amended soils although less than in control soils, thereby corroborating this low chlorpyrifos degradation rate in the presence of biochar. On the other hand, the physicochemical properties of chlorpyrifos (Log K_{OC} = 3.83, $S^{25°C}$ = 0.36 mg/L, Mackay et al. 2014) also indicate an adsorption of this pesticide onto the biochar surface.

Comparison of the ED_{50s} values for both biochar treatments suggest that the protective role of biochar depended on application dose, with the 10% (w/w) dose being the most effective for protecting the carboxylesterase activity against chlorpyrifos toxicity. Altogether, these preliminary results suggest that the addition of biochar to agricultural soils treated by organophosphorus pesticides may be an effective strategy for protecting the soil enzyme activity, although the presence of biochar may favor the persistence of the pesticides in soil.

Summary and Future Lines of Research

This chapter has discussed the effects of biochar on soil enzyme activities, with the aim of identifying the main variables that modulate the enzyme responses to biochar application. One significant finding emerging from the published studies is that the stimulation of soil enzyme production strongly depends on the type of biochar. In addition to the type of biomass used to produce biochar, the pyrolysis temperature determines the structure and chemical properties of biochar. Data suggest that biochar produced at low pyrolysis temperatures (< 450°C) increases the activity of soil enzymes such as dehydrogenase, phosphatases and β-glucosidase, whereas biochar produced at higher pyrolysis temperatures does not generally alter soil enzyme activity, and may even inhibit enzyme production relative to that in biochar-free soils. Surface functionality of biochar probably accounts for this temperature-dependent impact on soil enzyme activities.

Biochar has been demonstrated to immobilize a vast range of environmental contaminants ranging from metals to organic pollutants such as polycyclic aromatic hydrocarbons, dioxins, furans, polychlorinated biphenyls, pesticides and pharmaceuticals. However, the remediation potential of biochar in soils contaminated by organophosphorus pesticides has scarcely been investigated (Morillo and Villaverde 2017). The case study examined in this chapter reveals that biochar was able to retain the organophosphorus pesticide chlorpyrifos, increasing its persistence. This can be perceived as a drawback because the primary aim of any remediation action is to remove target pollutants. However, the measurement of soil carboxylesterase activity showed that the bioavailability of chlorpyrifos, and its highly toxic metabolite chlorpyrifos-oxon, was significantly lower in soils treated with 10% (w/w) biochar than in control (biochar free) soils. These results also suggest two novel strategies for the remediation of organophosphorus-contaminated soils. First, the activity of soil carboxylesterase may be used as a biomarker of organophosphorus contamination, particularly of oxon metabolites. These metabolites, are toxic to soil organisms and are also the least persistent, which hampers their detection by conventional analytical methods. In this respect, detection of carboxylesterase inhibition may be an indicator of the presence of oxon metabolites in soil. Second, the interaction between carboxylesterase and the oxon metabolite yields a stable enzyme-inhibitor complex (Fig. 1), thus representing an enzymatic bioremediation process whose efficiency will depend on the number of carboxylesterase molecules, the affinity of the oxon molecule for the active site of the enzyme, and the accessibility of the active site of esterases to the oxon metabolites.

Based on the research findings discussed in this chapter, we suggest that future work on the use of biochar to alleviate the impact of pesticides on soil biochemistry should focus on the following:

(1) The ecotoxicology of biochar. The most recent research shows that biochar alters soil enzyme activities, although the response (increase or inhibition) largely depends on the pyrolysis temperature at which the biochar is produced. The physicochemical characteristics of biochar and the molecular mechanisms underlying these enzymatic responses are not well understood. Future research should therefore focus on identifying the most suitable biochar properties for promoting soil biochemical performance. In this context, the impact of biochar itself on soil enzyme activity must be considered. In addition to the standardized ecotoxicity tests that use model plants and soil fauna, microbial processes and enzyme activities should form part of future guidelines for testing biochar toxicity pre- and post-authorization.

(2) Activation of biochar with extracellular enzymes. At present, biochar is often activated in a pre- or post-pyrolysis step as a way of increasing its remediation capacity. Methods of such activation include treatment of biochar with acid and alkaline solutions or with solutions containing metal oxides, clay minerals or organic compounds (e.g., chitosan), among other chemical treatments (Sizmur et al. 2017). Most of these methods, however, are not cost-effective and involve a great amount of research effort. In seeking to assess the impact of biochar on soil enzyme activities, several laboratories, including ours, have identified to the importance enriching biochar with extracellular enzymes during incubation with soil (Elzobair et al. 2016a, Sanchez-Hernandez 2018). These exciting findings lead to new opportunities for activating the biochar surface with extracellular

enzymes of bioremediation concern such as carboxylesterases, laccases and peroxidases. However, the most fascinating aspect of this functionality is that may occur naturally in soil, without any engineering being required. Interestingly biochar activation may be favored by the presence of earthworms (Sanchez-Hernandez 2018). We therefore, suggest a synergistic relationship between earthworms and biochar, whereby earthworms increase the production of soil enzymes, while biochar stabilizes and protects the enzymes. However, better knowledge of the molecular mechanisms underlying the sorption of extracellular enzymes onto the biochar surface, as well as how earthworms favor such activation (e.g., mucus secretion) is required in order to understand such functional cooperation.

Acknowledgements

We thank the Ministerio de Economía y Competitividad, Spanish Government (grant no. CTM2014-53915-R) for the financial support of this research.

References

Abdelhadi, S.O., C.G. Dosoretz, G. Rytwo, Y. Gerchman and H. Azaizeh. 2017. Production of biochar from olive mill solid waste for heavy metal removal. Bioresource Technol. 244(Pt 1): 759–767.

Ahmad, M., Y.S. Ok, B.Y. Kim, J.H. Ahn, Y.H. Lee, M. Zhang et al. 2016. Impact of soybean stover- and pine needle-derived biochars on Pb and As mobility, microbial community, and carbon stability in a contaminated agricultural soil. J. Environ. Manage 166: 131–139.

Asensio-Ramos, M., J. Hernández-Borges, L.M. Ravelo-Pérez and M.A. Rodríguez-Delgado. 2010. Evaluation of a modified QuEChERS method for the extraction of pesticides from agricultural, ornamental and forestal soils. Anal. Bioanal. Chem. 396(6): 2307–2319.

Babich, H., R.J.F. Bewley and G. Stotzky. 1983. Application of the "Ecological Dose" concept to the impact of heavy metals on some microbe-mediated ecologic processes in soil. Arch. Environ. Con. Tox. 12(4): 421–426.

Bailey, V.L., S.J. Fansler, J.L. Smith and H. Bolton. 2011. Reconciling apparent variability in effects of biochar amendment on soil enzyme activities by assay optimization. Soil Biol. Biochem. 43(2): 296–301.

Barron, M.G. and K.B. Woodburn. 1995. Ecotoxicology of chlorpyrifos. Rev. Environ. Contam. T. 144: 1–93.

Bastida, F., A. Zsolnay, T. Hernández and C. García. 2008. Past, present and future of soil quality indices: A biological perspective. Geoderma 147(3-4): 159–171.

Benavente, I., G. Gascó, C. Plaza, J. Paz-Ferreiro and A. Méndez. 2018. Choice of pyrolysis parameters for urban wastes affects soil enzymes and plant germination in a Mediterranean soil. Sci. Total Environ. 634: 1308–1314.

Bowles, T.M., V. Acosta-Martínez, F. Calderón and L.E. Jackson. 2014. Soil enzyme activities, microbial communities, and carbon and nitrogen availability in organic agroecosystems across an intensively-managed agricultural landscape. Soil Biol. Biochem. 68: 252–262.

Brown, H.M., M.M. Joshi, A.T. Van, T.H. Carski, J.J. Dulka, M.C. Patrick et al. 1997. Degradation of thifensulfuron methyl in soil: Role of microbial carboxyesterase activity. J. Agr. Food Chem. 45(3): 955–961.

Burns, R.G. and J.A. Edwards. 1980. Pesticide breakdown by soil enzymes. Pest Manag. Sci. 11(5): 506–512.

Burns, R.G., J.L. DeForest, J. Marxsen, R.L. Sinsabaugh, M.E. Stromberger, M.D. Wallenstein et al. 2013. Soil enzymes in a changing environment: Current knowledge and future directions. Soil Biol. Biochem. 58: 216–234.

Busch, D., C. Kammann, L. Grünhage and C. Müller. 2012. Simple biotoxicity tests for evaluation of carbonaceous soil additives: Establishment and reproducibility of four test procedures. J. Environ. Qual. 41(4): 1023.

Caldwell, B.A. 2005. Enzyme activities as a component of soil biodiversity: A review. Pedobiologia 49(6): 637–644.

Casey Laizure, S., V. Herring, Z. Hu, K. Witbrodt and R.B. Parker. 2013. The role of human carboxylesterases in drug metabolism: have we overlooked their importance. Pharmacotherapy 33(2): 210–222.

Casida, J.E. and G.B. Quistad. 2004. Organophosphate toxicology: safety aspects of nonacetylcholinesterase secondary targets. Chem. Res. Toxicol. 17(8): 983–998.

Chanda, S.M., S.R. Mortensen, V.C. Moser and S. Padilla. 1997. Tissue-specific effects of chlorpyrifos on carboxylesterase and cholinesterase activity in adult rats: an *in vitro* and *in vivo* comparison. Fund. Appl. Toxicol. 38(2): 148–157.

Cravo-Laureau, C., B. Lauga, C. Cagnon and R. Duran. 2017. Microbial responses to pollution—ecotoxicology: Introducing the different biological levels. pp. 45–62. *In*: Cravo-Laureau C., C. Gagnon, B. Lauga and R. Duran (eds.). Microbial Ecotoxicology. Springer International Publishing. Switzerland.

De Vriese, C., M. Hacquebard, F. Gregoire, Y. Carpentier and C. Delporte. 2007. Ghrelin interacts with Human Plasma Lipoproteins. Endocrinology 148(5): 2355–2362.

Demisie, W., Z. Liu and M. Zhang. 2014. Effect of biochar on carbon fractions and enzyme activity of red soil. Catena 121: 214–221.

Dick, R.P. 1997. Soil enzyme activities as integrative indicators of soil health. pp. 121–156. *In*: Pankhurst, C.E., B.M. Doube and V.V.S.R. Gupta (eds.). Biological Indicators of Soil Health. CABI Publishing, Wallingford, UK.

Dick, W.A. 2011. Development of a soil enzyme reaction assay. pp. 71–84. *In*: Dick, R.P. (ed.). Methods of Soil Enzymology. Soil Science Society of America, Madison, Wisconsin, USA.

Ding, Y., Y. Liu, S. Liu, Z. Li, X. Tan, X. Huang et al. 2016. Biochar to improve soil fertility. A review. Agron. Sustain. Dev. 36(2): 36.

Domene, X. 2016. A critical analysis of meso- and macrofauna effects following biochar supplementation. pp. 268–292. *In*: Ralebitso-Senior, T.K. and C.H. Orr (eds.). Biochar Application : Essential Soil Microbial Ecology. Elsevier. Amsterdam, Netherlands.

EBC. 2012. European Biochar Certificate–guidelines for a sustainable production of biochar. European Biochar Foundation (EBC), Arbaz, Switzerland. http://www.european-biochar.org/en/download. Version 6.3E of 14th August 2017, doi: 10.13140/RG.2.1.4658.7043.

European Commision. 2002. Guidance document on terrestrial ecotoxicology under Council Directive 91/414/EEC. Directorate E-Food Safety: plant health, animal health and welfare, international questions, Brussels.

Elzobair, K.A., M.E. Stromberger and J.A. Ippolito. 2016a. Stabilizing effect of biochar on soil extracellular enzymes after a denaturing stress. Chemosphere 142: 114–119.

Elzobair, K.A., M.E. Stromberger, J.A. Ippolito and R.D. Lentz. 2016b. Contrasting effects of biochar versus manure on soil microbial communities and enzyme activities in an Aridisol. Chemosphere 142: 145–152.

Gao, Y., P. Zhou, L. Mao, Y.-E. Zhi and W.-J. Shi. 2010. Assessment of effects of heavy metals combined pollution on soil enzyme activities and microbial community structure: modified ecological dose–response model and PCR-RAPD. Environ. Earth Sci. 60(3): 603–612.

Gascó, G., J. Paz-Ferreiro, P. Cely, C. Plaza and A. Méndez. 2016. Influence of pig manure and its biochar on soil CO_2 emissions and soil enzymes. Ecol. Eng. 95: 19–24.

Gershater, M.C. and R. Edwards. 2007. Regulating biological activity in plants with carboxylesterases. Plant Sci. 173(6): 579–588.

Gershater, M.C., I. Cummins and R. Edwards. 2007. Role of a carboxylesterase in herbicide bioactivation in Arabidopsis thaliana. J. Biol. Chem. 282(29): 21460–21466.

Getzin, L.W. and L. Rosefield. 1971. Partial purification and properties of a soil enzyme that degrades the insecticide malathion. Biochim. Biophys. Acta 235: 442–453.

Gianfreda, L. and P. Ruggiero. 2006. Enzyme activities in soil. pp. 257–311. *In*: Nannipieri, P. and K. Smalla (eds.). Nucleic Acids and Proteins in Soil. Springer -Verlag, Berlin Heidelberg, Germany.

Gianfreda, L. and M.A. Rao. 2008. Interactions between xenobiotics and microbial and enzymatic soil activity. Crit. Rev. Env. Sci. Tec. 38(4): 269–310.

Gianfreda, L., M.A. Rao, R. Scelza and M. de la Luz Mora. 2016. Role of enzymes in environment cleanup/remediation. pp. 133–155. Dhillon, G. and S. Kaur (eds.). *In*: Agro-Industrial Wastes as Feedstock for Enzyme Production. Academic Press, Elsevier. London, UK.

Guasch, H., B. Bonet, C. Bonnineau and L. Barral. 2017. Microbial biomarkers. pp. 251–281. *In*: Cravo-Laureau, C., C. Gagnon, B. Lauga and R. Duran (eds.). Microbial Ecotoxicology. Springer International Publishing, Switzerland.

Haanstra, L. and P. Doelman. 1991. An ecological dose-response model approach to short- and long-term effects of heavy metals on arylsulphatase activity in soil. Biol. Fert. Soils 11(1): 18–23.

Harms, H., D. Schlosser and L.Y. Wick. 2011. Untapped potential: exploiting fungi in bioremediation of hazardous chemicals. Nat. Rev. Microbiol. 9(3): 177–192.

Hatfield, M.J., R.A. Umans, J.L. Hyatt, C.C. Edwards, M. Wierdl, L. Tsurkan et al. 2016. Carboxylesterases: General detoxifying enzymes. Chem.-Biol. Interact. 259: 327–331.

Hmid, A., Z. Al Chami, W. Sillen, A. De Vocht and J. Vangronsveld. 2015. Olive mill waste biochar: a promising soil amendment for metal immobilization in contaminated soils. Environ. Sci. Pollut. Res. Int. 22(2): 1444–1456.

Hosokawa, M. 2008. Structure and catalytic properties of carboxylesterase isozymes involved in metabolic activation of prodrugs. Molecules 13(2): 412–431.

Hussain, S., T. Siddique, M. Saleem, M. Arshad and A. Khalid. 2009. Impact of pesticides on soil microbial diversity, enzymes, and biochemical reactions. pp. 159–200. *In*: D.L. Sparks (ed.). Advances in Agronomy, vol. 102. Academic Press, Elsevier, San Diego, USA.

Igalavithana, A.D., J. Park, C. Ryu, Y.H. Lee, Y. Hashimoto, L. Huang et al. 2017. Slow pyrolyzed biochars from crop residues for soil metal(loid) immobilization and microbial community abundance in contaminated agricultural soils. Chemosphere 177: 157–166.

International Biochar Initiative. 2015. Standardized product definition and product testing guidelines for biochar that is used in soil. IBI biochar standards, version 2.1, 23th November 2015, https://biochar-international.org/characterizationstandard.

Ippolito, J.A., K.A. Spokas, J.M. Novak, R.D. Lentz and K.B. Cantrell. 2015. Biochar elemental composition and factors influencing nutrient retention. pp. 139–163. *In*: Lehmann, J. and S. Joseph (eds.). Biochar for Environmental Management: Science, Technology and Implementation. Routledge, Taylor & Francis Group, New York, USA.

Jackson, C.J., J.G. Oakeshott, J.C. Sanchez-Hernandez and C.E. Wheelock. 2011. Carboxylesterases in the metabolism and toxicity of pesticides. pp. 57–75. *In*: Tetsuo Satoh and R.C. Gupta (eds.). Anticholinesterase Pesticides: Metabolism, Neurotoxicity, and Epidemiology. John Wiley & Sons, New Jersey, USA.

Kamita, S.G. and B.D. Hammock. 2010. Juvenile hormone esterase: biochemistry and structure. J. Pestic. Sci. 35(3): 265–274.

Kammann, C., J. Ippolito, N. Hagemann, N. Borchard, M.L. Cayuela, J.M. Estavillo et al. 2017. Biochar as a tool to reduce the agricultural greenhouse-gas burden—knowns, unknowns and future research needs. J. Environ. Eng. Landsc. 25(2): 114–139.

Kuppusamy, S., P. Thavamani, M. Megharaj, K. Venkateswarlu and R. Naidu. 2016. Agronomic and remedial benefits and risks of applying biochar to soil: Current knowledge and future research directions. Environ. Int. 87: 1–12.

Lee, J., K.-H. Kim and E.E. Kwon. 2017. Biochar as a catalyst. Renew. Sust. Energ. Rev. 77: 70–79.

Lehmann, J., M.C. Rillig, J. Thies, C.A. Masiello, W.C. Hockaday and D. Crowley. 2011. Biochar effects on soil biota—A review. Soil Biol. Biochem. 43(9): 1812–1836.

Lehotay, S.J., J. Tully, A.V. Garca, M. Contreras, H. Mol, V. Heinke et al. 2007. Determination of pesticide residues in foods by acetonitrile extraction and partitioning with magnesium sulfate: collaborative study. J. AOAC Int. 90(2): 485–520.

Lessard, I., S. Sauvé and L. Deschênes. 2014. Toxicity response of a new enzyme-based functional diversity methodology for Zn-contaminated field-collected soils. Soil Biol. Biochem. 71: 87–94.

Li, D., W.C. Hockaday, C.A. Masiello and P.J.J. Alvarez. 2011. Earthworm avoidance of biochar can be mitigated by wetting. Soil Biol. Biochem. 43(8): 1732–1737.

Lian, J., R. Nelson and R. Lehner. 2018. Carboxylesterases in lipid metabolism: from mouse to human. Protein Cell 9(2): 178–195.

Liao, N., Q. Li, W. Zhang, G. Zhou, L. Ma, W. Min et al. 2016. Effects of biochar on soil microbial community composition and activity in drip-irrigated desert soil. Eur. J. Soil Biol. 72: 27–34.

Luo, Y., M. Durenkamp, M. De Nobili, Q. Lin, B.J. Devonshire and P.C. Brookes. 2013. Microbial biomass growth, following incorporation of biochars produced at 350°C or 700°C, in a silty-clay loam soil of high and low pH. Soil Biol. Biochem. 57: 513–523.

Mackay, D., J.P. Giesy and K.R. Solomon. 2014. Fate in the environment and long-range atmospheric transport of the organophosphorus insecticide, chlorpyrifos and its oxon. pp. 35–76. *In*: Giesy, J.P. and K.R. Solomon (eds.). Ecological Risk Assessment for Chlorpyrifos in Terrestrial and Aquatic Systems in the United States: Rev. Environ. Contam. T., vol 231. Springer, Heidelberg, Germany.

Margesin, R. 2005. Determination of enzyme activities in contaminated soil. pp. 309–320. *In*: Margesin, R. and F. Schinner (eds.). Manual of Soil Analysis: Monitoring and Assessing Soil Bioremediation. Berlin, Germany: Springer-Verlag.

Marks, E.A.N., S. Mattana, J.M. Alcañiz and X. Domene. 2014. Biochars provoke diverse soil mesofauna reproductive responses in laboratory bioassays. Eur. J. Soil Biol. 60: 104–111.

Maxwell, D.M. 1992. The specificity of carboxylesterase protection against the toxicity of organophosphorus compounds. Toxicol. Appl. Pharmacol. 114(2): 306–312.

Mierzwa-Hersztek, M., K. Gondek and A. Baran. 2016. Effect of poultry litter biochar on soil enzymatic activity, ecotoxicity and plant growth. App. Soil Ecol. 105: 144–150.

Mijangos, I., I. Albizu, L. Epelde, I. Amezaga, S. Mendarte and C. Garbisu. 2010. Effects of liming on soil properties and plant performance of temperate mountainous grasslands. J. Environ. Manage 91(10): 2066–2074.

Mikhailov, A.T. and M. Torrado. 1999. Carboxylesterase overexpression in the male reproductive tract: a universal safeguarding mechanism. Reprod. Fert. Develop. 11(3): 133–146.

Moreno, J.L., C. Garcıa, L. Landi, L. Falchini, G. Pietramellara and P. Nannipieri. 2001. The ecological dose value (ED50) for assessing Cd toxicity on ATP content and dehydrogenase and urease activities of soil. Soil Biol. Biochem. 33(4-5): 483–489.

Morillo, E. and J. Villaverde. 2017. Advances technologies for the remediation of pesticide-contaminated soils. Sci. Total Environ. 586: 576–597.

Motulsky, H. and A. Christopoulos. 2003. Fitting models to biological data using linear and nonlinear regression: A practical guide to curve fitting. GraphPad Software Inc. San Diego, CA, USA.

Nannipieri, P., E. Kandeler and P. Ruggiero. 2002. Enzyme activities and microbiological and biochemical processes in soil. pp. 1–33. *In*: Burns, R.G. and R.P. Dick (eds.). Enzymes in the Environment: Activity, Ecology, and Applications. Marcel Dekker, New York, USA.

Nannipieri, P., C. Trasar-Cepeda and R.P. Dick. 2018. Soil enzyme activity: A brief history and biochemistry as a basis for appropriate interpretations and meta-analysis. Biol. Fert. Soils 54(1): 11–19.

Novak, J.M., J.A. Ippolito, T.F. Ducey, D.W. Watts, K.A. Spokas, K.M. Trippe et al. 2018. Remediation of an acidic mine spoil: Miscanthus biochar and lime amendment affects metal availability, plant growth, and soil enzyme activity. Chemosphere 205: 709–718.

Oliveira, F.R., A.K. Patel, D.P. Jaisi, S. Adhikari, H. Lu and S.K. Khanal. 2017. Environmental application of biochar: Current status and perspectives. Bioresource Technol. 246: 110–122.

Otero, S. and G. Kristoff. 2016. *In vitro* and *in vivo* studies of cholinesterases and carboxylesterases in Planorbarius corneus exposed to a phosphorodithioate insecticide: Finding the most sensitive combination of enzymes, substrates, tissues and recovery capacity. Aquat. Toxicol. 180: 186–195.

Paz-Ferreiro, J., G. Gascó, B. Gutiérrez and A. Méndez. 2012. Soil biochemical activities and the geometric mean of enzyme activities after application of sewage sludge and sewage sludge biochar to soil. Biol. Fert. Soils 48(5): 511–517.

Paz-Ferreiro, J., S. Fu, A. Méndez and G. Gascó. 2014. Interactive effects of biochar and the earthworm Pontoscolex corethrurus on plant productivity and soil enzyme activities. J. Soil Sediment. 14(3): 483–494.

Paz-Ferreiro, J. and S. Fu. 2016. Biological indices for soil quality evaluation: Perspectives and limitations. Degrad. Dev. 27(1): 14–25.

Pei, J., S. Zhuang, J. Cui, J. Li, B. Li, J. Wu et al. 2017. Biochar decreased the temperature sensitivity of soil carbon decomposition in a paddy field. Agr. Ecosyst. Environ. 249: 156–164.

Peiris, C., S.R. Gunatilake, T.E. Mlsna, D. Mohan and M. Vithanage. 2017. Biochar based removal of antibiotic sulfonamides and tetracyclines in aquatic environments: A critical review. Bioresource Technol. 246: 150–159.

Qambrani, N.A., M.M. Rahman, S. Won, S. Shim and C. Ra. 2017. Biochar properties and eco-friendly applications for climate change mitigation, waste management, and wastewater treatment: A review. Renew. Sust. Energ. Rev. 79: 255–273.

Qian, K., A. Kumar, H. Zhang, D. Bellmer and R. Huhnke. 2015. Recent advances in utilization of biochar. Renew. Sust. Energ. Rev. 42: 1055–1064.

Qian, T., Y. Wang, T. Fan, G. Fang and D. Zhou. 2016. A new insight into the immobilization mechanism of Zn on biochar: The role of anions dissolved from ash. Sci. Rep. 6: 33630.

Racke, K.D., D.A. Laskowski and M.R. Schultz. 1990. Resistance of chlorpyrifos to enhanced biodegradation in soil. J. Agr. Food Chem. 38(6): 1430–1436.

Racke, K.D. 1993. Environmental fate of chlorpyrifos. Rev. Environ. Contam. T. 131: 1–150.

Rao, M.A., R. Scelza, F. Acevedo, M.C. Diez and L. Gianfreda. 2014. Enzymes as useful tools for environmental purposes. Chemosphere 107: 145–162.

Renella, G., A.L.R. Ortigoza, L. Landi and P. Nannipieri. 2003. Additive effects of copper and zinc on cadmium toxicity on phosphatase activities and ATP content of soil as estimated by the ecological dose (ED50). Soil Biol. Biochem. 35(9): 1203–1210.

Riah, W., K. Laval, E. Laroche-Ajzenberg, C. Mougin, X. Latour and I. Trinsoutrot-Gattin. 2014. Effects of pesticides on soil enzymes: A review. Environ. Chem. Lett. 12(2): 257–273.

Ross, M.K., A. Borazjani, C.C. Edwards and P.M. Potter. 2006. Hydrolytic metabolism of pyrethroids by human and other mammalian carboxylesterases. Biochem. Pharmacol. 71(5): 657–669.

Ross, M.K. and J.A. Crow. 2007. Human carboxylesterases and their role in xenobiotic and endobiotic metabolism. J. Biochem. Mol. Toxic. 21(4): 187–196.

Ross, M.K., T.M. Streit and K.L. Herring. 2010. Carboxylesterases: Dual roles in lipid and pesticide metabolism. J. Pestic. Sci. 35(3): 257–264.

Sanchez-Hernandez, J.C., C. Mazzia, Y. Capowiez and M. Rault. 2009. Carboxylesterase activity in earthworm gut contents: Potential (eco)toxicological implications. Comp. Biochem. Physiol. 150(4): 503–511.

Sanchez-Hernandez, J.C., S. Martínez Morcillo, J. Notario del Pino and P. Ruiz. 2014. Earthworm activity increases pesticide-sensitive esterases in soil. Soil Biol. Biochem. 75: 186–196.

Sanchez-Hernandez, J.C. and M. Sandoval. 2017. Effects of chlorpyrifos on soil carboxylesterase activity at an aggregate-size scale. Ecotoxicol. Environ. Saf. 142: 303–311.

Sanchez-Hernandez, J.C., M. Sandoval and A. Pierart. 2017. Short-term response of soil enzyme activities in a chlorpyrifos-treated mesocosm: Use of enzyme-based indexes. Ecol. Indic. 73: 525–535.

Sanchez-Hernandez, J.C. 2018. Biochar activation with exoenzymes induced by earthworms: A novel functional strategy for soil quality promotion. J. Hazard. Mater. 350: 136–143.

Sanchez-Hernandez, J.C., J. Notario Del Pino, Y. Capowiez, C. Mazzia and M. Rault. 2018. Soil enzyme dynamics in chlorpyrifos-treated soils under the influence of earthworms. Sci. Total Environ. 612: 1407–1416.

Satoh, T. and M. Hosokawa. 1998. The mammalian carboxylesterases: from molecules to functions. Annu. Rev. Pharmacol. 38(1): 257–288.

Satyanarayana, T. and L.W. Getzin. 1973. Properties of a stable cell-free esterase from soil. Biochemistry 12(8): 1566–1572.

Singh, B. 2014. Review on microbial carboxylesterase: general properties and role in organophosphate pesticides degradation. Biochem. Biol. Mol. 2: 1–6.

Singh, B., J. Kaur and K. Singh. 2014. Microbial degradation of an organophosphate pesticide, malathion. Crit. Rev. Microbiol. 40(2): 146–154.

Singh, B.K., A. Walker and D.J. Wright. 2002. Degradation of chlorpyrifos, fenamiphos, and chlorothalonil alone and in combination and their effects on soil microbial activity. Environ. Toxicol. Chem. 21(12): 2600–2605.

Singh, B.K. and A. Walker. 2006. Microbial degradation of organophosphorus compounds. FEMS Microbiol. Rev. 30(3): 428–471.

Singh, B.K. 2009. Organophosphorus-degrading bacteria: ecology and industrial applications. Nat. Rev. Microbiol. 7(2): 156–164.

Sizmur, T., T. Fresno, G. Akgül, H. Frost and E. Moreno-Jiménez. 2017. Biochar modification to enhance sorption of inorganics from water. Bioresource Technol. 246: 34–47.

Song, D., J. Tang, X. Xi, S. Zhang, G. Liang, W. Zhou et al. 2018. Responses of soil nutrients and microbial activities to additions of maize straw biochar and chemical fertilization in a calcareous soil. Eur. J. Soil Biol. 84: 1–10.

Subedi, R., N. Taupe, I. Ikoyi, C. Bertora, L. Zavattaro, A. Schmalenberger et al. 2016. Chemically and biologically-mediated fertilizing value of manure-derived biochar. Sci. Total Environ. 550: 924–933.

Sun, Z., E.W. Bruun, E. Arthur, L.W. de Jonge, P. Moldrup, H. Hauggaard-Nielsen et al. 2014. Effect of biochar on aerobic processes, enzyme activity, and crop yields in two sandy loam soils. Biol. Fert. Soils 50(7): 1087–1097.

Tabatabai, M.A. and W.A. Dick. 2002. Enzymes in soil: Research and developments in measuring activities. pp. 567–596. *In*: Burns, R.G. and R.P. Dick (eds.). Enzymes in the Environment: Activity, Ecology, and Applications. Marcel Dekker, New York, USA..

Taketani, M., M. Shii, K. Ohura, S. Ninomiya and T. Imai. 2007. Carboxylesterase in the liver and small intestine of experimental animals and human. Life Sci. 81(11): 924–932.

Tan, Z., C.S.K. Lin, X. Ji and T.J. Rainey. 2017. Returning biochar to fields: A review. App. Soil Ecol. 116: 1–11.

Thies, J.E., M.C. Rillig and E.R. Graber. 2015. Biochar effects on the abundance, activity and diversity of the soil biota. pp. 327–389. *In*: Lehmann, J and S. Joseph (eds.). Biochar for Environmental Management: Science, Technology and Implementation. Routhledge, Taylor & Francis Group, London, UK.

Thomsen, R., H.B. Rasmussen, K. Linnet and C. Indices. 2014. *In vitro* drug metabolism by human carboxylesterase 1: focus on angiotensin-converting enzyme inhibitors. Drug Metab. Dispos. 42(1): 126–133.

Trasar-Cepeda, C., M.C. Leiros, S. Seoane and F. Gil-Sotres. 2000. Limitations of soil enzymes as indicators of soil pollution. Soil Biol. Biochem. 32(13): 1867–1875.

Vithanage, M., T. Bandara, M.I. Al-Wabel, A. Abduljabbar, A.R.A. Usman, M. Ahmad et al. 2018. Soil enzyme activities in waste biochar amended multi-metal contaminated soil; effect of different pyrolysis temperatures and application rates. Commun. Soil Sci. Plan. 49(5): 635–643.

Wallenstein, M.D. and R.G. Burns. 2011. Ecology of extracellular enzyme activities and organic matter degradation in soil: A complex community-driven process. pp. 35–55. *In*: Dick, R.P. (ed.). Methods of Soil Enzymology. Soil Science Society of America, Madison, Wisconsin, USA.

Wang, F., J. Yao, H. Chen, K. Chen, P. Trebse and G. Zaray. 2010. Comparative toxicity of chlorpyrifos and its oxon derivatives to soil microbial activity by combined methods. Chemosphere 78(3): 319–326.

Wang, X., W. Zhou, G. Liang, D. Song and X. Zhang. 2015. Characteristics of maize biochar with different pyrolysis temperatures and its effects on organic carbon, nitrogen and enzymatic activities after addition to fluvo-aquic soil. Sci. Total Environ. 538: 137–144.

Weber, K. and P. Quicker. 2018. Properties of biochar. Fuel 217: 240–261.

Weyers, S.L. and K.A. Spokas. 2011. Impact of biochar on earthworm populations: A review. Appl. Environ. Soil Sci. 1–12.

Wheelock, C.E., B.M. Phillips, B.S. Anderson, J.L. Miller, M.J. Miller and B.D. Hammock. 2008. Applications of carboxylesterase activity in environmental monitoring and toxicity identification evaluations (TIEs). Rev. Environ. Contam. T. 195: 117–178.

Wheelock, C.E. and Y. Nakagawa. 2010. Carboxylesterases—from function to the field: an overview of carboxylesterase biochemistry, structure–activity relationship, and use in environmental field monitoring. J. Pestic. Sci. 35(3): 215–217.

Williams, M., S. Martin and R.S. Kookana. 2015. Sorption and plant uptake of pharmaceuticals from an artificially contaminated soil amended with biochars. Plant Soil 395(1-2): 75–86.

Wu, F., Z. Jia, S. Wang, S.X. Chang and A. Startsev. 2013. Contrasting effects of wheat straw and its biochar on greenhouse gas emissions and enzyme activities in a Chernozemic soil. Biol. Fert. Soils 49(5): 555–565.

Xiao, X. and B. Chen. 2016. Interaction mechanisms between biochar and organic pollutants. pp. 225–257. *In*: Guo, M., Z. He and S.M. Uchimiya (eds.). Agricultural and Environmental Applications of Biochar: Advances and Barriers. Soil Science Society of America. Madison, WI, USA.

Xie, T., K.R. Reddy, C. Wang, E. Yargicoglu and K. Spokas. 2015. Characteristics and applications of biochar for environmental remediation: A review. Crit. Rev. Env. Sci. Tec. 45(9): 939–969.

Yang, D.I.N.G., L.I.U. Yunguo, L.I.U. Shaobo, X. Huang, L.I. Zhongwu, T.A.N. Xiaofei et al. 2017. Potential benefits of biochar in agricultural soils: A review. Pedosphere 27(4): 645–661.

Yoo, G. and H. Kang. 2012. Effects of biochar addition on greenhouse gas emissions and microbial responses in a short-term laboratory experiment. J. Environ. Qual. 41(4): 1193.

Zhu, X., B. Chen, L. Zhu and B. Xing. 2017. Effects and mechanisms of biochar-microbe interactions in soil improvement and pollution remediation: A review. Environ. Pollut. 227: 98–115.

Zumstein, M.T., D. Rechsteiner, N. Roduner, V. Perz, D. Ribitsch, G.M. Guebitz et al. 2017. Enzymatic hydrolysis of polyester thin films at the nanoscale: Effects of polyester structure and enzyme active-site accessibility. Environ. Sci. Technol. 51(13): 7476–7485.

Chapter 11

Dual Role of Vermicomposting in Relation to Environmental Pollution
Detoxification and Bioremediation

Juan C. Sanchez-Hernandez[1],* and *Jorge Domínguez*[2]

INTRODUCTION

In recent years, bioremediation has become common practice in the restoration of contaminated soils. Phytoremediation and biostimulation are the most popular strategies used for the bioremediation of contaminated soils, although they have some limitations. Biostimulation involves the addition of nutrients and other supplementary chemicals to increase native populations of microorganisms in an attempt to accelerate the biodegradation of environmental contaminants (Tyagi et al. 2011, Megharaj et al. 2011). Nutrients can be applied in the form of inorganic fertilizers, oleophilic fertilizers and organic waste (e.g., animal manure) (Tyagi et al. 2011); however, the application dose and nutrient ratios must be optimal to have significant impacts on the pollutant degradation rate (Megharaj et al. 2011).

In the last two decades, vermicomposting has emerged as an eco-friendly technology for improving the physicochemical and biological properties of degraded

[1] Laboratory of Ecotoxicology, Institute of Environmental Sciences, University of Castilla-La Mancha, 45071 Toledo, Spain.
[2] Departamento de Ecoloxía e Bioloxía Animal, Universidade de Vigo, 36310 Vigo, Spain.
Email: jdguez@uvigo.es
* Corresponding author: juancarlos.sanchez@uclm.es

soils as well as to increase crop yield (Edwards et al. 2010). The final product of vermicomposting, i.e., vermicompost, is a finely divided, porous peat-like material with several beneficial properties related to soil function: high water-holding capacity, high contents of humic substances and nutrients, well-established bacterial and fungal communities, and large loads of extracellular enzymes. Although vermicompost has several characteristics suitable for mitigating soil pollution, it has scarcely been investigated in relation to bioremediation.

This chapter provides up-to-date knowledge on two lines of active research in vermicomposting technology: (1) the use of vermicompost to remove hazardous chemicals from contaminated raw materials (detoxification), and (2) the impact of vermicompost for remediating contaminated soils (bioremediation). The chapter is divided into four sections. The first section provides an overview of the functional features of the vermicomposting process. The second section describes the main findings regarding the detoxification capacity of vermicomposting to degrade organic contaminants and to immobilize metals present in agroindustrial and municipal waste. The third section reports several important examples that demonstrate the potential use of vermicompost in the degradation and immobilization of pesticides and metals in soil. The microbiota and organic ligands (e.g., humic substances) present in vermicompost are the main features that account for this bioremediation capability. Moreover, recent studies suggest that vermicompost contains extracellular detoxifying enzymes (e.g., carboxylesterases, peroxidases and laccases), thus indicating possible new lines of research related to the bioremediation potential. The final section will suggest new lines of future research in the area of vermicomposting for bioremediation purposes.

Functional Processes of Vermicomposting

Intensive agriculture, which is characterized by conventional tillage, massive and repeated fertilization, high pesticide input and low plant diversity, has a negative impact on soil biodiversity (Tsiafouli et al. 2015, Steffen et al. 2015). In this context, organic or ecological agriculture has emerged as an alternative form of cropping that promotes natural soil processes and seeks to reduce the negative impacts on soil biota (Mäder et al. 2002). However, this type of sustainable agriculture does not reach the productive standards of intensive agriculture (Seufert et al. 2012), and novel eco-friendly strategies are therefore required to increase crop yield and maintain soil fertility and biodiversity. Although soil biological dynamics are fundamental for the balance and productivity of natural ecosystems, this fact is still ignored in managed ecosystems, such as agricultural ecosystems (Bender et al. 2016).

Soil biodiversity is lower in intensive agriculture than in organic agriculture or natural ecosystems (Tsiafouli et al. 2015). The reduced biodiversity implies a lower functional capacity to resist environmental stressors than in soil systems with a high biodiversity (Bender et al. 2016). Moreover, high-diversity communities are more productive and sustainable than low-diversity ones. Therefore, the loss of some species or functional groups in intensively managed agricultural systems will lead to greater maladaptation in ecosystem functioning than observed in natural ecosystems. Plant productivity depends on aboveground interrelationships of different functional groups, and soil microbial consortia provide key ecosystem functions (Putten et al. 2013). The importance of soil microbial communities for plant growth and the development of plant communities is demonstrated in studies in which the soil microbiome is manipulated (Chaparro et al. 2012), and in field studies involving the inoculation of different soils ranging from those in highly productive ecosystems to degraded soils (van de Voorde et al. 2012, Wubs et al. 2016). Therefore, soil biodiversity is a key concept in sustainable agriculture, as well as in the bioremediation of polluted soils.

The production and use of vermicompost, a known organic fertilizer that contains a well-established microbial community, may be a promising eco-friendly strategy for promoting soil fertility by supplying both nutrients and microbial communities to the soil. The most common organic materials used for vermicomposting are animal manures, biosolids generated in wastewater treatment plants, biosolids derived from industries dealing with any type of organic material (e.g., olive mill waste and dairy industry waste), the organic fraction of municipal solid waste, and a wide range of animal and plant residues.

Vermicomposting consists of the biooxidative decomposition of dead organic matter in a mesophilic environment ($< 30°C$) created by the cooperative actions of detritivore earthworm species (e.g., *Eisenia fetida, E. andrei*), microorganisms and, to a less extent, other members of the soil fauna decomposer community. During the vermicomposting process, the physical, microbiological and biochemical properties of the organic matter are greatly modified, thus stabilized (Aira et al. 2006). Earthworms are key drivers of the process because of the significant contribution that they make to the fragmentation of organic debris and dispersion of microorganisms. Likewise, earthworms play a significant role in microbial activity via comminution of organic matter and enhancement of the surface area available for microbial attack (Aira et al. 2007), and via grazing directly on microbiota (Aira et al. 2007, Domínguez et al. 2017). Through these and other specific activities, earthworms enhance the efficiency of the microbial communities and their turnover rates, thus substantially increasing the substrate decomposition rates. In addition to microbes, earthworms also affect other soil biota either directly, through the ingestion of protozoa and microfauna present in the detrital food webs, or indirectly, by modifying the availability of resources for these organisms (Edwards et al. 2010).

The nutritional, microbial and enzymatic enrichment of vermicompost is achieved by a long-term process including two successional stages: an active stage and a maturation stage.

(1) The active stage is characterized by the feeding activity of the earthworms, which contributes to modifying the physical and chemical properties of the feedstock as well as its microbial composition (Aira et al. 2006, 2007). During this stage, earthworms are continuously moving and burrowing into the substrate, thereby contributing to its oxygenation and homogenization which, in turn, accelerates its decomposition. On the other hand, the feeding activity of earthworms also alters the physicochemical and biological properties of the feedstock as it passes through their guts. During this gastrointestinal transit of organic matter, many digestive enzymes produced by the earthworms themselves and symbiont

microorganisms boost organic matter decomposition. These diverse extracellular enzymes, break down a wide variety of organic molecules, including cellulose and phenolic compounds. Because of the high content of organic carbon, these digestive enzymes become stable by binding to the organic complexes of casts (Domínguez et al. 2017, Sanchez-Hernandez and Domínguez 2017).

(2) The maturation stage starts once earthworms move away from the processed substrate, and new microbial communities take over the further decomposition of more recalcitrant molecules (Aira et al. 2007). During the maturation stage, the cast-associated processes most closely associated with the presence of material not processed by the transit through the earthworm gut and with physical modification of the egested material contribute to the further decomposition of the substrate. Therefore, vermicompost reaches an optimal stage in terms of its biological properties, marked by its capacity to promote plant growth and reduce plant diseases. The duration of both vermicompost stages are highly variable; thus while the active stage depends on the earthworm species, their population dynamics and the environmental conditions, the maturation stage depends on the efficiency of the active stage (Edwards et al. 2010).

Detoxification: Cleaning Contaminated Organic Waste

The use of vermicompost to promote soil fertility requires that this nutrient-rich and microbiologically active organic amendment meets a series of quality standards that involve microbiological and physicochemical parameters. Thus, vermicompost must be free of potentially hazardous contaminants such as metals, hydrocarbons, pesticides, pharmaceuticals, cosmetics and other contaminants typically found in agroindustrial and municipal organic residues. In this context, many studies have considered the capacity of the vermicomposting process to remove toxic metals and organic contaminants.

Metals

The use of the term "heavy metal" in environmental sciences continues to be a matter of intensive debate in the scientific literature (Pourret and Bollinger 2018), and its use is discouraged because the definition is vague and misleading (Duffus 2002, Hodson 2004, Hübner et al. 2010). Therefore, in this chapter, we have opted to use the term "metal" (or "metalloid" in the case of As, At, B, Ge, Po, Sb, Se, Si, and Te), as suggested by Chapman (2007).

The industrial and municipal waste commonly used in vermicomposting may contain important amounts of metals, and the concentrations in the final product (vermicompost) may pose a serious threat to soil function when used as a soil amendment. A recent review by Swati and Hait (2017) summarizes the range of metal concentrations in industrial, municipal and domestic organic waste as well as the standards for composts that are adopted in many countries. As pointed out by these authors, the standards for the environmentally safe use of compost are based on total metal concentrations, and the chemical speciation of metals in the environment is ignored. Chemical speciation can be defined as the formation of multiple chemical forms in which a metal exists in the environment (Wright and Welbourn 2002). Metal speciation is highly dependent on fluctuating environmental variables such as pH, organic matter content, exchange capacity, temperature and moisture, which ultimately affect metal bioavailability and toxicity in both the vermicompost itself and in the vermicompost-amended soil.

Chemical speciation affects the metal concentration in vermicompost. Changes in metal speciation affect bioaccumulation by earthworms and facilitate sorption of the metal on to organic ligands in the vermicompost. Many studies have demonstrated that vermicomposting significantly reduces the available fraction of several metals and metalloids such as As, Cu, Cd, Cr, Ni, Pb and Zn, irrespective of the earthworm species and type of raw material used (Singh and Kalamdhad 2013, Sahariah et al. 2015, Goswami et al. 2016, Lv et al. 2016, He et al. 2016). However, the reduction in metal availability does not necessarily indicate a lack of toxicity. Thus, toxicity testing with vermicompost should be adopted as a complementary measure prior to its use as a soil amendment. For example, the study by Vašíčková et al. (2016) clearly shows why toxicity assessment of vermicompost is highly recommended. These researchers compared the toxicity of As-contaminated sludges after three different types of treatment, i.e., composting, vermicomposting and mixing with soil. The results of ecotoxicity tests using *Folsomia candida* (reproduction test), *Enchytraeus crypticus* (reproduction test), and *Lactuca sativa* (root elongation test) demonstrated that vermicomposting As-contaminated sludge increased its toxicity relative to the other two procedures, although vermicomposting greatly reduced the available fraction of As. According to the authors, the unexpected toxicity may be due to ammonia, among other potentially toxic by-products generated during vermicomposting (Vašíčková et al. 2016).

Obviously metals cannot be mineralized as organic compounds. Therefore, the only rational way to reduce their concentrations in the vermicompost, relative to those in the feedstock, is for them to be accumulated in earthworms. Many studies have demonstrated the capacity of earthworms to accumulate metals (Swati and Hait 2017), which is favoured by the induction of metal-binding proteins such as metallothioneins (Goswami et al. 2016). Surprisingly, however, the same studies have documented a slight increase in the total metal concentrations of the vermicompost relative to the feedstock (Bakar et al. 2011, Yadav and Garg 2011, Lv et al. 2016). The marked differences in organic matter content and quality between the organic waste and the final vermicompost probably explain the increase in metal concentrations in the vermicompost. Vermicompost is a humic-rich material (Elvira et al. 1998, Goswami et al. 2016), and humic substances facilitate the formation of metal-humus complexes which, in turn, result in higher metal contents than in the initial organic residues (He et al. 2017). Indeed, the study by Goswami et al. (2016) suggests such a chemical relationship. These researchers observed higher metal concentrations in the vermicompost than in compost, with the former having average humic acid carbon contents ranging between 0.1 and 0.51%, whereas the contents in compost varied between 0.02 and 0.25%.

In summary, these examples illustrate that vermicompost obtained from metal-contaminated feedstock may increase the environmental risk of soil contamination due to over-supply of metals. In this context, novel vermicomposting strategies (e.g., periodic removal of composting earthworm population by fresh earthworm population) should be adopted to take advantage of the high capacity of earthworms to accumulate metals.

Organic Pollutants

Agroindustrial and municipal organic wastes usually contain a large variety of organic pollutants such as polycyclic aromatic hydrocarbons, polychlorinated and polybrominated biphenyls, pesticides, pharmaceuticals and personal care products. This chemical cocktail poses a challenge to vermicomposting because both earthworms and microflora may be negatively affected by exposure to mixtures of pollutants. Accordingly, monitoring of the toxic effects during vermicomposting of contaminated raw materials is recommended in order to evaluate the viability of the decomposing process and, if appropriate, to propose corrective measures. Most vermicomposting studies measure changes in the earthworm population (individual density and reproduction rate) and, occasionally, changes in microbial activity and biomass relative to the potential toxicity of the feedstock. For example, plant waste containing residues of the neonicotinoid imidacloprid had a significant impact on *E. fetida* during vermicomposting for 15 weeks. Although a pesticide concentration of 2 mg/kg did not kill the earthworms, they did not produce cocoons (Fernández-Gómez et al. 2011). Similarly, degradation of the antibiotic oxytetracycline and its metabolites was monitored during the decomposition of chicken manure by a first thermophilic composting phase (20 d), followed by a second vermicomposting (7 wk) phase with the earthworm species *E. fetida* (Ravindran and Mnkeni 2017). The highest rate of degradation of this pharmaceutical took place during the composting phase (30–80% reduction compared to the initial concentration), whereas degradation rates of 15–40% occurred in the vermicomposting phase. The high temperature typically reached during composting (50–70°C) probably accounted for oxytetracycline degradation. However, it is not clear whether a single vermicomposting phase would have been as effective as composting for degrading this antibiotic. Although high temperatures are not generated during vermicomposting, the higher microbial activity may have a similar impact on oxytetracycline persistence. Villalobos-Maldonado et al. (2015) suggested that the cooperation between earthworms and microorganisms may have the same outcome as composting. These researchers demonstrated that vermicomposting reduced the initial concentration of the recalcitrant contaminant decachlorobiphenyl by 80–95% during incubation for 3 months with *E. fetida* and a mixture of peat moss and rabbit excrement as organic feedstock.

Although the aforementioned studies have demonstrated that vermicomposting reduces the concentration of organic pollutants, the toxic effects on earthworms, microorganisms and extracellular enzyme activities, which are the main drivers of organic matter decomposition, have received little attention. Moreover, understanding the ability of earthworms and microorganisms to adapt to feedstocks containing environmental contaminants demands future research. This toxicological data would help in the selection of the most appropriate conditions (earthworm species, organic waste mixture, pre-treatment of raw material, and so on) in the vermicomposting of contaminated-feedstocks.

Bioremediation: Cleaning Contaminated Soils

The aim of vermicomposting is to obtain a stabilized, nutrient-rich material for use as a soil amendment. Most studies concerning vermicomposting technology have provided knowledge about the following issues: the impact of the feedstock characteristics on the chemical and microbiological properties of the final vermicompost, the functional role of earthworms in the oxidative decomposition of organic matter, and the effects of vermicompost on plant growth (Edwards et al. 2010). By comparison, the application of vermicomposting in the field of bioremediation has received less attention, even though vermicompost is a nutrient- and microbial-rich material. The high contents of organic matter and nutrients undoubtedly promote soil microbial proliferation (biostimulation). Likewise, vermicompost is a microbiologically active substrate, and as such, the addition of vermicompost is a means of inoculating the soil with non-native microorganisms that may participate in the degradation of organic contaminants (bioaugmentation). In the following two subsections, we will identify the main effects of vermicompost in soil contaminated by metals and organic pollutants.

Metals

In previous sections, we have highlighted that one of the main impacts of the vermicomposting process on the chemical speciation of metals is the reduced availability, probably due to the formation of metal-humic complexes. However, the immediate question arising is whether vermicompost has the same effect on metals present in soil when used as an amendment.

Although the chemical characteristics of vermicompost suggest that it will be an excellent sorbent substrate for metals, thereby reducing their bioavailability and toxicity in soil, very few studies have examined the potential applications of vermicompost in the remediation of metal-contaminated soils. For example, Zhu et al. (2017) suggested that vermicompost can efficiently retain toxic metals from the soil solution. In this study, a series of kinetic assays of adsorption and desorption were performed with different Pb^{+2} and Cd^{+2} solutions to compare the metal binding capacity of both cow manure and the vermicompost derived from this waste material. Although the adsorption isotherm kinetics of these metals differed, vermicompost was better for retaining both metals. The functional capacity of vermicompost was suggested to be the main cause of binding metals from water. Use of Fourier transform infrared spectroscopy revealed that the cow manure-derived vermicompost displayed functional groups such as –OH (aliphatic alcohol), –COOH (aromatic compounds), and bonds such as C=O and C–O (carbonates and aliphatic alcohol) and P–O (phosphates), thus explaining the high functional capacity of vermicompost to bind Pb^{+2} and Cd^{+2} (Zhu et al. 2017). In a similar study, Singh and Kaur (2015) studied the potential of a cow manure-derived vermicompost to remove metals from industrial effluents. Using a simulated bio-filter, these authors demonstrated that vermicompost retained Cu^{+2} and Zn^{+2} from beverage plant, paper mill and distillery effluents. With the same aim, He et al. (2017) examined the kinetics of Pb^{+2} and Cd^{+2} adsorption on a vermicompost generated from sewage sludge. This vermicompost displayed a high capacity to adsorb both metals (individually or together), although the maximal adsorption capacity was lower for Cd^{+2}, probably because of competition with Pb^{+2} for the binding sites. Together, these examples suggest that vermicompost may be used as a suitable and complementary filter-like support for the removal of potentially toxic metals from contaminated water. However, it remains to be elucidated whether such an environmental service can also be achieved in metal-contaminated soils. Moreover, in order to gain a deeper insight into the use of vermicompost for remediating metal-contaminated soils, some aspects related to the type and surface functionality of vermicompost must be investigated: the mechanism of chemical interaction between metal and vermicompost, the impact of environmental variables, such as soil pH, moisture and dissolved organic carbon on the capacity of vermicompost to sorb metals, as well as the effect of aging on the capacity of vermicompost to retain metals.

Vermicompost can act as a metal adsorbent. For example, in a laboratory experiment using metal-contaminated soils, Hoehne et al. (2016) examined the capacity of vermicompost to immobilize Cd, Cr, and Pb. Using a sequential extraction procedure, these researchers measured the different chemical forms in which these metals could be distributed in the soil. Moreover, the chemical analysis was accompanied by a phytoremediation experiment in which three-black oat (*Avena strigosa*) was used to verify whether vermicompost altered the chemical speciation of metals, thus favoring their phytoextraction. Doses of 50, 75 and 100% vermicompost in metal-contaminated soil increased the bioavailable fraction of Cd and Cr, which was corroborated by the metal concentrations measured in the three-black oat. Although the application of vermicompost facilitated the phytoextraction, marked metal-specific differences that depended on the dose of vermicompost were observed. The vermicompost doses of 25% (for Cr- and Pb-contaminated soils) and 50% (for Cd-contaminated soil) were suggested to be suitable for mobilizing metals from soil and thus improving phytoextraction (Hoehne et al. 2016). By contrary, the study by Wang et al. (2018) showed that the available Cd fraction in soil contaminated by this metal was slightly lower in the presence of vermicompost, biochar or a mixture of both amendments. However, all treatments were regularly irrigated with water of different pH during 2 months in an attempt to simulate the impact of acid rain on the adsorptive capacity of these amendments to immobilize Cd. Although the results of the study clearly showed that vermicompost and biochar reduced the available fraction of Cd, the complex experimental design did not allow the researchers to determine whether the vermicompost itself was able to alter the chemical speciation of Cd in non-water saturated soils. Despite these promising findings demonstrating the qualities of vermicompost as an efficient adsorbent of metals, the understanding of the impact of vermicompost on historically metal contaminated soils remains limited.

In the aforementioned studies, the test soils were experimentally spiked with metals (generally as metal salts); however, aging of metals in the soil may affect the capacity of vermicompost to immobilize metals. For example, Abbaspour and Golchin (2011) suggested that this type of aging effect was critical for the use of vermicompost to remediate contaminated soils. The addition of manure-derived vermicompost to a soil collected from a Pb-Zn mine area had little impact on metal speciation after an incubation period of 6 months. Similarly, Fernández-Gómez et al. (2012) concluded that the addition of two different types of vermicompost to a soil under *Trifolium repens* (planted for purposes of phytoremediation) did not improve the capacity of this plant species to remove Ni, Pb and Cd from soil, even in the presence of arbuscular mycorrhizal fungi. However, the addition of vermicompost had a beneficial impact on the growth of *T. repens* and soil enzyme activity (Fernández-Gómez et al. 2012), suggesting that vermicompost may be used to alleviate the toxicity of contaminated soils on other biological components with a more significant role in bioremediation such as plants and microorganisms.

Organic Pollutants

Vermicompost has attractive properties as a substrate for the bioremediation of contaminated soils. Three of these properties play an important role in the bioremediation capacity: the high organic matter content, microbial abundance and diversity, and the existence of pollutant-detoxifying exoenzymes.

Vermicompost is an organic carbon-rich substrate. Although the total amount of organic matter in the final vermicompost is lower than in the original feedstock, the content of humic substances are higher. These humic substances facilitate the formation of metal-humic complexes, as well as the adsorption of organic contaminants with a high partitioning coefficient between soil organic carbon and the soil solution (Log K_{OC}). The functional significance of this interaction was highlighted by Fernández-Bayo et al. (2007), who evaluated the capacity of spent grape marc-derived vermicompost to adsorb the insecticide imidacloprid from soil. The adsorption of imidacloprid was higher in the vermicompost-amended soils than in vermicompost-free soils, and was higher in soils with a low organic carbon content. The absence of a direct relationship between imidacloprid adsorption and the organic carbon content of soil led these researchers to confirm the importance of vermicompost in this adsorptive process, which was attributed to the lignocellulosic nature of the vermicompost (Fernández-Bayo et al. 2007). Indeed, a comparison of the potential of three types of vermicompost (elaborated from spent grape marc, biosolid vinasse and olive-mill waste) to bind the herbicide diuron showed that the vermicompost with the highest lignin content, i.e., spent grape marc-derived vermicompost, displayed the highest adsorption capacity (Fernández-Bayo et al. 2009). Similarly, the sorptive properties of vermicompost also explained the reduction in the leaching potential of diuron, imidacloprid and the metabolites of these in an experimental soil column with vermicompost on top (Fernández-Bayo et al. 2015). Nevertheless, the intrinsic properties of contaminants determine the efficacy of vermicompost as an adsorptive substrate. For example, PAHs are also efficiently retained by vermicompost, although the sorption process seems to depend on the number of benzene rings in the molecule (Dores-Silva et al. 2018), probably due to the positive relationship between the adsorption capacity (Log K_{oc}) and the number of benzene rings in the PAH molecule.

Vermicompost is also a microorganism-rich substrate, and the pollutant biodegradation rate is therefore expected to increase in vermicompost-amended soils. The bioaugmentation effect has been addressed in several studies. For example, Di Gennaro et al. (2009) demonstrated that the microbial communities present in soils historically contaminated with PAHs changed after addition of a vermicompost generated from olive-mill waste. The naphthalene dioxygenase activity was higher in these soils than in vermicompost-free soils, thus indicating why naphthalene was degraded at a higher rate in the vermicompost-amended soils than in vermicompost-free contaminated soils. Perhaps more interestingly, spiking the vermicompost with naphthalene induced expression of biodegradation indicator genes in the native microbiota of the vermicompost (Di Gennaro et al. 2009), suggesting that it is possible to stimulate potential contaminant degraders in the vermicompost before it is used for bioaugmentation purposes (Castillo et al. 2014, Castillo Diaz et al. 2016).

Vermicompost contains a significant fraction of extracellular enzymes involved both in the nutrient cycling (e.g., phosphatases, β-glucosidases, cellulase, protease and ureases) and in the metabolism of organic pollutants (e.g., laccases, peroxidases and carboxylesterases). During vermicomposting, the microbiota changes greatly relative to that in the original raw material (Aira et al. 2007, Gómez-Brandón et al. 2011), and the final vermicompost contains a high diversity of bacteria and fungi (Anastasi et al. 2005). Changes in the activity of many enzymes is also observed during vermicomposting as a consequence of microbial foraging. Therefore, vermicompost contains a high load of extracellular enzymes, which are stabilized by organic matter. Recent studies performed in our laboratory have reported the existence of carboxylesterase activity in vermicompost derived from different types of organic waste (Sanchez-Hernandez and Domínguez 2017, Domínguez et al. 2017). The enzyme activity is involved in the metabolism of synthetic pyrethroid and anticholinesterase (organophosphorus and carbamate) insecticides in animals (Sogorb and Vilanova 2002, Wheelock et al. 2005), plants (Gershater and Edwards 2007, Gershater et al. 2007), microorganisms (Bornscheuer 2002, Singh 2014) and soil (Sanchez-Hernandez et al. 2017, Sanchez-Hernandez et al. 2018). In the particular case of organophosphorus pesticides, carboxylesterase-mediated detoxification involves the formation of a stable enzyme-pesticide complex by the direct interaction between the organophosphorus molecule and the active site of the enzyme (Sogorb and Vilanova 2002). Some laboratory experiments with vermicompost carboxylesterase suggest that this enzyme irreversibly binds the organophosphorus chlorpyrifos-oxon, thus acting as a molecular scavenger of this class of pesticides (Domínguez et al. 2017, Sanchez-Hernandez and Domínguez 2017).

In order to assess the enzymatic bioremediation capacity of vermicompost, we performed a laboratory study to explore the stability of vermicompost carboxylesterase activity in response to physical stress (desiccation and heat shock) applied to the vermicompost. We also assessed the potential of these esterases to bind organophosphorus pesticides in dried vermicompost. In this preliminary study, the carboxylesterase activity was measured in suspensions of cow manure-derived vermicompost in water (1:50, w/v, rotating mixing for 30 min at 25°C). The esterase activity decreased slightly over two weeks during which the vermicompost was air-dried, but the enzyme lost between 48% and 66% of the initial activity during longer (21 days) air-desiccation treatments (Fig. 1A). In addition, heating vermicompost to 40°C (24 h), 50°C (24 h) and 90°C (6 h) showed the high stability of this esterase activity, which was probably due to an interaction with the high organic matter content of vermicompost (Fig. 1B). Indeed, it is well known that a high proportion of the total enzyme activity in the soil, for instance, corresponds to extracellular enzymes that are stabilized by association with soil organo-mineral complexes (Nannipieri et al. 1996). This association protects and stabilizes the enzymes against physicochemical (e.g., temperature- and protease-induce degradation) and biological stressors (e.g., microbial foraging). Because vermicompost is rich in both organic matter and microorganisms, it can also be assumed that enzyme activities in vermicompost are mainly extracellular in location and microbial in origin (released as exoenzymes). On the other hand, soil carboxylesterases are generally glycoproteins, and it has been postulated that the stability of the enzyme in the soil matrix is due to the interaction between the carbohydrate residue of the enzyme and the organic matter and clays (Satyanarayana and Getzin 1973). A comparable interaction could be assumed in the case of vermicompost, which is rich in organic matter.

Early studies indicated a high level of carboxylesterase activity in vermicompost (Sanchez-Hernandez and Domínguez 2017, Domínguez et al. 2017). The existence of these enzymes suggests that vermicompost provides binding sites for organic contaminants of chemical nature (e.g., humic substances) and of enzymatic nature. Indeed, the carboxylesterase activity of cow manure-derived vermicompost was sensitive to *in vitro* inhibition by chlorpyrifos-oxon, dichlorvos and paraoxon-methyl in a dose-dependent manner (Fig. 2). The sensitivity to organophosphorus insecticides was confirmed by spiking both wet and air-dried vermicompost with 10 µg/g chlorpyrifos-oxon; the carboxylesterase activity in both types of vermicompost was significantly inhibited (58% of control activity) 2 days after the spiking (Fig. 3). Interestingly, in these preliminary trials, the response of vermicompost carboxylesterase to desiccation, heat stress and pesticide exposure strongly depended on the substrate used in the enzyme assay. Therefore, and as suggested by others (Wheelock et al. 2005), the use of more than one model substrate is highly recommended for studying the dynamics of vermicompost carboxylesterase activity against environmental stressors.

Together, these findings suggest that vermicompost contains a significant level of carboxylesterase activity that can bind organophosphorus pesticides, thus boosting the bioremediation capacity of vermicompost in relation to organic contaminants. In this context, the addition of vermicompost to the topsoil could act as a molecular and biochemical barrier to reduce contaminant transportation and thus facilitate its

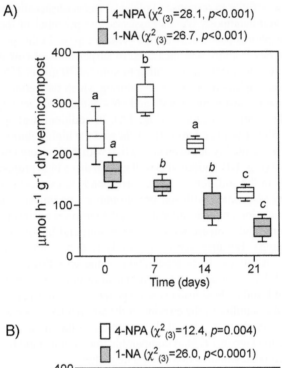

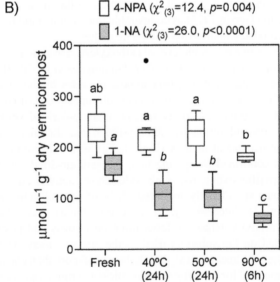

Fig. 1. Stability of carboxylesterase activity in air-dried vermicompost derived from cow manure (graph A) and treated for thermal denaturing of proteins (graph B). The enzyme activity was assayed in suspensions of vermicompost in water with two different substrates, i.e., 4-nitrophenyl acetate (4-NPA) and 1-naphthyl acetate (1-NA), as described in Sanchez-Hernandez et al. (2017). Tukey box plots indicate the median, the 25th and 75th percentiles (box edges), and the range (whiskers) of 10 samples. Significant differences between treatments are indicated by different letters (normal typeface for 4-NPA and italics for 1-NA) after Kruskal-Wallis test followed by a post hoc Mann-Whitney test ($p < 0.05$).

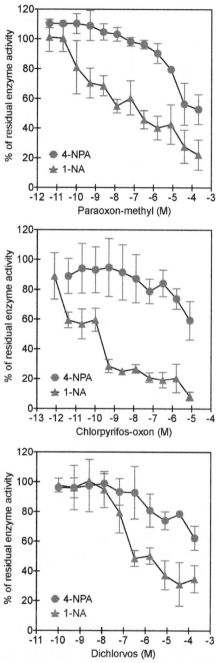

Fig. 2. *In vitro* dose-response relationships between carboxylesterase inhibition and the molar concentration of organophosphorus insecticides (30-min incubation at 20°C). Symbols represent the mean and standard deviation of three independent incubations. Substrates for carboxylesterase assay as in Fig. 1.

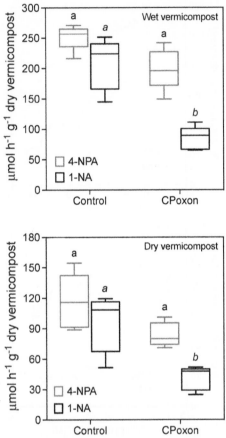

Fig. 3. Effect of chlorpyrifos-oxon (CPoxon) treatment (10 µg/g dry mass; incubated thermostatized chamber for two days at 15°C and continuous dark) on the activity of carboxylesterase in wet (82% H_2O, w/v) and dry (8% H_2O, w/v) vermicompost. Tukey box plots indicate the median, the 25th and 75th percentiles (box edges), and the range (whiskers) of 5 samples. Significant differences between treatments are indicated by different letters (normal typeface for 4-NPA and italics for 1-NA) after Kruskal-Wallis test followed by a post hoc Mann-Whitney test ($p < 0.05$).

biodegradation. Although significant advances have been made in different research laboratories, the use of vermicompost for the bioremediation of contaminated soils, or as a preventive measure to avoid the impact of pesticides on soil function remains to be validate.

Concluding Remarks and Knowledge Gaps

Vermicomposting is a low cost, simple and eco-friendly technology for generating value-added products (vermicompost) that can be used as fertilizers and as reactive substrates for the effective removal of soil contaminants. In this context, vermicomposting technology provides a dual role in protecting the environment from chemical stressors. First, the joint function of microorganisms and earthworms during the decomposition of organic matter provides a unique scenario for the removal of pollutants from agroindustrial and municipal waste (detoxification capability). Vermicomposting alters the chemical speciation of metals and thus decreases their bioavailability. In addition, the accumulation of metals by earthworms contributes to reducing the concentration of metals in the final vermicompost. This bioaccumulation process also occurs with organic contaminants present in the agroindustrial and municipal waste used as feedstock. However, in the case of organic pollutants, biodegradation is the main route of dissipating pollutants. Second, vermicompost is a microbiologically and enzymatically active substrate with a large abundance of organic ligands (e.g., humic substances), both characteristics of which are ideal for exploitation in the biostimulation and bioaugmentation of contaminated soils.

Nevertheless, the dual role of vermicomposting demands future research, particularly when new families of contaminants challenge the current technology for reclaiming wastewater and solid waste. Nowadays, the so-called emerging pollutants (pharmaceuticals, personal care products, flame retardant chemicals, etc.), as well as nanomaterials and micro(nano)plastics are commonly detected in sewage sludges (Mahon et al. 2017, Vicent et al. 2013, Hurley and Nizzetto 2018, Boix et al. 2016). In the following subsections, we suggest two lines of research aimed at increasing the understanding of the potential applications of vermicomposting, and vermicompost, in environmental pollution.

Toxicity of Vermicomposting Feedstock to Earthworms and Microorganisms

To date, most vermicompost studies have dealt with changes in earthworm population dynamics (number of adults, juvenile and cocoons per m^3 substrate) and in microbial communities, in relation to the physicochemical properties of feedstock. However, very few studies have considered the toxicity of the raw material, particularly agroindustrial and municipal waste. Particular attention has been given to the impact of metals on earthworms, which is often assessed in terms of bioaccumulation (Yadav and Garg 2011). The induction of metal-binding proteins (e.g., metallothioneins) in *E. fetida* and *Lampito mauritii* during vermicomposting, thus favoring the accumulation of metals in the earthworm tissues, has been described (Goswami et al. 2014, Goswami et al. 2016). Apart from these studies, no data are available on the molecular mechanisms involved in toxicity and tolerance of earthworms to contaminated feedstock. The lack of information is more evident in relation to the microbiota involved in vermicomposting.

Fungi are well represented during the vermicomposting process. For example, Anastasi et al. (2005) isolated up to 142 entities in vermicompost obtained from a mixture of 70% dung from cows, poultry and other animals and 30% plant debris with *Lumbricus rubellus* as the composting earthworm species, compared to 118 entities isolated from composting plant debris for 6 months. In addition, a laboratory experiment by Aira et al. (2006) demonstrated that fungal communities increased during vermicomposting of fresh pig slurry using *E. fetida*. Because fungi are important in the bioremediation of contaminated soils, they should be considered key biological entities in the decomposition of organic pollutants present in raw material, or in the immobilization of metals. Indeed, fungi have multiple systems for immobilizing or accumulating metals (e.g., complexation with glomalin, sorption to cell wall chitin and chitosan, binding to siderophores, storage in vacuoles, complexation with metallothioneins and phytochelatins, formation of organometals, among others), and for degrading organic contaminants (e.g., exoenzymes) (Harms et al. 2011). The role of fungi in removing contaminants from feedstock during vermicomposting should be investigated in the coming years as a possible detoxifying function of the process. In this respect, many ecotoxicological approaches and methods (ranging from biomarker measurements to standardized toxicity testing) could be exploited in the assessment of the feedstock toxicity during vermicomposting.

Vermicomposting in Microplastic Research

The terms "microplastic" (0.1 μm–5 mm size) and "nanoplastic" (< 0.1 μm) refer to particles of synthetic and semisynthetic materials, which are generated by fragmentation and degradation of polymer-based materials or are engineered as supplements in cosmetic products and other goods (Ng et al. 2018). Nowadays, micro(nano)plastics are pollutant entities of global concern in all environmental compartments and ecosystems. Municipal wastewater treatment plants are an important source of micro(nano)plastics in the environments. Sewage sludge is often applied to farmland and probably represents the major source of micro(nano)plastics to agricultural soils (Mahon et al. 2017). Because sewage sludge is often used as a feedstock in vermicomposting, the questions arise as to whether vermicomposting could reduce the occurrence of micro(nano)plastics in the final vermicompost and whether vermicomposting could be used as potentially means of producing plastic degraders. The previously mentioned study by Anastasi et al. (2005), which examined the bacterial and fungal diversity in both compost and vermicompost, opens an exciting line of work in the field of vermicomposting for environmental protection. Many of the microorganisms identified in compost and vermicompost have been characterized as potential degraders of synthetic polymers (Shah et al. 2008, Sivan 2011, Bhardwaj et al. 2013, Lambert et al. 2014, Ojha et al. 2017, Osman et al. 2018). Moreover, the levels of carboxylesterase activity, which seems to play an important role in biodegrading polyester polymers (Bhardwaj et al. 2013, Zumstein et al. 2017), are particularly high in vermicompost (Sanchez-Hernandez and Domínguez 2017, Domínguez et al. 2017).

Acknowledgements

We thank the Ministerio de Economía y Competitividad, Spanish Government (grants no. CTM2011-25788/TECNO and CTM2014-53915-R), and the Regional Government of Castilla-La Mancha (grant no. PEII-2014-001-P) for the financial support of this research.

References

Abbaspour, A. and A. Golchin. 2011. Immobilization of heavy metals in a contaminated soil in Iran using di-ammonium phosphate, vermicompost and zeolite. Environ. Earth Sci. 63(5): 935–943.

Aira, M., F. Monroy and J. Domínguez. 2006. *Eisenia fetida* (Oligochaeta, Lumbricidae) activates fungal growth, triggering cellulose decomposition during vermicomposting. Microb. Ecol. 52: 738–747.

Aira, M., F. Monroy and J. Domínguez. 2007. Earthworms strongly modify microbial biomass and activity triggering enzymatic activities during vermicomposting independently of the application rates of pig slurry. Sci. Total Environ. 385: 252–261.

Anastasi, A., G.C. Varese and V. Filipello Marchisio. 2005. Isolation and identification of fungal communities in compost and vermicompost. Mycologia 97(1): 33–44.

Bakar, A.A., N.Z. Mahmood, J.A. Teixeira da Silva, N. Abdullah and A.A. Jamaludin. 2011. Vermicomposting of sewage sludge by *Lumbricus rubellus* using spent mushroom compost as feed material: Effect on concentration of heavy metals. Biotechnol. Bioproc. E. 16(5): 1036–1043.

Bender, S.F., C. Wagg and M.G.A. van der Heijden. 2016. An underground revolution: biodiversity and soil ecological engineering for agricultural sustainability. Trends Ecol. Evol. 31(6): 440–452.

Bhardwaj, H., R. Gupta and A. Tiwari. 2013. Communities of microbial enzymes associated with biodegradation of plastics. J. Polym. Environ. 21(2): 575–579.

Boix, C., M. Ibáñez, D. Fabregat-Safont, E. Morales, L. Pastor, J.V. Sancho et al. 2016. Behaviour of emerging contaminants in sewage sludge after anaerobic digestion. Chemosphere 163: 296–304.

Bornscheuer, U.T. 2002. Microbial carboxyl esterases: classification, properties and application in biocatalysis. FEMS Microbiol. Rev. 26(1): 73–81.

Castillo Diaz, J.M., L. Delgado-Moreno, R. Núñez, R. Nogales and E. Romero. 2016. Enhancing pesticide degradation using indigenous microorganisms isolated under high pesticide load in bioremediation systems with vermicomposts. Bioresource Technol. 214: 234–241.

Castillo, J.M., R. Nogales and E. Romero. 2014. Biodegradation of 3,4 dichloroaniline by fungal isolated from the preconditioning phase of winery wastes subjected to vermicomposting. J. Hazard Mater. 267: 119–127.

Chaparro, J.M., A.M. Sheflin, D.K. Manter and J.M. Vivanco. 2012. Manipulating the soil microbiome to increase soil health and plant fertility. Biol. Fert. Soils 48(5): 489–499.

Chapman, P.M. 2007. Heavy metal—music, not science. Environ. Sci. Technol. 41, 6C.

Di Gennaro, P., B. Moreno, E. Annoni, S. García-Rodríguez, G. Bestetti and E. Benitez. 2009. Dynamic changes in bacterial community structure and in naphthalene dioxygenase expression in vermicompost-amended PAH-contaminated soils. J. Hazard Mater. 172(2-3): 1464–1469.

Domínguez, J., J.C. Sanchez-Hernandez and M. Lores. 2017. Vermicomposting of winemaking by-products. pp. 55–78. *In*: Galanakis, G. (ed.). Handbook of Grape Processing By-Products. Elsevier, London, UK.

Dores-Silva, P.R., J.A.O. Cotta, M.D. Landgraf and M.O.O. Rezende. 2018. Soils impacted by PAHs: Would the stabilized organic matter be a green tool for the immobilization of these noxious compounds. J. Environ. Sci. Health B 53(5): 313–318.

Duffus, J.H. 2002. Heavy metals a meaningless term. Pure Appl. Chem. 74(5): 793–807.

Edwards, C.A., N.Q. Arancon and R.L. Sherman. 2010. Vermiculture Technology: Earthworms, Organic Wastes, and Environmental Management. CRC Press, Taylor & Francis Group, Boca Raton, Fl, USA.

Elvira, C., L. Sampedro, E. Benitez and R. Nogales. 1998. Vermicomposting of sludges from paper mill and dairy industries with *Eisenia andrei*: A pilot-scale study. Bioresource Technol. 63(3): 205–211.

Fernández-Bayo, J.D., R. Nogales and E. Romero. 2007. Improved retention of imidacloprid (Confidor) in soils by adding vermicompost from spent grape marc. Sci. Total Environ. 378(1-2): 95–100.

Fernández-Bayo, J.D., R. Nogales and E. Romero. 2009. Assessment of three vermicomposts as organic amendments used to enhance diuron sorption in soils with low organic carbon content. Eur. J. Soil Sci. 60(6): 935–944.

Fernández-Bayo, J.D., R. Nogales and E. Romero. 2015. Winery vermicomposts to control the leaching of diuron, imidacloprid and their metabolites: role of dissolved organic carbon content. J. Environ. Sci. Health B 50(3): 190–200.

Fernández-Gómez, M.J., E. Romero and R. Nogales. 2011. Impact of imidacloprid residues on the development of Eisenia fetida during vermicomposting of greenhouse plant waste. J. Hazard Mater. 192(3): 1886–1889.

Fernández-Gómez, M.J., M. Quirantes, A. Vivas and R. Nogales. 2012. Vermicomposts and/or arbuscular mycorrhizal fungal inoculation in relation to metal availability and biochemical quality of a soil contaminated with heavy metals. Water Air Soil Poll. 223(5): 2707–2718.

Gershater, M.C. and R. Edwards. 2007. Regulating biological activity in plants with carboxylesterases Plant Sci. 173(6): 579–588.

Gershater, M.C., I. Cummins and R. Edwards. 2007. Role of a carboxylesterase in herbicide bioactivation in *Arabidopsis thaliana*. J. Biol. Chem. 282(29): 21460–21466.

Gómez-Brandón, M., M. Aira, M. Lores and J. Domínguez. 2011. Changes in microbial community structure and function during vermicomposting of pig slurry. Bioresource Technol. 102(5): 4171–4178.

Goswami, L., S. Sarkar, S. Mukherjee, S. Das, S. Barman, P. Raul et al. 2014. Vermicomposting of tea factory coal ash: Metal accumulation and metallothionein response in *Eisenia fetida* (Savigny) and *Lampito mauritii* (Kinberg). Bioresource Technol. 166: 96–102.

Goswami, L., S. Pratihar, S. Dasgupta, P. Bhattacharyya, P. Mudoi, J. Bora et al. 2016. Exploring metal detoxification and accumulation potential during vermicomposting of tea factory coal ash: sequential extraction and fluorescence probe analysis. Sci. Rep. 6: 30402.

Harms, H., D. Schlosser and L.Y. Wick. 2011. Untapped potential: Exploiting fungi in bioremediation of hazardous chemicals. Nat. Rev. Microbiol. 9(3): 177–192.

He, X., Y. Zhang, M. Shen, G. Zeng, M. Zhou and M. Li. 2016. Effect of vermicomposting on concentration and speciation of heavy metals in sewage sludge with additive materials. Bioresource Technol. 218: 867–873.

He, X., Y. Zhang, M. Shen, Y. Tian, K. Zheng and G. Zeng. 2017. Vermicompost as a natural adsorbent: Evaluation of simultaneous metals (Pb, Cd) and tetracycline adsorption by sewage sludge-derived vermicompost. Environ. Sci. Pollut. Res. Int. 24(9): 8375–8384.

Hodson, M.E. 2004. Heavy metals—geochemical bogey men. Environ. Pollut. 129(3): 341–343.

Hoehne, L., C.V.S. de Lima, M.C. Martini, T. Altmayer, D.T. Brietzke, J. Finatto et al. 2016. Addition of vermicompost to heavy metal-contaminated soil increases the ability of black oat (*Avena strigosa* Schreb) plants to remove Cd, Cr, and Pb. Water Air Soil Poll. 227(12).

Hübner, R., K.B. Astin and R.J.H. Herbert. 2010. 'Heavy metal'—time to move on from semantics to pragmatics. J. Environ. Monitor. 12(8): 1511.

Hurley, R.R. and L. Nizzetto. 2018. Fate and occurrence of micro(nano)plastics in soils: Knowledge gaps and possible risks. Curr. Opin. Environ. Sci. Health 1: 6–11.

Lambert, S., C. Sinclair and A. Boxall. 2014. Occurrence, degradation, and effect of polymer-based materials in the environment. Rev. Environ. Contam. Toxicol. 227: 1–53.

Lv, B., M. Xing and J. Yang. 2016. Speciation and transformation of heavy metals during vermicomposting of animal manure. Bioresource Technol. 209: 397–401.

Mahon, A.M., B. O'Connell, M.G. Healy, I. O'Connor, R. Officer, R. Nash et al. 2017. Microplastics in sewage sludge: Effects of treatment. Environ. Sci. Technol. 51(2): 810–818.

Megharaj, M., B. Ramakrishnan, K. Venkateswarlu, N. Sethunathan and R. Naidu. 2011. Bioremediation approaches for organic pollutants: a critical perspective. Environ. Int. 37(8): 1362–1375.

Nannipieri, P., P. Sequi and P. Fusi. 1996. Humus and enzyme activity. pp. 293–328. *In*: Piccolo, A. (ed.). Humic Substances in Terrestrial Ecosystems. Elsevier. Science B.V., Amsterdam, Netherlands.

Ng, E.-L., E. Huerta Lwanga, S.M. Eldridge, P. Johnston, H.-W. Hu, V. Geissen et al. 2018. An overview of microplastic and nanoplastic pollution in agroecosystems. Sci. Total Environ. 627: 1377–1388.

Ojha, N., N. Pradhan, S. Singh, A. Barla, A. Shrivastava, P. Khatua et al. 2017. Evaluation of HDPE and LDPE degradation by fungus, implemented by statistical optimization. Sci. Rep. 7: 39515.

Osman, M., S.M. Satti, A. Luqman, F. Hasan, Z. Shah and A.A. Shah. 2018. Degradation of polyester polyurethane by *Aspergillus* sp. strain S45 isolated from soil. J. Polym. Environ. 26(1): 301–310.

Pourret, O. and J.C. Bollinger. 2018. Heavy metal—What to do now: To use or not to use. Sci. Total Environ. 610-611: 419–420.

Putten, W.H., R.D. Bardgett, J.D. Bever, T.M. Bezemer, B.B. Casper, T. Fukami et al. 2013. Plant–soil feedbacks: The past, the present and future challenges. J. Ecol. 101(2): 265–276.

Ravindran, B. and P.N.S. Mnkeni. 2017. Identification and fate of antibiotic residue degradation during composting and vermicomposting of chicken manure. Int. J. Environ. Sci. Te. 14(2): 263–270.

Sahariah, B., L. Goswami, K.H. Kim, P. Bhattacharyya and S.S. Bhattacharya. 2015. Metal remediation and biodegradation potential of earthworm species on municipal solid waste: A parallel analysis between *Metaphire posthuma* and *Eisenia fetida*. Bioresour. Technol. 180: 230–236.

Sanchez-Hernandez, J.C. and J. Domínguez. 2017. Vermicompost derived from spent coffee grounds: assessing the potential for enzymatic bioremediation. pp. 369–398. *In*: Galanakis, G. (ed.). Handbook of Coffee Processing By-Products. Elsevier, London, UK.

Sanchez-Hernandez, J.C., M. Sandoval and A. Pierart. 2017. Short-term response of soil enzyme activities in a chlorpyrifos-treated mesocosm: Use of enzyme-based indexes. Ecol. Indic. 73: 525–535.

Sanchez-Hernandez, J.C., J. Notario Del Pino, Y. Capowiez, C. Mazzia and M. Rault. 2018. Soil enzyme dynamics in chlorpyrifos-treated soils under the influence of earthworms. Sci. Total Environ. 612: 1407–1416.

Satyanarayana, T. and L.W. Getzin. 1973. Properties of a stable cell-free esterase from soil. Biochemistry 12(8): 1566–1572.

Shah, A.A., F. Hasan, A. Hameed and S. Ahmed. 2008. Biological degradation of plastics: A comprehensive review. Biotechnol. Adv. 26(3): 246–265.

Singh, B. 2014. Review on microbial carboxylesterase: general properties and role in organophosphate pesticides degradation. Biochem. Mol. Biol. 2: 1–6.

Singh, J. and A.S. Kalamdhad. 2013. Effect of *Eisenia fetida* on speciation of heavy metals during vermicomposting of water hyacinth. Ecol. Eng. 60: 214–223.

Singh, J. and A. Kaur. 2015. Vermicompost as a strong buffer and natural adsorbent for reducing transition metals, BOD, COD from industrial effluent. Ecol. Eng. 74: 13–19.

Sivan, A. 2011. New perspectives in plastic biodegradation. Curr. Opin. Biotechnol. 22(3): 422–426.

Sogorb, M.A. and E. Vilanova. 2002. Enzymes involved in the detoxification of organophosphorus, carbamate and pyrethroid insecticides through hydrolysis. Toxicol. Lett. 128(1-3): 215–228.

Swati, A. and S. Hait. 2017. Fate and bioavailability of heavy metals during vermicomposting of various organic wastes—A review. Process. Saf. Environ. 109: 30–45.

Tyagi, M., M.M. da Fonseca and C.C. de Carvalho. 2011. Bioaugmentation and biostimulation strategies to improve the effectiveness of bioremediation processes. Biodegradation 22(2): 231–241.

van de Voorde, T.F.J., W.H. van der Putten and T.M. Bezemer. 2012. Soil inoculation method determines the strength of plant–soil interactions. Soil Biol. Biochem. 55: 1–6.

Vašíčková, J., B. Maňáková, M. Šudoma and J. Hofman. 2016. Ecotoxicity of arsenic contaminated sludge after mixing with soils and addition into composting and vermicomposting processes. J. Hazard Mater. 317: 585–592.

Vicent, T., G. Caminal, E. Eljarrat and D. Barceló. 2013. Emerging Organic Contaminants in Sludges: Analysis, Fate and Biological Treatment. Springer-Verlag Berlin Heidelberg, New York, USA.

Villalobos-Maldonado, J.J., R. Meza-Gordillo, N.A. Mancilla-Margalli, T.R. Ayora-Talavera, M.A. Rodríguez-Mendiola, C. Arias-Castro et al. 2015. Removal of decachlorobiphenyl in vermicomposting process amended with rabbit manure and peat moss. Water Air Soil Poll. 226(5).

Wang, Y., Y. Xu, D. Li, B. Tang, S. Man, Y. Jia et al. 2018. Vermicompost and biochar as bio-conditioners to immobilize heavy metal and improve soil fertility on cadmium contaminated soil under acid rain stress. Sci. Total Environ. 621: 1057–1065.

Wheelock, C.E., G. Shan and J. Ottea. 2005. Overview of carboxylesterases and their role in the metabolism of insecticides. J. Pestic. Sci. 30(2): 75–83.

Wright, D.A. and P. Welbourn. 2002. Environmental Toxicology. Cambridge University Press, Cambridge, UK.

Wubs, E.R.J., W.H. van der Putten, M. Bosch and T.M. Bezemer. 2016. Soil inoculation steers restoration of terrestrial ecosystems. Nat. Plants 2(8): 16107.

Yadav, A. and V.K. Garg. 2011. Industrial wastes and sludges management by vermicomposting. Rev. Environ. Sci. Bio. 10(3): 243–276.

Zhu, W., W. Du, X. Shen, H. Zhang and Y. Ding. 2017. Comparative adsorption of Pb^{2+} and Cd^{2+} by cow manure and its vermicompost. Environ. Pollut. 227: 89–97.

Zumstein, M.T., D. Rechsteiner, N. Roduner, V. Perz, D. Ribitsch, G.M. Guebitz et al. 2017. Enzymatic hydrolysis of polyester thin films at the nanoscale: Effects of polyester structure and enzyme active-site accessibility. Environ. Sci. Technol. 51(13): 7476–7485.

Part 3
Biological Methodologies for Monitoring Bioremediation

Chapter 12

Soil Enzyme Activities for Soil Quality Assessment

Liliana Gianfreda and *Maria A. Rao**

INTRODUCTION

Whatever is the condition of a soil (whether healthy or disturbed, polluted or restored, fertile or unfertile), it is of paramount importance to assess its quality to manage soil resources for sustainable future use. Indeed, soil quality evaluation is a tool to improve soil management and land use system. Therefore, two aspects of this matter need to be analyzed: to well identify and define soil quality and to find reliable, reproducible, low-cost and environmentally friendly indicators capable of monitoring the status of a soil and its changes. A soil quality indicator should be a property easily measurable, sensitive and quickly responsive to changing its value, even due to small changes of soil parameters and properties, induced by natural and/ or anthropogenic alterations. A limitation of any selected indicator is to establish the threshold, above or below which, its response is indicative of a significant change occurring or having occurred in soil.

Soil quality assessment is much more important when a remediation program has to be implemented for restoring a polluted environment. Indeed, to assess the results of restoration of polluted sites it is mandatory not only to measure the remaining amount of pollutants still present in the restored environment, or their transformation in non-toxic end products, but also to evaluate the quality of sites in terms of their safety and regain of their original properties. Moreover, it is important to monitor whether and how soil biological functions have been affected by, during and after the process.

Dipartimento di Agraria, Università di Napoli Federico II, Portici (Napoli), Italy.
* Corresponding author: mariarao@unina.it

As claimed by Rao et al. (2014a) "to maintain increasing levels of production and at the same time to remediate the adverse effects of environmental pollution, good monitoring and reliable indicators of soil quality are required. Consequently, it is mandatory to have a number of properties that supply vital information concerning structure, function and composition of the soil system. In addition, they should behave as early and sensitive indicators of soil ecological stress or restoration and should optimally differentiate the extent of soil pollution and/or soil remediation". Enzymes may be helpful as biosensors of pollution, as well as bioindicators because they may be utilized to reveal the presence of pollutants and quantify their concentrations as well as to evaluate the health and quality status of the polluted/restored target environment.

This chapter will try to give a brief overview on: (i) the basic concepts of soil heath and soil indicators, (ii) enzyme activities suitable to detect soil changes and examples related mainly to polluted soils, and (iii) numerical indexes and some models based on enzyme activities. Future perspectives will be also addressed to stimulate the research on the understanding of soil enzyme dynamic in the context of environmental challenges in the coming years (e.g., global climate change, biofuel cropping).

Soil Quality

Soil quality, often used as synonymous of soil health and vice versa, is a complex topic and its definition depends on the different perspectives on the utilization and value of soils. Because of its long genesis, soil is a vital natural resource, non-renewable on a human time scale and its quality is susceptible of chemical, physical and biological alterations by natural events and anthropogenic activities (Gianfreda and Bollag 1996, Gianfreda and Ruggiero 2006). Therefore, soil quality can change and deteriorate by poor management, natural adverse events, pollution by indiscriminate use of agrochemicals. Each easily available, technique restoring key soil functions will contribute to improve and/or stabilize it.

Depending on the utilization of soils, several definitions of soil quality have been attempted. All refer to the capability of soils to perform useful services for human life and to influence land and ecosystems. In the early 1997, a broad definition of soil quality was given by the Soil Science Society of America (SSSA): "The ability of a specific type of soil to function within natural or managed ecosystem boundaries, to sustain plant and animal productivity, maintain or improve air quality and water to support human health and livable" (Karlen et al. 1997). Previously, Parr et al. (1992) and Doran and Parkin (1994) claimed that soil quality is "the capability of soil to produce safe and nutritious crops in a sustained manner over the long-term, and to enhance human and animal health, without impairing the natural resource base or harming the environment" (Parr et al. 1992) and "the capacity of a soil to function within ecosystem boundaries to sustain biological productivity, maintain environmental quality, and promote plant and animal health" (Doran and Parkin 1994). A simpler definition of soil quality, related above all to anthropic demand, is to evaluate how useful is soil for the functions humans desire to perform (Schjønning et al. 2003).

Indicators of Soil Quality

Several static and dynamic soil properties and processes may behave as indicators of soil quality. Static soil properties are the intrinsic characteristics (soil texture, mineralogy, and classification) of a given site affected by geologic history and climatic conditions. Moreover, factors like topography, hydrology, and climate influence productivity and environmental quality of a soil. The measurement of these properties is crucial when a polluted site has to be managed. Indeed, the first step of any remediation program is the exact characterization of the site along with the individualization of nature, quantity and complexity of pollutants in the site. Dynamic soil properties change their values over relatively short time periods (e.g., months, years, and decades). Obviously, these properties are of pivotal importance when assessing dynamic soil quality. Indeed, they often change rapidly, and even considerably, in response to climate and anthropogenic variations. Soil enzyme activities, organic matter content, microbial activity and biomass, meso- and macro-fauna communities, are only some examples of dynamic soil properties (Franzlubbers and Haney 2006).

Indicators can be also classified as physical, chemical or biological indicators. Physical indicators include depth, bulk density, porosity, aggregate stability, texture and compaction whereas chemical indicators include pH, salinity, organic matter content, phosphorus availability, cation exchange capacity, nutrient cycling, as well as contaminants such as potentially toxic elements, organic pollutants, radioactive substances, etc. These indicators are important when soil-plant-related organisms, nutrient availability, water for plants, and mobility of contaminants are of concern.

Micro- and macro-organisms (earthworms, nematodes, termites, ants, microbial biomass, fungi, actinomycetes, or lichens), their activities and ecological functions are also considered biological indicators (Garcia-Ruiz et al. 2009). Other, nonetheless important, biological indicators are metabolic processes (soil respiration and associated-parameters as C-biomass, N-biomass, potentially mineralizable-N, C/N ratios, and so on) or metabolic products of organisms such as enzymes, which are important catalysts in the degradation of substrates and mineralization of organic matter.

To be a useful indicator, a property must respond to some general criteria and requirements (Table 1) (Doran and Safley 1997, Doran and Zeiss 2000, Dale and Beyler 2001). Indeed, it has to be efficient in (a) monitoring chemical, physical or biological and biochemical changes of a soil as the result of anthropic and environmental effects; (b) making evident and quantifiable these changes; (c) providing a quick response to the occurring changes.

Usually, a minimum data set (MDS) for soil quality assessment is considered to reliably evaluating the capacity of soil to perform some defined functions. It has also to monitor changes of soil quality over time, following changed environmental conditions, management practices or pollution events. Due to different soil properties, use or conditions, the elements included in a given MDS are not universal and their choice depends on the specific soil and its utilization (Andrews et al. 2004).

As underlined by Karaca et al. (2011), soil quality indicators to include in a MDS should meet criteria listed in Table 1. They should be easily measurable and

Table 1. Criteria and requirements for a good indicator of soil quality.

Criteria	Requirements
Easily and cheaply measurable in a wide range of different soils, possible to do control assays.	Representative, reproducible, accessible, economic, reliable (*low values of systematic errors*), operative, compatible and adaptable to different types of soils.
Sensitive but not labile.	Sensitive to environmental changes to be measured and non-labile (*i.e., its changes correspond to changes scientifically valid*).
Standardized and validated at a general scientific level.	Measurable with standard units, enabling unambiguous comparisons between soils.

interpretable, responsive to changes in soil function, integrate the effects of physical, biological and chemical properties and processes, cost-effective, valid under different management and environmental conditions, and available to a large number of users with different knowledge. In addition, when selecting indicators of soil quality, time and expense needed for a correct and unfailing evaluation should be considered.

Moreover, when assessing changes in soil quality, selected soil indicators should be able to evaluate the resistance of soils to degradation and their resilience to regain initial quality levels after a period of their declining. In both aspects, dynamic soil indicators can be helpful because the extent of their change and/or their fast or slow return to the initial level may give useful information on the resistance of soils to disturbance and their resilience after a period of poor management. In the first case, if the selected property acting as dynamic indicator is low resistant to disturbance, long-lasting and harmful variations of soil functionality will be measured. By contrast, a high resistant property will less change to indicate a strong functional capability of soils even after several disturbing events (Zaborowska et al. 2015).

Biological Properties and Enzymatic Activities as Soil Indicators

Microorganisms and enzymatic activities may behave as efficient biological indicators or bioindicators of soil quality because they seem to meet most of requirements of a suitable soil indicator (Table 1). Indeed, they are capable of revealing the biological and biochemical changes of soil caused by anthropogenic and environmental factors (such as climatic changes, addition of fertilizers, presence of pollutants, agricultural management or land use) and providing a fast-measurable response to such changes.

Among soil biological properties, soil enzyme activities have been suggested as suitable indicators of soil quality dynamic. They are a measure of the soil microbial activity. Therefore, they are strictly related to the nutrient cycles and transformations, are influenced by natural and anthropogenic factors (Gianfreda and Bollag 1996), rapidly respond to ecosystem variations and changes in soil use and management and, being easily measurable, they may provide an early warning tool to follow environmental modifications (Rao et al. 2014a,b and references therein). Their high degree of variability is, however, one of the main limitations to their use as indicators of soil quality because contradictory results in different studies might result (Cardoso et al. 2013). Likewise, extracellular enzymes may be differently bound to inorganic and organic components of soil, where they can long persist although microbial activity and community change. Therefore, this fraction of soil enzyme activity, usually so-called soil-bound or stabilized enzymes, may interfere when these biochemical endpoints are used as indicators of short-term environmental changes associated to microbial activity. Moreover, current methodologies for activity measurement provide potential rather than real soil enzyme activities. Therefore, it could be difficult to use absolute soil enzyme activity values as a powerful tool for understanding soil biological processes, particularly when there are not standardized protocols for measuring potential enzyme activities that facilitate data interpretation among laboratories (Deng et al. 2017).

Several reports are available in literature on enzyme activities as bioindicators of soils affected by natural or anthropogenic events. In particular, soil enzyme activities seem to be good candidates for examining the quality of polluted environments and their response to the remediation/restoration process (see reviews by Dick 1994, 1997, Gianfreda and Bollag 1996, Nannipieri et al. 2002, 2012, Gianfreda and Ruggiero 2006, Dawson et al. 2007, Gianfreda and Rao 2008, Karaca et al. 2011, Burns et al. 2013, Rao et al. 2014a,b). Although single enzymes like dehydrogenase, β-glucosidase, phosphatase or urease have been identified as indicators of soil quality (Martinez-Salgado et al. 2010), according to the MDS requirements, more than one enzyme should be analyzed, along with either physical and chemical properties, to have a good and precise picture of soil quality and its changes (Lessard et al. 2014). Often kinetics of soil enzyme activities has been utilized as index of soil quality to compare different ecosystems (Moscatelli et al. 2009, Kujur and Kumar Patel 2014) or after reclamation in a chronosequence coal mine overburden spoil (Maharana and Patel 2013). The absolute values of kinetic parameters such as V_{max} and Michaelis constant K_m as well as the catalytic efficiency (V_{max}/K_m) along with the activity of some enzymes were used to monitor the evolution of microbiological activity in soil, and hence they were considered early and useful index of soil quality.

Several are the possible mechanisms justifying the variations of enzyme activities in soil perturbed by pollutants (Gianfreda and Rao 2011a,b, Rao et al. 2014a,b). Environmental contaminants may act on soil enzyme activities by direct, either reversible or irreversible, and indirect effects. Direct reversible or irreversible actions may influence the catalytic active site of the enzyme as well as the protein conformation. Indirect effects are the result of the toxic action of pollutants on the growth and activity of soil microorganisms. Therefore, altered production of enzymes will occur by considerable changes in size, structure, and functionality of the microbial community. In this context, few examples, selected among those prevalently related to polluted or restored soils and recently published, will be reported and commented in the following sections.

Enzyme Activities as Bioindicators of Soil Polluted by Potentially Toxic Elements

Contamination by potentially toxic elements (PTEs) from anthropogenic activities may be a widespread phenomenon in many urban or industrial areas. Particularly, mining is generally the main cause of high PTE concentrations in the environment, with serious consequences for agriculture. As an example, Chile is one of the countries with a high mining activity in the world, and consequently this highly contaminating activity has led a legacy of soil PTE pollution in many agroecosystems (Verdejo et al. 2015, Corradini et al. 2017). In this scenario, it becomes necessary to develop and implement assessment tools, based on soil quality indicators, for monitoring PTE bioavailability and toxicity in the Chilean agricultural soils. Monitoring of soil enzyme activities in these heavily impacted agricultural soils is a suitable approach in the decision making related to soil remediation and agricultural soil management.

PTEs may interact even irreversibly with soil microbial communities and their functions. However, PTEs are naturally present in the environment and are essential for living organisms, because they perform important biological functions acting as cofactors of enzymes or participating in metabolic pathways. Nevertheless, at high concentrations, PTEs become dangerous to the environment and organisms, affecting microbial and enzymatic diversity (Gianfreda and Rao 2016). Since soil can be considered the natural and preferred sink for contamination, measurement of soil enzyme activities may be appropriate for discriminating among different levels of pollution, resulting useful indicators for detecting PTE pollution in soils. PTE bioavailability usually governs their impact on soil microbial processes, and consequently on their activities (Giller et al. 1998). It depends on several properties of soil (e.g., pH, organic matter content, clay and iron oxide content, plant exudates) and on the nature and chemical speciation of PTEs (i.e., by its different chemical forms present in soil, being the free form the only one available to microorganisms and therefore the main responsible of affecting microbial processes) (D'Ascoli et al. 2006).

Zaborowska et al. (2016) measured the activities of six enzymes (dehydrogenases, urease, acid phosphatase, alkaline phosphatase, arylsulfatase, and catalase) along with the counts of twelve microorganism groups in soils exposed at different times (25 and 50 d) to aqueous solutions of Sn^{2+}, Co^{2+} and Mo^{5+}, which were applied at different doses (0, 25, 50, 100, 200, 400, and 800 mg kg^{-1} soil dw). Results indicated a PTE-specific sensibility of enzymes in the order dehydrogenase > urease > arylsulfatase > alkaline phosphatase > acid phosphatase > catalase, thus suggesting both dehydrogenase and urease activities as the most reliable soil health indicators. The activities of the same six enzymes were also monitored at increasing levels of Cd and in the presence of additives, able to mitigate the effect of PTEs. Additives were used to evaluate soil restoration and to calculate indicators of soil resistance (RS) and soil resilience (RL) (Zaborowska et al. 2015). In this case, dehydrogenase and urease were also the most sensitive to Cd effect and additive function (Zaborowska et al. 2015). The activity of dehydrogenase was still the most sensitive biochemical endpoint to the toxicity of PTEs, especially Cu, Zn and Cd, in four representative paddy fields distributed at different towns in China (northern Hunan Province). The enzyme responses were due to long-term irrigation with the nearby stream water contaminated by mining wastewater, thus supporting the use of dehydrogenases as eco-indicator of soil pollution in the study areas (Hu et al. 2014).

The usefulness of urease (and invertase) activity was also confirmed for estimating the progress of the rehabilitation processes at a Zn, Cd and Pb ore mining and the surrounding processing area (Ciarkowska et al. 2014). Results indicated, however, that the response of this enzymatic activity was strongly dependent on the content and decomposition of organic matter in the soil. The influence of organic matter and other soil properties, in general, on soil enzyme activities was confirmed by studies performed in PTE-contaminated soils characterized by various properties (Xian et al. 2015). In these studies, the experiments were conducted to determine the joint effects of PTEs and soil properties on enzyme activities. Results showed that arylsulfatase was the most sensitive soil enzymatic activity and could be utilized as indicator of the enzymatic toxicity of PTEs under various soil properties. Even in this case, soil organic matter was the dominant factor affecting the activity of the enzyme in soil (Xian et al. 2015).

Soil enzyme activities could be significant indicators of PTE toxicity and bioavailability assessment in soil (Sang-Hwan et al. 2009, Ruggiero et al. 2011, Xian et al. 2015, Mahbub et al. 2016). For instant, no or negligible effect by Hg on soil enzyme activities was measured in an agricultural soil for long time heavily polluted by inorganic contamination and with a high Hg concentration (Ruggiero et al. 2011). Possibly, microflora and consequently its activity, as expressed by enzymes, adapted to Hg contamination as this metalloid was present in non-bioavailable forms. Characterization of the site in terms of total Hg concentration and its speciation indicated that Hg was present mainly in insoluble forms, thus being less bioavailable (Cattani et al. 2008, Ruggiero et al. 2011).

Studies performed on three soil types, i.e., neutral, alkaline and acidic, experimentally contaminated with nine different concentrations of Hg (0, 5, 10, 50, 100, 150, 200, 250, 300 mg kg^{-1}) showed that significant inhibition by PTEs on soil microbial activities, as assessed by dehydrogenase and nitrification rate, occurred (Mahbub et al. 2016). As the effective concentrations of Hg that reduced 10% (EC10) of enzyme activity were lower than the available safe limits for inorganic Hg contamination, authors concluded that the existing guideline values are inadequate (Mahbub et al. 2016).

In experimental field studies performed on soil contaminated by PTEs (Cd, Cr, Cu, Pb, Ni, and Zn) from sewage sludge application, the activity of urease and arginine ammonification increased with increasing sludge in soil as respect to the control soil. Both soil enzyme activities were not inhibited by DTPA-extractable Cd, Cr, Cu, Pb, Ni, and Zn in the sludge treated soil. These results indicated that the recycling of sewage sludge, applied at a rate of 200 ton/ha for three successive growth seasons, avoided accumulation of PTEs in soil to such levels to hinder soil biochemical activity (Òlkmnb et al. 2016).

Hagmann et al. (2015) demonstrated long-term adaptation of soil microbial communities to PTE contamination. The activity of alkaline phosphatase, cellobiohydrolase, and L-leucine-amino-peptidase was used as bioindicator of soil pollution. These enzymes were measured across four plots in an un-remediated, urban brownfield site in Jersey City (New Jersey, USA). The soils of the site developed over the last 150 yrs through the dumping of urban fill from New York City as well as industrial rail use. This study area, abandoned and fenced in the late 1960s, had a gradient of PTE concentration, including As, Pb, Cr, Cu, Zn, and V. Unexpectedly and differently on results reported in literature, enzyme activities were highest at soils with the greatest PTE contents, thus indicating that soil microbial communities adapted to PTE contamination after long-term exposure. Taken together these studies evidenced that the use of a single enzyme activity as bioindicator of soil pollution may be an inappropriate strategy. Therefore, it is highly recommended the use of multiple enzyme activities covering more than one biogeochemical cycle for a more comprehensive assessment of pollution impact on soil quality (Lessard et al. 2014).

The resort to a "core enzyme", i.e., a group of enzymes, oxidoreductases activities (rather than a specific enzyme activity), was introduced by Yang et al. (2016) to "simply and universally indicate the PTE pollution degrees of different environments". In that study, PTE concentrations and enzyme activities (oxidoreductases and hydrolases) of soil samples collected from 47 sites exposed to long-term mining activities were examined. Pb, Cd, Zn and As were the primary PTE pollutants and polyphenoloxidase significantly correlated with almost all the PTE ($p < 0.05$), this oxidoreductase being the most sensitive among the target soil enzymes. Furthermore, hydrolase activities involved in C, N, P and S cycles were suggested as supplement to better correlate PTE indication in various polluted environments.

Similarly, Wahsha et al. (2017) used a consortium of different microbial enzymes as an early warning tool for the assessment and monitoring of soil quality by PTE pollution. Investigations were performed on soils collected from an abandoned mine area in northeast Italy and containing different PTEs (Cr, Cu, Fe, Mn, Ni, Pb, and Zn). Tested enzymes (arylsulfatase, leucine aminopeptidase, β-glucosidase, alkaline phosphatase, and chitinase) were all negatively affected by PTEs and their activities varied proportionally with the concentrations of PTEs in soils. Results, clearly, demonstrated that "the assessment of changes in the activities of microbial community can precede some detectable changes in soil chemical and physical properties, and strongly support the utility of using enzymatic activity of soil microbial communities as an early warning tool for monitoring soil pollution with PTEs" (Wahsha et al. 2017).

Enzyme Activities as Bioindicators of Soil Polluted by Organic Contaminants

A large number of organic substances such as petroleum hydrocarbons, halogenated and nitro aromatic compounds, phthalate esters, solvents, and pesticides with severe polluting and hazardous properties are released into the environment by natural and anthropogenic activities. Many of these chemicals may affect soil microbial diversity, resulting in a change of soil biological functionality. The activity of soil enzymes produced by a particular group of microorganisms and involved in a given biological function may be also altered with consequent negative effects on the correct performance of nutrient biogeochemical cycles (Gianfreda et al. 2016). Therefore, even in the case of soil polluted by organic compounds, enzymatic activities may assist in the assessment of soil health.

One of most common sources of soil organic contamination is the use of reclaimed wastewater and sludge from wastewater treatment plants for soil irrigation and fertilization, respectively. Usually, wastewaters are rich of organic pollutants (and PTEs) and may severely affect soil biological functions and its quality. For example, municipal waste water contains minerals and nutrients such as nitrogen and phosphorus as well as other organic compounds that can produce beneficial effects on soil microbial communities and, in turn, promote soil fertility. Nevertheless, the transfer of metals, pharmaceuticals and even pathogens could be not excluded and directly affects soil microbial diversity and activity (Ibekwe et al. 2018).

As an example, Dindar et al. (2015) evaluated the activities of dehydrogenase, alkaline phosphatase, urease, and β-glucosidase as quality indicators of a soil amended with municipal wastewater sludge at different application rates (50, 100 and 200 t dry sludge ha^{-1}). Increases by 9 up to 70% of the four enzymatic activities were found, thus indicating a beneficial effect of the sludge on soil biochemical properties. In particular, sludge amendment strongly increased urease activity, resulting this enzyme as the best suited quality indicator for measuring potential changes in sludge-amended soil. However, this potential positive effect on soil enzymatic activities can become adverse in those situations of continuous industrial wastewater input. Microbial activity, as assessed by basal soil respiration and *in situ* enzyme activities, was deleteriously affected by long-term (20 yrs) industrial waste effluent (IWE) pollution (Subrahmanyam et al. 2016). In that field study, some genetic studies such as community composition profiling of soil bacteria using 16S rRNA gene amplification, and denaturing gradient gel electrophoresis (DGGE) method, revealed that a shift of bacterial community towards more tolerant phyla occurred in IWE-affected soils. Authors suggested that specific bacterial phyla along with soil enzyme activities could be used as relevant biological indicators for long-term pollution assessment of soil quality (Subrahmanyam et al. 2016).

Olive oil mill wastewater (OMW), a by-product derived from olive oil production, is usually used for fertigation in the Mediterranean area. This organic waste contains an elevated, heterogeneous load of different phenolic compounds with high levels of toxicity and antibacterial activity. Although its use is quite limited to countries of the Mediterranean basin, soils in that area are subjected to potential pollution by phenols, also because OMW is massively used in those countries affected by severe water-deficiency. Accordingly, chemical and biological properties of these soils, and therefore their microbial and enzymatic activities, may be severely affected, even irreversibly. For example, temporary and permanent changes in several chemical and biochemical soil properties occurred when a soil, representative of lands with dramatic water insufficiency and very poor in organic matter content, was amended in a laboratory model system with different OMW amounts, applied as such and after the removal of its toxic phenolic compounds (dephenolized OMW) (Piotrowska et al. 2006, 2011). In that laboratory study, tested soil enzymatic activities (hydrolases and oxido-reductases) were very sensitive to the presence of phenolic compounds as assessed by the lower inhibitory effect measured with dephenolized OMW. The results showed the capability of soil enzyme activities to give information on the intensity and on kind and duration of the effects of pollutants on the metabolic activity of soil.

Soil pollution by polycyclic aromatic hydrocarbons (PAHs) is another issue of current concern in agriculture. Combustion of fossil fuel and biomass, atmospheric deposition of PAH-bound particles, flooding, wastewater irrigation, and accidental spills are among the main sources of PAH contamination in agricultural soils (Sun et al. 2018). However, contamination of soils by these organic contaminants appears to be more significant in urban areas, road traffic being the primary source of PAHs. In the last decade, there has been a progressive growth of urban agriculture worldwide, which is defined as the production of crop and livestock within and around cities and towns (Zezza and Tasciotti 2010). Undoubtedly, this kind of sustainable agriculture demands specific and sensitive bioindicators of soil pollution to guarantee the quality of urban soil to support safe crops. The measurement of soil enzyme activities could be suitable indicators of soil deterioration by PAHs and other petroleum-derived hydrocarbons. For example, the activities of eight enzymes (dehydrogenase, arylsulfatase, urease, phosphatase, invertase, β-glucosidase, *o*-diphenol oxidase, and fluorescein diacetate hydrolase) were measured in an agricultural soil and some uncultivated soils from various parts of Europe (Andreoni et al. 2004, Gianfreda et al. 2005). The agricultural soil has received repeated flooding over by uncontrolled urban and industrial wastes, whereas uncultivated soils had a different hydrocarbon-pollution history by organic contaminants (phenanthrene and other PAHs). Highly productive, not subjected to any flooding event, and uncultivated, unpolluted soils served as control, respectively (Andreoni et al. 2004, Gianfreda et al. 2005). Results indicated that enzyme activities of the agricultural soil flooded by wastes were mainly higher than in the control soil. Uncultivated polluted soils showed very low values or a full absence of enzymatic activity with a direct negative relationship between degree of pollution and enzyme activity levels. Even in this case, organic pollution caused a bacterial diversity as assessed by DGGE profiles of the 16S rDNA genes in heavily polluted soils, thus confirming the capability of biological properties, including enzymes, to be sensitive to the extent of soil pollution (Andreoni et al. 2004, Gianfreda et al. 2005).

The negative impact of PAH and/or phenol pollution on soil biological and enzymatic activities and the consequent use of these latter as indicators of soil quality, even after restoration processes, is supported by findings available in literature, that for brevity are not commented here (Diez et al. 2006, Dawson et al. 2007, Scelza et al. 2007, 2008, Pérez-Leblic et al. 2012).

Enzyme activities have been also used to detect soil quality in the presence of pesticides (Schaffer 1993, Gianfreda and Rao, 2008, 2011, 2016). Usually, divergent and contradictory research findings are reported in literature. This is ascribed to several factors like chemical nature of pesticides, their concentration, structure and nature of microbial community, type and conditions of soil (Utobo and Tewari 2015). For instance, acetamiprid and carbofuran differently affected urease activity depending on the amount of the two insecticides applied to the soil (Mohiddin et al. 2015). Moreover, urease activity seemed more sensitive to the carbamate insecticide because its stimulation was higher at lower concentrations of carbofuran than acetamipirid. These results support that it is not possible to predict the impact of a given pesticide on soil enzyme activities. Likewise, this study shows that it is highly risky to draw conclusions about pesticide impact on soil function by the measuring of a single enzyme activity. The study by Jyot et al. (2015) is a clear example of what important is to use multiple enzyme activities in the assessment of soil pollution by agrochemicals. In that study, no effect on urease activity was found when thiamethoxam was used as seed treatment. By contrast, this neonicotinoid insecticide had an inhibitory effect on the activity of both dehydrogenase and phosphatase, thus indicating that thiamethoxam might have a significant impact on the biochemical performance of soil by an enzyme-specific effect (Jyot et al. 2015). Similar results were obtained with the herbicide butachlor that caused a different response in the activity of β-glucosidase, alkaline phosphatase and urease in an agroecosystem under rice-wheat summer fallow crop rotation (Singh and Ghoshal 2013).

The activity of dehydrogenases, catalase, urease, acid phosphatase, and alkaline phosphatase to Falcon® 460EC exposure (an emulsifiable concentrate containing the active ingredients of the triazole fungicides tebuconazole and triadimenol) was measured in sandy loam soil at different times after fungicide treatment (Baćmaga et al. 2016). Falcon® 460EC was applied at the dose recommended by the manufacturer, 30-fold, 150-fold and 300-fold higher than the recommended dose. Microbial colony development index and the eco-physiological index were measured together with soil enzymes as bioindicators. The fungicide inhibited the activity of dehydrogenases, catalase, urease, acid phosphatase, and alkaline phosphatase and modified the biological diversity of the analyzed groups of soil microorganisms. The most dramatic changes occurred at the highest fungicide dose, although dehydrogenase was the most resistant to the fungicide toxicity. Still in this case, analyzed microbiological and biochemical parameters could be considered reliable indicators of the fungicide's toxic effects on soil quality (Baćmaga et al. 2016). Similar inhibitory effects were observed with the new fungicide boscalid, which inhibited the activity of four soil enzymes (i.e., phenol-oxidase, peroxidase, phosphatase, and β-glucosidase) over a period of 60 days, although some positive effects on soil phenol oxidase and peroxidase occurred in a long-term scale (Dan et al. 2014).

Numerical Indexes and Models Based on Enzymatic Activities

Several enzyme activities and other microbial properties well reflect the healthy status of a soil. As discussed in the previous sections, soil quality may be estimated by observing or measuring different properties or processes. Many of these indicators can be rationally used to develop soil quality indexes. However, individual soil biochemical properties may be ineffective for revealing the overall state of a soil because they are environmentally fluctuating variables. Moreover, they usually reflect local or regional conditions where they were developed, so very often they cannot be applied at a global scale (Bastida et al. 2008). However, the wide variation occurring in individual soil properties can be compensated by integrating several properties in a numerical value, i.e., an index capable of differentiating between soils with different quality features or affected by environmental stressors such as pollution or land use.

Another important issue linked to enzyme activities as bioindicators is their ecological role in chemico-biological processes of the soil. Indeed, complex enzymatic reactions take place within living (mainly microbial) cells and in the soil solution. A continual interplay of mineralization and immobilization processes brought by the activities of intracellular and extracellular enzymes affects the availability of nutrients in the soil, affecting ultimately the crop production. To identify and quantify the cause-effect relationship between the activity level of a given enzyme and the biochemical fate of a nutrient, and consequently to have a picture of how these processes affect soil quality, several simulation models have been developed. They examine the influence of competition, nutrient availability, and spatial structure on microbial growth and enzyme synthesis (Gianfreda et al. 2011, and references herein).

Numerical Indexes

The main goal of a valuable soil quality index should be its capability to condense in a single and simple numerical index the achieved information. With this scope, the definition of index related to soil quality requires:

- the choice of the quality criteria, to which the index values refer, and the selection of reliable properties;
- the definition of reference criteria;
- the selection of reliable properties sensitive to changes in management practices and environmental stresses, by which to build up the index;
- the application of an appropriate numerical algorithm to the development of the index.

The numerical index should be sensitive, accurate and capable of quantifying and evaluating changes in soil quality in a short time-scale (Bastida et al. 2008). Empirical relationships, often based on multiple regression models and combining different soil properties, have been developed as useful indexes of soil biochemical quality (Table 2). In some cases, they considered just one single enzyme activity or category of enzyme activities as hydrolases and oxidases. In other cases, more complex indexes combine biological and biochemical properties such as enzyme activities, microbial C- and N-biomass.

Table 2 lists a large number of indexes until now developed. Usually, their validity has to be tested by applying them to soils characterized by different properties and quality status. Details on each of indexes reported in Table 2, their application and effectiveness in evaluating soil quality are summarized in Gianfreda and Ruggiero (2006) and Rao and Gianfreda (2014b). All these indexes include enzyme activities as reliable properties of the soil. Interestingly, what stands out in the Table 2 is that the most represented enzyme activities are dehydrogenase and phosphatase (9 cases), and β-glucosidase and urease (8 times). This observation seems to support that these enzymes, that are involved in the P-, C- and N-cycling, are mostly related to soil quality.

The usefulness of quality indexes, developed by canonical discriminant analysis (CDA), and validated on published data sets dealing with enzymatic activities in different soils and characterized by different alteration events (Puglisi et al. 2006), has been recently confirmed in the studies by Innangi et al. (2017) and Igalavithana et al. (2017).

Innangi et al. (2017) applied the soil alteration AI 3 index (Puglisi et al. 2006, No. 1 in Table 2) to agricultural soils exposed for long term to olive pomace used as a soil amendment. Olive pomace could have detrimental effects by salts and its high content of phenolic substances. However, the results of that study indicated

Table 2. Indexes of soil quality integrating soil enzyme activities (as modified from Rao et al. 2014b).

No.	Index (*)	Purpose	References
1	Soil Alteration Index AI 1 = -21.30 arylsulphatase AR $+ 35.2$ β-glucosidase $- 10.20$ phosphatase $- 0.52$ urease $- 4.53$ invertase $+ 14.3$ dehydrogenase $+ 0.003$ phenoloxidase. Soil Alteration Index AI 2 = 36.18 β-glucosidase $- 8.72$ phosphatase $- 0.48$ urease $- 4.19$ invertase. Soil Alteration Index AI 3 = 7.87 β-glucosidase $- 8.22$ phosphatase $- 0.49$ urease	Effects on the quality of agricultural soils contaminated with industrial and municipal wastes, organic fertilisation or irrigation with poor quality water under different crops (*Ficus carica*, maize, tomato, etc.)	Puglisi et al. (2006)
2	Enzymatic Activity Number (EAN) = 0.2 (0.15 dehydrogenase + catalase $+ 1.25 \times 10^{-5}$ phosphatase $+ 4 \times 10^{-5}$ protease $+ 6 \times 10^{-4}$ protease)	Effect in cultivated and forest soils and pastures. Effect of soil management on its quality	Beck (1984)
3	Soil Quality Index (SEI) = 0.26 dehydrogenase $+ 0.05$ alkaline phosphatase $- 0.25$ β-glucosidase $- 0.18$ arylsulphatase $+ 0.02$ urease	Assessment of agricultural urban soil's quality	Igalavithana et al. (2017)
4	Biological Index of Fertility (BIF) = (1.5 dehydrogenase $+ 100$ k catalase · catalase)/2	Effect in untilled management systems (natural grassland and orange grooves) comparing to tilled systems in south-eastern Sicily (Italy)	Stefanic (1984)
5	Biochemical Index of Soil Fertility (B) = $C_{org} + N_{total}$ + dehydrogenase + phosphatase + protease	Effect of organic and mineral fertilisation	Koper and Piotrowska (2003)
6	Organic C = -0.4008 arylsulphatase $+ 0.4153$ dehydrogenase $+ 0.4033$ phosphatase $+ 0.4916$ β-glucosidase	Evaluation of soil in different states of degradation	De la Paz-Jiménez et al. (2002)
7	Total N = 0.38×10^{-3} microbial biomass-C $+ 1.4 \times 10^{-3}$ mineral N$+ 13.6 \times 10^{-3}$ phosphatase $+ 8.9 \times 10^{-3}$ β-glucosidase $+ 1.6 \times 10^{-3}$ urease	Evaluation of soils under climax vegetation	Trasar-Cepeda et al. (1998)
8	Total N = 0.44 available phosphate $+ 0.017$ WHC $+ 0.410$ phosphatase $- 0.567$ urease $+ 0.001$ microbial biomass-C $+ 0.419$ β-glucosidase $- 0.980$	Valid for Mollisol. Evaluation of forests soils under natural vegetation without human intervention	Zornoza et al. (2007)
9	Organic C content = 4.247 available phosphate $+ 8.185$ β-glucosidase $+ 7.949$ Urease $+ 17.333$	Valid for Entisol. Evaluation of forests soils under natural vegetation without human intervention	Zornoza et al. (2007)
10	Microbial Degradation Index (MID) = $[0.89 \, (1/1+ (\text{dehydrogenase}/4.87)^{-2.5})] + [0.86 \, (1/1+ (\text{water-soluble carbohydrates}/11.09)^{-2.5})] + [0.84 \, (1/1+ (\text{urease}/1.79)^{-2.5})] + [0.72 \, (1/1+ (\text{respiration}/18.01)^{-2.5})]$	Assessment of semiarid degraded soils	Bastida et al. (2006)

Table 2 contd. ...

...Table 2 contd.

No.	Index (*)	Purpose	References
11	Relative Soil Stability Index (RSSI) $= \dfrac{\int_{day\ 2}^{day\ 15} \text{EAperturbed}(t)dt}{\int_{day\ 2}^{day\ 15} \text{EAcontrol}(t)dt}\,100$	Effects of the herbicide 2,4-dichlorofenoxiacetic acid on soil functional stability	Becaert et al. (2006)
12	Biological Quality Index (BQI) Total C (calculated) $= -2.924 + 0.037$ hot water extractable C -0.096 cellulase $+0.081$ dehydrogenase $+ 0.009$ respiration	Variation in relation to the ecosystem degradation	Armas et al. (2007)
13	Soil Quality Index including microbial biomass C, respiration, dehydrogenase activity, seeds germination and earthworms (**)	Evaluation of recuperation of hydrocarbon contaminated soils by nutrient applications, surfactants or soil agitation	Dawson et al. (2007)
14	GMean (geometric mean) index	High values indicate high microbial functional diversity. Evaluation of soil quality in chlorpyrifos-treated soils	Hinojosa et al. (2004), Lessard et al. (2014), Sanchez-Hernandez et al. (2017)
15	WMean (weighted mean) index	High values indicate high microbial functional diversity. Evaluation of soil quality in chlorpyrifos-treated soils	Lessard et al. (2014)
16	T-SQI (treated-soil quality index)	Assessment of soil quality in soil intentionally treated with organic or inorganic amendments to increase its biological activity. Measurement of increase or inhibition of changes caused by an environmental stressor on soil enzyme activities	Mijangos et al. (2010), Sanchez-Hernandez et al. (2017)
17	IBRv2 (integrated biological response) index	Index used for assessing animal's health inhabiting contaminated sites. It integrates the response of multiple biomarkers to evaluate the health status of organisms living in contaminated environments. Soil enzyme activities have been used as biological responses in this index instead of animal biomarkers	Sanchez et al. (2012), Sanchez-Hernandez et al. (2017)

C_{org} (organic carbon content, %); N_{total} (total nitrogen content, %); EA = enzyme activity (enzymes measured = arylsulphatase, β-glucosidase, urease, protease, acid and alkaline phosphatase). (*) For the units in which each enzyme activity or soil properties were expressed see references. (**) For the development of the Soil Quality Index see Andrews et al. (2004).

a sharp improvement in quality in the amended soil compared to its non-amended counterpart, and this effect was well revealed by the trend of the AI 3 index (lower values corresponding to better quality), thus confirming the validity of this index as useful tool to measure soil quality in organic amended soils (Innangi et al. 2017).

According to the approach by Puglisi et al. (2006), Igalavithana et al. (2017) developed using CDA a numerical soil enzyme index (SEI) based on the activities of five soil enzymes (dehydrogenase, alkaline phosphatase, β-glucosidase, arylsulfatase, and urease) (No. 3 in Table 2). This index was applied to agricultural urban soils, belonging to five different textural classes, clay loam, loam, loamy sand, sandy loam, and silt loam. Sandy loam soils had the highest frequency (n = 27) and were used to develop the numerical index of soil quality. The remaining nine soils with different textural classes were used to validate the index. The SEI well discriminated between the two soil groups. Moreover, it was also validated using published data sets including contaminated and non-contaminated soils, and altered and unaltered soils mainly due to management practices (Igalavithana et al. 2017).

Investigations performed with a combination of three herbicides (diflufenican, mesosulfuron-methyl and iodosulfuron-methyl-sodium) on the activity of dehydrogenase, alkaline phosphatase, arylsulfatase, and β-glucosidase in a sandy loam soil confirmed the validity of dehydrogenase and urease as valuable indicators of soil quality. Indeed, both enzymes were inhibited by the herbicide mixture at its highest applied amount, urease being the most resistant and dehydrogenases the least resistant. A direct negative correlation was found between the values of the biochemical soil fertility indicator (BA21) and the high doses of the herbicide (Bacmaga et al. 2015).

The potential utility and versatility of dehydrogenase in assessing changes in soil quality is further supported by a different approach proposed by Veum et al. (2014). Across a continuum of long-term agricultural practices in Missouri (USA), the authors measured several physical and chemical properties, the activities of dehydrogenase and phenol oxidase, and collected ^{13}C nuclear magnetic resonance (^{13}C NMR), diffuse reflectance Fourier transform (DRIFT) spectra of soil organic matter and visible, near-infrared reflectance (VNIR) spectra of whole soil. They used the SMAF (Soil Management Assessment Framework) soil index, that incorporates several biological, physical, chemical, and nutrient data scoring curves to score and rank all the 12 investigated plots for overall soil quality. Dehydrogenase and phenol oxidase activities as well as the organic matter composition strongly correlated, both positively and negatively, with the SMAF index, thus highlighting the robust link among microbial community and its function, common indicators of soil quality, and properties of soil organic matter.

Sanchez-Hernandez et al. (2017) used a more complex approach to assess the environmental risk of pesticides. They performed a mesocosm study where an Andisol was exposed to two doses (4.8 and 24 kg a.i. ha^{-1}) of chlorpyrifos. After 14 days of pesticide treatment, the activity of the hydrolases carboxylesterase, acid phosphatase, β-glucosidase, urease, protease and the oxidoreductase activity of dehydrogenase and catalase were measured. A strong inhibition of almost all tested activities occurred. The four enzymatic indexes (geometric mean, weighted mean, treated-soil quality index [T-SQI] and integrated biological response [IBRv2] index) (No. 14 – 17 in Table 2), that these authors used, were significantly lower in the chlorpyrifos-sprayed soils compared with controls. Moreover, a significant agreement was found between the T-SQI and IBRv2 scores. The authors concluded that the unknown IBRv2 index in soil biochemistry could be very useful in the assessment of soil pollution by pesticides and future research should be performed to support its potential use as index of soil quality (Sanchez-Hernandez et al. 2017).

Models

The assessment of outcomes from soil enzyme activities or their derived indexes is complicated by the complexity of enzyme location in the soil. It is widely accepted that different intracellular and extracellular enzymatic components, deriving from different origins, contribute quantitatively and qualitatively to the overall enzymatic activity of soil (Gianfreda and Ruggiero 2006, Burns et al. 2013). As mentioned before, a significant component of this plethora of soil enzymes is made by enzymes differently bound to soil inorganic and organic components. The interactions of extracellular enzymes with soil components will affect substantially their behavior and consequently their ecological role.

Moreover, as stated by Nannipieri et al. (2012), current soil enzyme assays provide a measurement of potential enzyme activities, indicative of overall enzyme concentrations. Therefore, the real rates of enzymatic reactions under natural situations remain poorly understood. The limit of existing soil enzymology methodologies may further invalidate the use of enzyme activities as soil quality indicators.

More or less complex models were developed to describe precisely the ecological role of enzymes in soil, to enhance the understanding of *in situ* enzyme activities and consequently to confirm their validity as soil quality indicators. However, in several of these models, the classic enzymatic assays do not allow to discriminate between free and stabilized enzymes, when the presence of these latter fraction can mask a quick response of microbial enzyme allocation determined by new or altered nutrient inputs.

Gianfreda et al. (2011) and Rao et al. (2014b) detailed each of the many investigations performed on this topic, but the main points concluded by various authors can be summarized as follow:

– Complex enzymatic reactions occurring within living (mainly microbial) cells and in the soil solution involve inorganic and organic derivatives of the more essential macronutrients C, N, P and S through a continual interplay of mineralization and immobilization processes. Although nutrients availability and enzyme activities in soil are closely correlated, the cause-effect relationship between the activity level of a given enzyme and the biochemical fate of a nutrient is hardly identified and quantified.

– According to Allison (2005), *cheaters*, i.e., microbes that do not synthesize enzymes but utilize products and compete with the enzyme-producing microorganisms, influence the ecological role of extracellular enzymes in soil. A reduced efficiency of the secreted enzymes for the growth and survival of their originating cells may result.

- As claimed before, several extracellular enzymes may persist in soil as a stable and active (even if reduced) enzymatic pool.
- By a simulation model involving the influence of competition, nutrient availability, and spatial structure on microbial growth and enzyme synthesis, it could be concluded that cheaters are favored by conditions in which the cost of enzyme is high, whereas lower rates of enzyme diffusion favored producers. Moreover, nitrogen supply may limit enzyme production because enzymes with a high content of N-units require high amounts of N sources (Allison 2005).
- According to the economic theory, the production of enzymes is limited by scarce resources and should increase when simple nutrients are insufficient and complex nutrients are abundant. Indeed, "microorganisms produce enzymes according to 'economic rules', but a substantial pool of mineral stabilized or constitutive enzymes mediates this response" (Allison and Vitousek 2005).
- As proposed by Sinsabaugh et al. (2008), an elemental stoichiometric evaluation may exist between the more common measured soil enzymatic activities and all microbial communities. Moreover, by extending the stoichiometry theory at ecological level, it can be concluded that the relative activities of soil enzymes, involved in mineralizing organic C, N, and P, are stoichiometrically and energetically constrained by microbial biomass growth. A biogeochemical equilibrium model, combining the enzyme kinetics and community growth, when multiple resources are limited, with elements of metabolic and ecological stoichiometry theory, may help in understanding the role of enzymes in soil microbial metabolism (Sinsabaugh and Shah 2012).

Conclusions and Future Perspectives

Findings so far commented have demonstrated the suitability of enzyme activities as valid indicators of soil quality. Their involvement in more complex expressions (indexes) and models devoted to well describe biogeochemical importance of soil processes, support the use of soil enzyme activities as reliable tool for the assessment of soil health.

Despite of the great diversity of soil quality indexes, they have never been used on larger scales, nor even in similar climatological or agronomic conditions. Several are the causes: poor standardization of some methodologies; some methods may be technically expensive for some laboratories; spatial scale problems (soil heterogeneity); poor definition of soil natural conditions (climate and vegetation); poor definition of soil function to be tested for soil quality (Bastida et al. 2008).

Although quite recently a book has collected and discussed the most common enzyme activity assays in soil (Dick 2011), standardized, universally accepted methodologies for measuring the activity of soil enzymes are still not available. Moreover, the experimental conditions of enzyme assays (e.g., incubation pH and temperature, assay duration, type and concentration of substrate, presence of cofactors, treatment of the soil before the assay, etc.) are quite different among laboratories. Therefore, inter-comparison exercises of enzyme activity outcomes among laboratories or environmental scenarios is a difficult task. Nowadays, a key research priority in soil enzymology is to optimize the procedures for enzymatic activity determination in order to obtain the best values and indexes according to intrinsic soil properties. This standardization procedure should be mandatory if soil enzyme activities are used as bioindicators of soil quality.

Another difficulty in the use of soil enzyme activities as indicators of soil quality is still linked to the methodology. What measured under laboratory conditions does not represent the real biochemical processes of the soil, but just the potential soil enzyme activity. Efforts have been made to develop rapid, simple and easily adaptable methods for assaying soil enzyme activities using alternative methodologies like hydrodynamic voltammetry (Sazawa and Kuramitz 2015) or the microplate-scale fluorimetric assay as a method improving the effectiveness and the efficiency of measuring universal soil quality indicators using enzymes (Trap et al. 2012). However, these new and promising methodologies are not exempt from criticism. In particular, the fluorimetric assay may present many failing aspects as demonstrated by the discrepancy found when comparing the results obtained with spectroscopic and fluorimetric based approaches (Rao et al. 2014b).

The main goal in the assessment of soil pollution and remediation should be to develop one absolute indicator capable of evaluating soil quality under different soil management systems and under various climatic conditions and geographic regions. This goal seems difficult to achieve because of the intrinsic variability of biological properties and several site-specific factors that affect soil enzyme activity. Emerging, innovative approaches based on molecular biology could, however, overcome many of bottlenecks still present in soil enzymology. As underlined by Rao et al. (2014b), genomic studies recognizing functional genes coding for given enzymes can identify the genetic potential of microorganisms to produce enzymes. Transcriptomic studies of mRNA may help in assessing microbial regulation of enzyme production. Proteomic studies may evaluate the main features (pool sizes, diversity, and microbial source) of soil enzymes producers. Results of these studies may help in understanding the role of enzymes and their microbial sources in the complex biogeochemical cycles of nutrients. Therefore, a clearer picture will achieve on the role and involvement of enzyme activities in soil. More precise and valuable enzyme indicators for the assessment of soil quality may derive.

References

Allison, S.D. 2005. Cheaters, diffusion, and nutrients constrain decomposition by microbial enzymes in spatially structured environments. Ecol. Lett. 8: 626–635.

Allison, S.D. and P.M. Vitousek. 2005. Responses of extracellular enzymes to simple and complex nutrient inputs. Soil Biol. Biochem. 37: 937–944.

Armas, C.M., B. Santana, J.L. Mora, J.S. Notario, C.D. Arbelo and A. Rodríguez-Rodríguez. 2007. 2007. A biological quality index for volcanic Andisols and Aridisols (Canary Islands, Spain): Variations related to the ecosystem development. Sci. Total Environ. 378: 238–244.

Andreoni, V., L. Cavalca, M.A. Rao, G. Nocerino, S. Bernasconi, E. Dell'Amico, M. Colombo and L. Gianfreda. 2004. Bacterial communities and enzymatic activities of PAHs polluted soils. Chemosphere 57: 401–412.

Andrews S.S., D.L. Karlen and C.A. Cambardella. 2004. The soil management assessment framework: A quantitative soil quality evaluation method. Soil Sci. Soc. Am. J. 68: 1945–1962.

Armas, C.M., B. Santana, J.L. Mora, J.S. Notario, C.D. Arbelo and A. Rodríguez-Rodríguez. 2007. A biological quality index for volcanic Andisols and Aridisols (Canary Islands, Spain): variations related to the ecosystem development. Sci. Total Environ. 378: 238–244.

Baćmaga, M., A. Borowik, J. Kucharski, M. Tomkiel and J. Wyszkowska. 2015. Microbial and enzymatic activity of soil contaminated with a mixture of diflufenican + mesosulfuron-methyl + iodosulfuron-methyl-sodium. Environ. Sci. Poll. Res. 22: 643–656.

Baćmaga, M., J. Wyszkowska and J. Kucharski. 2016. The effect of the Falcon 460 EC fungicide on soil microbial communities, enzyme activities and plant growth. Ecotoxicology 25: 1575–1587.

Bastida, F., J.L. Moreno, T. Hernández and C. García. 2006. Microbiological degradation index of soils in a semiarid climate. Soil Biol. Biochem. 38: 3463–3473.

Bastida, F., A. Zsolnay, T. Hernández and C. García. 2008. Past, present and future of soil quality indices: A biological perspective. Geoderma 147: 159–171.

Bécaert, V., R. Samson and L. Deschênes. 2006. Effect of 2,4-D contamination on soil functional stability evaluated using the relative soil stability index (RSSI). Chemosphere 64: 1713–1721.

Beck, T.H. 1984. Methods and application of soil microbiological analysis at the Landensanstalt fur Bodenkultur und Pflanzenbau (LBB) for determination of some aspects of soil fertility. pp. 13–20. *In:* Nemes, M.P., S. Kiss, P. Papacostea, C. Stefanic and M. Rusan (eds.). Proc. 5th Symp. Soil Biol. Rumanian Nat. Soc. Soil Sci. Bucharest, Rumania.

Burns, R.G., J.L. DeForest, J. Marxsen, R.L. Sinsabaugh, M.E. Stromberger, M.D. Wallenstein et al. 2013. Soil enzymes in a changing environment: Current knowledge and future directions. Soil Biol. Biochem. 58: 216–234.

Cardoso, E.J.B.N., R.L.F. Vasconcellos, D. Bini, M.Y.H. Miyauchi, C.A. dos Santos, P.R.L. Alves et al. 2013. Soil health: Looking for suitable indicators. What should be considered to assess the effects of use and management on soil health? Sci. Agric. 70: 274–289.

Cattani, I., S. Spalla, G.M. Beone, A.A. Del Re, R. Boccelli and M. Trevisan. 2008. Characterization of mercury species in soils by HPLC-ICP-MS and measurement of fraction removed by diffusive gradient in thin films. Talanta 74: 1520–1526.

Ciarkowska, K., K. Sołek-Podwika and J. Wieczorek. 2014. Enzyme activity as an indicator of soil-rehabilitation processes at a zinc and lead ore mining and processing area. J. Environ. Manage. 132: 250–256.

Corradini, F., A. Correa, M.S. Moyano, P. Sepúlveda and C. Quiroz. 2017. Nitrate, arsenic, cadmium, and lead concentrations in leafy vegetables: expected average values for productive regions of Chile. Arch. Agron. Soil Sci. 64: 299–317.

Dale, V.H. and S.C. Beyeler. 2001. Challenges in the development and use of ecological indicators. Ecol. Indic. 1: 3–10.

D'Ascoli, R., M.A. Rao, P. Adamo, G. Renella, L. Landi, F.A. Rutigliano et al. 2006. Impact of river overflowing on trace element contamination of volcanic soils in south Italy: part II. Soil biological and biochemical properties in relation to trace element speciation. Environ. Pollut. 144: 317–326.

Dawson, J.J.C., E.J. Godsiffe, I.P. Thompson, T.K. Ralebitso-Senior, K.S. Killham and G.I. Paton. 2007. Application of biological indicators to assess recovery of hydrocarbon impacted soils. Soil Biol. Biochem. 39: 164–177.

De la Paz-Jiménez, M., A.M. De la Horra, L. Pruzzo and R.M. Palma. 2002. Soil quality: A new index base microbiological and biochemical parameters. Biol. Fertil. Soils 35: 302–306.

Deng, S., R. Dick, C. Freeman, E. Kandeler and M.N. Weintraub. 2017. Comparison and standardization of soil enzyme assay for meaningful data interpretation. J. Microbiol. Meth. 133: 32–34.

Dick, W.A. and M.A. Tabatabai. 1993. Significance and potential uses of soil enzymes. pp. 95–125. *In*: Metting, F.B. (ed.). Soil Microbial Ecology: Application in Agricultural and Environmental Management. Marcel Dekker, New York.

Dick, R.P. 1994. Soil enzymes activities as indicators of soil quality. pp. 107–124. *In*: Doran, J.W., D.C. Coleman, D.F. Bezdicek and B.A. Stewart. (eds.). Defining Soil Quality for a Sustainable Environment. SSSA Special Publication no. 35, Madison, WI, USA.

Dick, R.P. 1997. Soil enzyme activities as integrative indicators of soil health. pp. 121–156. *In*: Pankhurst, C.E., B.M. Doube and V.V.S.R. Gupta (eds.). Biological Indicators of Soil Health. CAB International, Wallingford USA.

Dick, R.P. (ed.). 2011. Methods in Soil Enzymology, Soil Science Society of America, Madison, WI.

Diez, M.C., F. Gallardo, G. Saavedra, M. Cea, L. Gianfreda and M. Alvear. 2006. Effect of pentachlorophenol on selected soil enzyme activities in a Chilean andisol. J. Soil Sc. Plant Nutr. 6: 40–51.

Dindar, E., F.O.T. Şa1ban and H.S. Başkaya. 2015. Evaluation of soil enzyme activities as soil quality indicators in sludge-amended soils. J. Environ. Biol. 36: 919–926.

Doran, J.W. and T.B. Parkin. 1994. Defining and assessing soil quality. pp. 3–21. *In:* Doran, J.W., D.C. Coleman, D.F. Bezdicek and B.A. Stewart (eds.). Defining Soil Quality for a Sustainable Environment. Soil Science Society of America Special Publication 35, ASA-SSSA, Madison, WI.

Doran, J.W. and M. Saffley. 1997. Defining and assessing soil health and sustainable productivity. pp. 1–28. *In*: Pankhurst, C.E., B.M. Doube and V.V.S.R. Gupta (eds.). Biological Indicators of Soil Health. CAB International, Wallingford, UK.

Doran, J.W. and M.R. Zeiss. 2000. Soil health and sustainability: managing the biotic component of soil quality. Appl. Soil Ecol. 15: 3–11.

Franzlubbers, A.J. and R.L. Haney. 2006. Assessing soil quality in organic agriculture. USDA Agricultural Research Service. Critical Issue Report: Soil Quality. The Organic Center.

García-Ruiz, R., V. Ochoa, B. Viñegla, M.B. Hinojosa, R. Peña-Santiago, G. Liébanas et al. 2009. Soil enzymes, nematode community and selected physico-chemical properties as soil quality indicators in organic and conventional olive oil farming: Influence of seasonality and site features. Appl. Soil Ecol. 41: 305–314.

Gianfreda, L. and J.-M. Bollag. 1996. Influence of natural and anthropogenic factors on enzyme activity in soil. pp. 123–194. *In:* Stotzky, G. and J.-M. Bollag (eds.). Soil Biochemistry Vol. 9. Marcel Dekker, New York.

Gianfreda, L., M.A. Rao, A. Piotrowska, G. Palombo and C. Colombo. 2005. Soil enzyme activities as affected by anthropogenic alterations: intensive agricultural practices and organic pollution. Sci. Total Environ. 241: 265–279.

Gianfreda, L. and P. Ruggiero. 2006. Enzyme activities in soil. pp. 257–311. *In*: Nannipieri, P. and K. Smalla (eds.). Nucleic Acids and Proteins in Soil. Soil Biology Vol. 8. Springer Verlag, Berlin, Heidelberg.

Gianfreda, L. and M.A. Rao. 2008. Interaction between xenobiotics and microbial and enzymatic soil activity. Crit. Rev. Environ. Sci. Tech. 38: 269–310.

Gianfreda, L. and M.A. Rao. 2011. The influence of pesticides on soil enzymes. pp. 293–312. *In*: Shukla, S. and A. Varma (eds.). Soil Enzymology, Soil Biology vol. 22. Springer Verlag, Berlin, Heidelberg.

Gianfreda, L., M.A. Rao and M. Mora. 2011. Enzymatic activity as influenced by soil mineral and humic colloids and the impact on biogeochemical processes. pp. 5.1–5.24. *In*: Huang, P.M., Y. Li and M.E. Summer (eds.). Handbook of Soil Sciences, Section E Soil Physical, Chemical, and Biological Interfacial Interactions. 2nd edition. Taylor and Francis Group, LLC, Boca Raton.

Gianfreda, L. and M.A. Rao. 2016. Soil microbial and enzymatic diversity as affected by the presence of xenobiotics. pp. 153–170. *In*: Varma, A. and M. Zaffar Hashmi (eds.). Xenobiotics in Soil: Monitoring and Remediation. Soil Biology. Vol. 49. Springer Verlag, Germany.

Giller, K.E., E. Witter and S.P. McGrath. 1998. Toxicity of heavy metals to microorganisms and microbial processes in agricultural soils: A review. Soil Biol. Biochem. 30: 1389–1414.

Gülser, F. and E. Erdoğan. 2008. The effects of heavy metal pollution on enzyme activities and basal soil respiration of roadside soils. Environ. Monit. Assess. 145: 127–133.

Hagmann, D.F., N.M. Goodey, C. Mathieu, J. Evans, M.F.J. Aronson, F. Gallagher et al. 2015. Effect of metal contamination on microbial enzymatic activity in soil. Soil Biol. Biochem. 91: 291–297.

Hinojosa, M.B., R. García-Ruíz, B. Viñegla and J.A. Carreira. 2004. Microbiological rates and enzyme activities as indicators of functionality in soils affected by the Aznalcóllar toxic spill. Soil Biol. Biochem. 36: 1637–1644.

Hu, X.-F., Y. Jiang, Y. Shu, X. Hu, L. Liu and F. Luo. 2014. Effects of mining wastewater discharges on heavy metal pollution and soil enzyme activity of the paddy fields. J. Geochem. Explor. 147: 139–150.

Ibekwe, A.M., A. Gonzalez-Rubio and D.L. Suarez. 2018. Impact of treated wastewater for irrigation on soil microbial communities. Sci. Total Environ. 622–623: 1603–1610.

Igalavithana, A.D., M. Farooq, K.-H. Kim, Y.-H. Lee, M.F. Qayyum, M.I. Al-Wabel et al. 2017. Determining soil quality in urban agricultural regions by soil enzyme-based index. Environ. Geochem. Health https://doi.org/10.1007/s10653-017-9998-2.

Innangi, M., E. Niro, R. D'Ascoli, T. Danise, P. Proietti, L. Nasini et al. 2017. Effects of olive pomace amendment on soil enzyme activities. Appl. Soil Ecol. 119: 242–249.

Jyot, G., K. Mandal and B. Singh. 2015. Effect of dehydrogenase, phosphatase and urease activity in cotton soil after applying thiamethoxam as seed treatment. Environ. Monit. Assess. 187: 298. https://doi.org/10.1007/s10661-015-4432-7.

Karaca, A., S.C. Cetin, O.C. Turgay and R. Kizilkaya. 2011. Soil enzymes as indication of soil quality. pp. 119–148. *In*: Shukla, G. and A. Varma (eds.). Soil Biology. Vol. 22, Soil Enzymology. Springer Verlag, Berlin, Heidelberg.

Karlen, D.L., M.J. Mausbach, J.W. Doran, R.G. Cline, R.F. Harris and G.E. Schuman. 1997. Soil quality: A concept, definition, and framework for evaluation. Soil Sci. Soc. Am. J. 61: 4–10.

Koper, J. and A. Piotrowska. 2003. Application of biochemical index to define soil fertility depending on varied organic and mineral fertilization. Electron. J. Polish Agric. Univ. 6. http://www.ejpau.media.pl/volume6/issue1/agronomy/art-06.html.

Kujur, M. and A. Kumar. 2014. Patel Kinetics of soil enzyme activities under different ecosystems: An index of soil quality. Chilean J. Agric. Res. 74: 96–104.

Leirós, M.C., C. Trasar-Cepeda, F. García-Fernández and F. Gil-Sotres. 1999. Defining the validity of a biochemical index of soil quality. Biol. Fert. Soils 30: 140–146.

Lessard, I., S. Sauvé and L. Deschênes. 2014. Toxicity response of a new enzyme-based functional diversity methodology for Zn-contaminated field-collected soils. Soil Biol. Biochem. 71: 87–94.

Mahbub, K.R., K. Krishnan, M. Megharaj and R. Naidu. 2016. Mercury inhibits soil enzyme activity in a lower concentration than the guideline value. Bull. Environ. Contam. Toxicol. 96: 76–82.

Martinez-Salgado, M.M., V. Gutiérrez-Romero, M. Jannsens and R. Ortega-Blu. 2010. Biological soil quality indicators: A review. pp. 319–328. *In*: Mendez-Vilas, A. (ed.). Current Research, Technology and Education Topics in Apples Microbiology and Microbial Biotechnology. Formatex, Badajoz, Spain.

Mijangos, I., I. Albizu, L. Epelde, I. Amezaga, S. Mendarte and C. Garbisu. 2010. Effects of liming on soil properties and plant performance of temperate mountainous grasslands. J. Environ. Manage. 91: 2066–2074.

Mohiddin, G.J., V. Rangaswamy, B. Manjunatha, D. Rueda, P.F. Salas, M.C.V. Aboleda et al. 2015. Activity of soil urease as influenced by acetamiprid and carbofuran. Pharm. Lett. 7: 147–152.

Moscatelli, M.C., A. Lagomarsino, A.M.V. Garzillo, A. Pignataro and S. Grego. 2012. β-Glucosidase kinetic parameters as indicators of soil quality under conventional and organic cropping systems applying two analytical approaches. Ecol. Indic. 13: 322–327.

Nannipieri, P., E. Kandeler and P. Ruggiero. 2002. Enzyme activities and microbiological and biochemical processes in soil. pp. 1–33. *In*: Burns, R.G. and R.P. Dick (eds.). Enzymes in the Environment: Activity, Ecology and Applications. Marcel Dekker, New York.

Nannipieri, P., L. Giagnoni, G. Renella, E. Puglisi, B. Ceccanti, G. Masciandaro et al. 2012. Soil enzymology: Classical and molecular approaches. Biol. Fertil. Soils 48: 743–762.

Òlkmnb, A., H.E. Elsalam, D.I. Saleh, M. El-Sharnouby, S.F. Mahmoud and E.-H. Yasser. 2016. Soil enzymes activities as bio-indicators for soil contamination by heavy metals from sewage sludge application. Res. J. Pharmac. Biol. Chem. Sci. 7: 3005–3011.

Parr, J.F., R.I. Papendick, S.B. Hornick and R.E. Meyer. 1992. Soil quality: Attributes and relationship to alternative and sustainable agriculture. Am. J. Altern. Agric. 7: 5–11.

Pérez-Leblic, M.I., A. Turmero, M. Hernández, A.J. Hernández, J. Pastor, A.S. Ball et al. 2012. Influence of xenobiotic contaminants on landfill soil microbial activity and diversity. J. Environ. Manage. 95: 285–290.

Perucci, P. 1992. Enzyme-activity and microbial biomass in a field soil amended with municipal refuse. Biol. Fertil. Soils 14: 54–60.

Piotrowska, A., G. Iamarino, M.A. Rao, R. Scotti and L. Gianfreda. 2006. Short-term effects of olive mill waste water (OMW) on chemical and biochemical properties of a semiarid Mediterranean soil. Soil Biol. Biochem. 38: 600–610.

Piotrowska, A., M.A. Rao, R. Scotti and L. Gianfreda. 2011. Changes in soil chemical and biochemical properties following amendment with crude and dephenolized olive mill waste water (OMW). Geoderma 161: 8–17.

Puglisi, E., A.A.M. Del Re, M.A. Rao and L. Gianfreda. 2006. Development and validation of numerical indexes integrating enzyme activities of soils. Soil Biol. Biochem. 38: 1673–1681.

Rao, M.A., R. Scelza, F. Acevedo, M.C. Diez and L. Gianfreda. 2014a. Enzymes as useful tools for environmental purposes. Chemosphere 107: 145–162.

Rao, M.A., R. Scelza and L. Gianfreda. 2014b. Soil enzymes. *In*: Gianfreda, L. and M.A. Rao (eds.). Enzymes in Agricultural Sciences. OMICS eBooks Group.

Ruggiero, P., R. Terzano, M. Spagnuolo, L. Cavalca, M. Colombo, V. Andreoni et al. 2011. Hg bioavailability and impact on bacterial communities in a long-term polluted soil. J. Environ. Monit. 13: 145–156.

Sanchez, W., T. Burgeot and J.-M. Porcher. 2012. A novel integrated biomarker response calculation based on reference deviation concept. Environ. Sci. Pollut. Res. 20: 2721–2725.

Sanchez-Hernandez J.C., M. Sandoval and A. Pierart. 2017. Short-term response of soil enzyme activities in achlorpyrifos-treated mesocosm: Use of enzyme-based indexes. Ecol. Indicat. 73: 525–535.

Sazawa, K. 2015. Hydrodynamic voltammetry as a rapid and simple method for evaluating soil enzyme activities. Sensors (Switzerland) 15: 5331–5343.

Scelza, R., M.A. Rao and L. Gianfreda. 2007. Effects of compost and phenanthrene-degrading cultures on the properties of an artificially phenanthrene-contaminated soil. Soil Biol. Biochem. 39: 1303–1317.

Scelza, R., M.A. Rao and L. Gianfreda. 2008. Response of an agricultural soil to pentachlorophenol (PCP) contamination and the addition of compost or dissolved organic matter. Soil Biol. Biochem. 40: 2162–2169.

Schaffer, A. 1993. Pesticide effects on enzyme activities in the soil ecosystem. pp. 273–340. *In*: Bollag, J.-M. and G. Stotzky (eds.). Soil Biochemistry. Vol. 8. Marcel Dekker, New York.

Schjønning, P., S. Elmholt and B.T. Christensen. 2003. Managing Soil Quality: Challenges in Modern Agriculture. CABI Publishing, Wallingford, UK.

Singh, A. and N. Ghoshal. 2013. Impact of herbicide and various soil amendments on soil enzymes activities in a tropical rainfed agroecosystem. Eur. J. Soil Biol. 54: 56–62.

Sinsabaugh, R., C. Lauber, M. Weintraub, B. Ahmed, S. Allison et al. 2008. Stoichiometry of soil enzyme activity at global scale. Ecol. Lett. 11: 1252–1264.

Sinsabaugh, R.L. and J.J.F. Shah. 2012. Ecoenzymatic stoichiometry and ecological theory. Annu. Rev. Ecol. Evol. Syst. 43: 313–343.

Stefanic, F., G. Ellade and J. Chirnageanu. 1984. Researches concerning a biological index of soil fertility. pp. 35–45. *In:* Nemes, M.P., S. Kiss, P. Papacostea, C. Stefanic and M. Rusan (eds.). Proc. Fifth Symp. Soil Biology. Rumanian Natl. Soc. Soil Sci., Bucharest, Rumania.

Subrahmanyam, G., J.-P. Shen, Y.-R. Liu, G. Archana and L.-M. Zhang. 2016. Effect of long-term industrial waste effluent pollution on soil enzyme activities and bacterial community composition. Environ. Monit. Assess. 188: 112.

Sun, J., L. Pan, D.C.W. Tsang, Y. Zhan, L. Zhu and X. Li. 2018. Organic contamination and remediation in the agricultural soils of China: A critical review. Sci. Total Environ. 615: 724–740.

Trap J., W. Riah, M. Akpa-Vinceslas, C. Bailleul, K. Laval and I. Trinsoutrot-Gattin. 2012. Improved effectiveness and efficiency in measuring soil enzymes as universal soil quality indicators using microplate fluorimetry. Soil Biol. Biochem. 45: 98–101.

Trasar-Cepeda, C., C. Leiros, F. Gil-Sotres and S. Seoane. 1998. Towards a biochemical quality index for soils: An expression relating several biological and biochemical properties. Biol. Fertil. Soils 26: 100–106.

Utobo, E.B. and L. Tewari. 2015. Soil enzymes as bioindicators of soil ecosystem status. Appl. Ecol. Environ. Res. 13: 147–169.

Verdejo, J., R. Ginocchio, S. Sauvé, E. Salgado and A. Neaman. 2015. Thresholds of copper phytotoxicity in field-collected agricultural soils exposed to copper mining activities in Chile. Ecotoxicol. Environ. Saf. 122: 171–177.

Veum, K.S., K.W. Goyne, R.J. Kremer, R.J. Miles and K.A. Sudduth. 2014. Biological indicators of soil quality and soil organic matter characteristics in an agricultural management continuum. Biogeochemistry 117: 81–99.

Xian, Y., M. Wang and W. Chen. 2015. Quantitative assessment on soil enzyme activities of heavy metal contaminated soils with various soil properties. Chemosphere 139: 604–608.

Xiong, D., Y. Li, Y. Xiong, X. Li, Y. Xiao, Z. Qin et al. 2014. Influence of boscalid on the activities of soil enzymes and soil respiration. Eur. J. Soil Biol. 61: 1–5.

Yang, J., F. Yang, Y. Yang, G. Xing, C. Deng, Y. Shen et al. 2016. A proposal of "core enzyme" bioindicator in long-term Pb-Zn ore pollution areas based on topsoil property analysis. Environ. Poll. 213: 760–769.

Zaborowska, M., J. Wyszkowska and J. Kucharski. 2015. The possibilities of restoring the enzymatic balance of soil contaminated with cadmium. Int. J. Environ. Pollut. 58: 197–214.

Zaborowska, M., J. Kucharski and J. Wyszkowska. 2016. Biological activity of soil contaminated with cobalt, tin, and molybdenum. J. Environ. Monit. Assess. 188: 398.

Zezza, A. and L. Tasciotti. 2010. Urban agriculture, poverty, and food security: empirical evidence from a sample of developing countries. Food Policy 35: 265–273.

Zornoza, R., J. Mataix-Solera, C. Guerrero, V. Arcenegui, F. García-Orenes, J. Mataix-Beneyto et al. 2007. Evaluation of soil quality using multiple lineal regression based on physical, chemical and biochemical properties. Sci. Total Environ. 378: 233–237.

Chapter 13

Biomarkers in Soil Organisms
Their Potential use in the Assessment of Soil Pollution and Remediation

Antonio Calisi,[1] *Maria Elena Latino,*[1] *Angelo Corallo,*[1]
Annalisa Grimaldi,[2] *Chiara Ferronato,*[3]
Livia Vittori Antisari[3] and *Francesco Dondero*[4,*]

INTRODUCTION

Soil is defined as "the top layer of the earth's crust, formed by mineral particles, organic matter, water, air and living organisms" (ISO 1996). Therefore, soil represents the interface between geosphere, atmosphere and hydrosphere, and it is a key natural resource for environmental, economic, social and cultural development. This environmental compartment is the habitat for many organisms that distribute between above- and belowground systems, contributing thereby to the ecological relationships between both systems. However, human activities have largely exploited soil resources, causing a slow but constant degradation of soil quality (Lal 2005). In particular, soil pollution has strongly increased during the last century as a consequence of: (1) industrialization, that has boosted depositions of atmospheric contaminants onto the soil; (2) the growing use of chemical products directly or

[1] Department of Engineering for Innovation, University of Salento, Via Prov.le per Monteroni, 73100, Lecce, Italy.
[2] Department of Biotechnology and Life Science, University of Insubria, Via H.J. Dunant 3, 21100, Varese, Italy.
[3] Department of Agricultural Science, University of Bologna, Viale Fanin 40, 40127, Bologna, Italy.
[4] Department of Science and Technological Innovation, Università del Piemonte Orientale, Viale Teresa Michel 11, 15121, Alessandria, Italy.
* Corresponding author: francesco.dondero@uniupo.it

indirectly conferred to lands; and (3) intensive agricultural practices using high amounts of pesticides and fertilizers (Aelion 2004, Ashraf 2014).

Beside point-sources of soil contamination, atmospheric deposition (dry and wet) is another significant route of soil pollution. Dry deposition is characterized by pollutants transferred simply by gravity and interception of topsoil, while wet deposition is characterized by pollutant desorption/resorption or washout during meteoric events (Lagzi et al. 2014). When toxic compounds reach the soil surface, they can be subjected to different processes such as infiltration, immobilization, transformation and accumulation. The growing use of toxic substances has triggered the alteration of soil, reducing its fertility and increasing contamination of crops and groundwater as well as a general disturbance of ecosystem services (Abbasi et al. 2013). Soil pollution can be distinguished in diffuse and point-source according to the source and effects of pollution. Diffuse contamination of soil is caused by the systematic introduction of low amounts of organic and inorganic chemicals from different sources (industrial, municipal, and agricultural activities). Soil diffuse pollution is critical since it can affect large extension of lands, so identification of contamination sources is generally a difficult task. Conversely, point-source contamination is limited to well-defined hot spot (e.g., industrial area, municipal and industrial effluents or landfills) (APAT 2004). Regardless of the nature of soil pollution (diffuse or point-source), contaminated soils elicit changes in the microbiome and soil fauna structure that modify, in turn, the quantity and quality of soil organic matter and nutrient cycles (Ashraf et al. 2014). Likewise, the presence of pollutants in soils above certain levels may lead to their progressive accumulation through the terrestrial food web (Nesheim 2015).

In the last two decades, soil pollution has been a topic of intense research. However, traditional approaches such as those merely based on chemical analyses and the establishment of threshold values are far from being applicable to soil pollution assessment. As reported by some authors, bioavailability of pollutants is the most critical parameter since it can be shaped by soil properties such as pH, texture, water-holding capacity, moisture and organic matter content, greatly affecting the species sensitivity to pollutants (Bradham et al. 2006, Spurgeon et al. 2006, Criel et al. 2008). Accordingly, methods other than the chemical analysis of environmental contaminants are required for a better environmental risk assessment of soil pollution based in the concept of bioavailability. For example, Chapman (1990) proposed the *Sediment Quality Triad* framework for sediment toxicity assessment. This approach consists of three evaluation stages: chemical analysis to determine the level of sediment contamination (chemistry), toxicity testing for exposure-effect assessment (ecotoxicology), and macroinvertebrate community structure (ecology). Nowadays, the emerging strategies for evaluating ecosystem alterations are mainly based on effect-based approaches integrating chemical analysis with a series of prognostic biological markers (Ashraf et al. 2014). The use of biological responses to contaminant exposure, i.e., biomarkers, can help to identify pollutant bioavailability and their potential adverse effects (Forbes et al. 2006). Soil invertebrates are in direct contact with both soil and pore-water (Kammenga et al. 2000), so they are excellent bioindicators of soil pollution when used in combination with biomarker responses to contaminant bioavailability (Hyne and Maher 2003, Weeks et al. 2004).

The aim of this chapter is to provide an overview on the use of biomarkers for soil monitoring and assessment of remediation programs, with particular emphasis in the agroecosystem. The chapter is subdivided into five sections. The first two sections make a distinction between the concepts such a bioindicator, indicator or sentinel species, and biomarker, which are often confused in the scientific literature. The third and fourth sections will provide an overview on what kind of information can be obtained from biomarkers and the most common biomarkers in soil ecotoxicology. The last section will summarize the main concluding remarks on this chapter and provide some lines of future research in the field of ecotoxicological biomarkers.

Soil (Bio)Indicator Species

The term *bioindicator* refers to living organisms, whose presence or absence in a specific site (quantified by community metrics as species abundance and diversity), may be used as an endpoint to assess environmental disturbance (Iserentant and De Sloover 1976, Sanchez-Hernandez 2006). Likewise, *indicator* or *sentinel* species is that organism that through a series of molecular, biochemical and physiological responses can inform us on the environment quality (Amiard Triquet et al. 2013, Dondero and Calisi 2015). Examples of indicator species are collembola (Fountaine and Hopkin 2005, Filser et al. 2013), earthworms (Sanchez Hernandez 2006, Lionetto et al. 2012) or snails (Kammenga et al. 2000, Dallinger et al. 2001). A good indicator species should show a reliable relationship between pollutant exposure and biological responses. However, the quality and intensity of these responses may change in relation to pollutant dose, i.e., exposure time and concentration.

In addition, a frequent term in the ecotoxicological literature is *bioaccumulator* (plant or animal species). This term defines those organisms inhabiting environments with a high pollution level, so they are resistant to pollution, so high concentrations of contaminants can be found in their tissues. For example, Lagadic et al. (2000) define bioaccumulators as "organisms able to survive in the presence of a pollutant absorbed from environmental matrices, allowing a qualification and quantification". Indeed, a great volume of literature has demonstrated a good relationship between contaminant concentrations in the environment and concentrations in bioaccumulator.

Other authors have established similar classification of the organisms used for ecotoxicological purposes. For example, Gherardt (2002) and Dondero and Calisi (2015) distinguish three categories:

1. *Bioindicator*, a species that provides information on pollution in its habitat either through its relative abundance or sub-individual responses at the biochemical, physiological and behavioural levels.

2. *Ecological indicator*, a species that describes the structure, composition and functioning of the biological community, and ecosystem quality in response to stressful or disturbance situations.

3. *Biodiversity indicator*, it refers to taxa whose richness in species can be useful to estimate the biodiversity of a broader biological community.

Regardless the terminology we adopt, a good candidate to be used for assessing environmental pollution should have the following characteristics (Hopkin 1993, Edwards et al. 1995, Doran and Parkin 1996, Gerarardt 2002, Chaffai 2014, Dondero and Calisi 2015):

- Sensitivity to ecosystem changes;
- Representing a key species or having a key role in the ecosystem;
- Wide distribution in the study area;
- Reduced capability of long-range dispersion;
- Long life-cycle;
- Genetic uniformity;
- Known genomic data (sequenced genome);
- Easiness of sampling, sorting, laboratory breeding and manipulation.

On the other hand, a good knowledge of the organism, i.e., its anatomy, physiology and ecology as well as the natural variability of biological response is ideally required (Lowe and Kendal 1990, Niemi and McDonald 2004). These are important variables to a better comprehension of pollution effects, and to distinguish between normal physiological variations and the actual toxic effects of pollutants.

The organisms used as indicator species of soil deterioration appertain at a wide range of life forms such as microorganism (bacteria, fungi, algae), micro-invertebrates (nematodes, mites, springtails), macro-invertebrates (isopods, earthworms, snails, spider, insects) and plants. Particularly, earthworms have received much attention in the last two decades due to their capability for contaminant bioaccumulation and continue exposure to soil contaminants through their gastrointestinal tract and tegument. Earthworm species are generally classified into three categories: epigeic (*Eisenia* spp.), anecic (*Lumbricus terrestris*) and endogeic (*Allolobophora chlorotica*) species. Whereas epigeic and anecic species feed the litter of the top-soil, endogeic species ingest large amounts of soil (geophagus), so selection of the most suitable species for assessing soil pollution will depend on the scope of the monitoring program (contaminants accumulated on soil surface or distributed in the bulk soil). Chapter 9 in this book provides a detailed description about these three ecological groups of earthworms and their potential use in terrestrial ecotoxicology. Nevertheless, and whenever possible, the use of a combination of earthworm species belonging to different ecological habits would represent a more robust strategy for ecological risk assessment of soil pollution.

Biomarkers: An Overview

A biomarker is defined as a variation, induced by a contaminant or contaminant mixture, at the physiological, biochemical, cellular or molecular level of a process, structure or function, that can be measured in a biological system (Moore et al. 2004, Dondero and Calisi 2015).

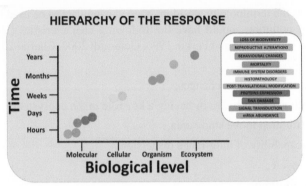

Fig. 1. Biological response hierarchy of stress.

When an organism is exposed to a toxic compound, it may cause a variety of defects or damages at different levels of biological complexity ranging from the molecular till to organism (Sánchez-Bayo et al. 2011). Due to the hierarchical architecture of living organisms, early responses may take place at the protein post-translational modification, mRNA relative abundances or enzyme activity. Depending on the exposure time and the stressor intensity, these early effects can be transduced and detected at higher levels of biological organization such as organelles, cells, tissues and organs. In fact, if ecotoxicology looks at the organism level (i.e., survival) and ecology to population dynamics and community structure, biomarkers rely on previous and less resilient steps of this hierarchy (Fig. 1).

Yet, in response to adverse pollutant impacts, organisms develop a series of adaptive responses aimed to keep homeostasis or join new steady state levels similar to the original conditions. Indeed, homeostatic responses tend to decrease the toxic effect of pollutants through the activation of detoxification systems or excretion mechanisms. Therefore, the set of responses against chemical stress represents the potential biomarkers commonly used in environmental biomonitoring (Kammenga et al. 2000, Hyne and Maher 2003, Weeks et al. 2004, Moore et al. 2004, Dondero and Calisi 2015).

Beside the use of biomarkers for assessing contaminant exposure, the understanding of the dose-response relationship is critical to use biomarkers for a predictive purpose (Dondero and Calisi 2015). Therefore, the elaboration of a dose-response curve based on laboratory experiments, in which biomarker responses are measured in the test organism exposed to serial concentrations of the test chemical, should be a mandatory exercise to gain knowledge on biomarker sensitivity and specificity. In the natural environment, however, it is difficult to establish this kind of single dose-response relationship. By contrary, it is more likely that multiple dose-response relationships occur, each corresponding to different combinations and interactions (e.g., antagonism, synergism, potentiation) of environmental stressors (Peakall and Shugart 1993).

Nowadays, there exist a vast range of biomarkers covering responses at multiple levels of biological organization. However, most of researchers agree in the following biomarker classification:

Table 1. Biomarker classification.

Criterium	Type of Biomarkers		
Impact on individual health	*Exposure:* Not reflecting health state impairment, just exposure or bioavailability of environmental contaminants.	*Effect:* Related to adverse effects. Having direct implications with the health status of the individual.	*Susceptibility:* Related to intrinsic or acquired ability to develop pathological stress.
Identification of chemical stressor	*Generic:* Its response is unspecific to multiple stressors.	*Specific:* Its response identifies a specific class of contaminants.	

According to the Table 1, a first classification is made on the basis of the type and level of interaction between the contaminant and the organism. In this way, biomarkers are classified as:

- Biomarkers of exposure. They can detect solely exposure to contaminants but do not necessarily reflect an impairment of the health state of the organism (Hagger et al. 2006). Most of biomarkers belong to this category, and some examples are the induction of metallothioneins, changes in the activity of antioxidant enzymes, changes in the reduced-to-oxidized glutathione concentrations, and inhibition of butyrylcholinesterase activity.

- Biomarkers of effect. They quantitatively measure the occurrence of adverse effects elicited by pollutant exposure and have direct implications on the health state of the organism (Regoli 2001). Some examples are the inhibition of brain acetylcholinesterase activity, and the DNA damage.

- Biomarkers of susceptibility. They inform on the intrinsic or acquired ability to develop a pathological stress syndrome from chemical exposure (Van der Oost et al. 2003, Wheelock et al. 2008). An example of this group is the carboxylesterase activity, which is implied in the biochemical mechanism of pesticide resistance in some pest species.

Biomarkers are also classified according to the specificity level of their response (Lagadic 2000, Conti 2008):

Specific biomarkers are those detecting the response to a specific class of pollutants. For example, the assessment of metallothioneins can be used as biomarkers to metal exposure and, occasionally, it can inform on the specific type of metal, e.g., Cd or Cu (Dondero et al. 2005, Banni et al. 2007). Another well-known example is the inhibition of acetylcholinesterase activity by organophosphate pesticides (Lionetto et al. 2011).

General biomarkers reflect, however, a generic response to different stressors. They are useful in a first tier of exposure/effect assessment. Some examples of general biomarkers are the measurement of lysosomal membrane stability, the induction of cytochrome P450-dependent monooxygenases, and changes in the activity of antioxidant enzymes (Kammenga et al. 2000, Svendsen et al. 2004, Dondero and Calisi 2015).

Use of Biomarkers

Biomarkers are useful tools in environmental biomonitoring (Kammenga et al. 2000, Weeks et al. 2004, Fontanetti et al. 2011, Chaffai 2014, Dondero and Calisi 2015). Although they do not provide a quantitative assessment of contaminant levels in exposed organisms, biomarkers allow the assessment of health status of individuals, occasionally providing a predictive perspective at population level (biomarkers related to behavior and reproduction). The goal of using biomarkers is the detection of the transition from homeostatic compensation to disease or viceversa (Peakall and Shugart 1993). In this approach, it is fundamental to measure the natural variability of the biomarkers in order to correctly identify changes attributable to chemical stress. For example, many enzymes undergo changes in their activity according to the hormonal state, age and sex of the individual, or environmental factors such as season and temperature.

Traditionally, biomarkers have been used as predictors of adverse effects at population and community levels. But very few examples exist in the ecotoxicological literature that had demonstrated such as functional link (Forbes et al. 2006). Nowadays, biomarkers are used in ecotoxicology as a measurement of contaminant bioavailability, and to gain understanding on the mode of toxic action of pollutants in the environment. This means that biomarkers should have ecological relevance. For example, changes of enzyme activities related to the detoxification of contaminants do not necessarily imply ecological consequences such as the reduction of biodiversity at the contaminated site. In this context, particularly useful are biomarkers directly implicated on survival and reproduction of individuals (Sanchez-Hernandez 2006), and those that measure genotoxicity in gametes (Dondero and Calisi 2015).

The selection of biomarkers in a biomonitoring program is also a critical point. Many authors agree that a multi-biomarker scheme encompassing several levels of biological organization represents the most effective advance. The use of a biomarker battery may provide an integrated response accounting for the various effects of pollutants and, further, it may inform on the mode of toxic action of environmental stressors. Therefore, biomarkers may be applied in three levels of biomonitoring programs:

1. The first stage is the identification of the hazard. Biomarker are used at this stage because the presence and composition of the chemical contamination is rather unknown. The hazard is usually assessed with the use of general and/or effect biomarkers targeting simple (sub)-cellular components such as lysosomal-membrane integrity and detoxifying enzymes.

2. The second stage is the early warning effects. This approach is performed through the use of specific biomarkers allowing the identification of a class of pollutants, and the intensity of the contamination. Biomarker of exposure such as metallothionein, acetylcholinesterase inhibition, cytochrome P450 enzymatic activity (CYP450s) are usually considered at this stage.

3. The third stage is risk prediction. In this case, biomarkers should give a clear indication on the potential long-term consequence at population level. Biomarkers of susceptibility (i.e., carboxylesterase activity) and mutagenicity biomarkers (such as micronuclei frequency) are some examples.

Types of Biomarkers

Beside the classification of biomarkers attending to their role in biomonitoring of polluted sites, or the toxicological meaning of their response, biomarkers are also classified according to the nature of their response (molecular, biochemical, cellular, genetic, and so on). Following the classification by Mc Carthy and Shugart (1990), biomarkers may belong to the following responses:

- Biochemical responses
- Genotoxic responses
- Cytological and histological responses
- Behavioural responses
- "Omics" responses

Biochemical Responses

In this category, we find many enzyme activities involved with the detoxification and toxic mechanisms of pollutants (Kammenga et al. 2000). Some examples of biochemical biomarkers are the CYP450s, metallothioneins, heat shock proteins (HSPs), esterases, multidrug resistance proteins (usually belonging to the ABC transporter superfamily), and several antioxidant enzymes.

In the particular case of the CYP450s, the most important reaction catalysed by this complex multienzymatic system is the oxidation of organic lipophilic compounds such as polycyclic aromatic hydrocarbons (PAHs), aromatic pesticides, polychlorinated (PCBs) and polybrominated biphenyls (PBBs), among others, to increase their water solubility, therefore facilitating their excretion or conjugation with endogenous compounds (e.g., glutathione). The CYP450 system includes substrate-inducible and substrate-specific enzymes. In fact, many studies have demonstrated the induction of specific CYP450s such as the CYP4501A1 isoform in organisms exposed to organic lipophilic chemicals (Conti 2008). Accordingly, CYP4501A1 is one of the most used biomarkers in vertebrates such as fish and birds (van der Oost et al. 2003). However, an accurate selection of the indicator species is critical for the correct use of CYP450 as a biochemical biomarker in the case of terrestrial invertebrates, since the classical enzyme activities commonly used in vertebrates (e.g., ethoxyresorufin *O*-deethylase or EROD, pentoxyresorufin *O*-deethylase or PROD, etc.) may not be workable in several invertebrates (Achazi et al. 1998, Brown et al. 2004). For example, assessment of CYP450 induction in the earthworm *Eisenia fetida* exposed to soil contaminated with benzo(a)pyrene is more evident when the methoxyresorufin *O*-deethylase (MROD) activity is used as reporter (Saint-Denies et al. 1999). Furthermore, there is a clear indication that metals may change the activity of CYP450 system. In a recent work, Cao et al. (2017) demonstrated that CYP450 activity of *E. fetida* as well as the activity of CYP3A4 isoform was modulated by some metals with complex response dynamics depending on the time and dose. Although these studies may give some insight on the capability of earthworms to adapt to polluted soils, they complicate the interpretation of contaminant exposure and effect.

Metallothioneins are ubiquitous cytosolic cysteine-rich proteins (20–30% of cysteine residues), with a low molecular weight (6,000–7,000 Da), and a high capacity for binding essential metal ions (Cu or Zn) and toxic metals (Cd or Hg). In this latter case, metallothioneins may act as molecular scavengers, reducing the toxicity of these metals. These proteins are genetically inducible by the substrate and their cytosolic increase represents a specific and sensitive biomarker of metal exposure (Viarengo et al. 1999). Nevertheless, the evaluation of RNA relative abundances may provide further advantage over protein determination (Dondero et al. 2005, Banni et al. 2007). Like CYP450s, metallothioneins are one of the most used biomarkers in the exposure and effect assessment of pollution (Kammenga et al. 2000, Sanchez and Porcher 2009, Khati et al. 2012, Dondero and Calisi 2015). The most common indicator species to detect metal exposure through metallothionein induction is earthworm because they are efficient bioaccumulator of metals, and yet metallothionein genes are promptly activated (Vijver et al. 2003, Calisi et al. 2011a, Le Roux et al. 2016).

The heat shock proteins (HSP)70 are ubiquitous cytosolic proteins that play a very important role in the folding as well as maturation process of other proteins (Calabrese et al. 2005). Furthermore, these proteins are involved in physicochemical stress such as the protection of cells from denaturation (Feder and Hoffman 1999). Many studies have demonstrated that these proteins respond (viz. increase) to a large number of toxic substances such as metals, PCBs, PAHs, pesticides in different soil organisms (Kohler et al. 1992, 1999, Eckwert et al. 1997, Nadeau et al. 2001). For example, Kohler et al. (1999) demonstrated that the isopods HSP70 can act as a biomarker of chronic exposure and effect to pentachlorophenol and g-hexachlorocyclohexane. On the other hand, Nadeau et al. (2001) demonstrated that the induction of Hsp70 gene(s) in earthworms can represent a good wide-spectrum biomarker of exposure and effect since known toxicants altered gene expression in tissues of these animals, by contrast only a slight accumulation of HSP70 protein was observed.

The two major esterases used in soil pollution assessment are acetylcholinesterase and carboxylesterase activities (Sanchez Hernandez 2010, 2011). Acetylcholinesterase activity is the enzyme involved in the hydrolysis of the neurotransmitter acetylcholine at the chemical synapsis, terminating thereby the nervous transmission signal and preventing continuous depolarization or excitation of the postsynaptic cell (Lionetto et al. 2011, 2013). Acetylcholinesterase activity is inhibited by organophosphorus compounds and, to a less extent, by methyl carbamates. In fact, the chemical bond between the active site of the enzyme and the latter pesticides is weak in comparison with that generated by the organophosphorus. Acetylcholinesterase activity can be easily measured for the assessments of soil pesticides pollution in many terrestrial invertebrates using the Ellman's reaction and thionated substrates (Ribeiro et al. 1999, Stanek et al. 2006, Rault et al. 2007, Santos et al. 2010, Ferreira et al. 2010, Mazzia et al. 2011, Lionetto et al. 2012, Calisi et al. 2013, Smina et al. 2016). In earthworm, acetylcholinesterase has the main cholinesterase activity in the pre-clitellar part of the organism whose main role is in functioning the dorsal brain (Rault 2007, Calisi et al. 2011b). Acetylcholinesterase activity inhibition can show higher interspecies differences, therefore particular attention should be given to bioindicator selection. In general, epigeic earthworms such as *Eisenia* spp. have displayed a good degree of inhibition for both organophosphorus and methyl carbamates, yet lasting for weeks after the exposure. Other species such as *Allobophora chlorotica* can be less prone responding at environmentally relevant concentrations, therefore biomonitoring of pesticide exposure in agriculture systems should include also alternative species such as *Aporrectodea caliginosa* (Jouni et al. 2018). Carboxylesterases catalyze are important biomarker of pesticide exposure because of their implication in the detoxification of organophosphorus and synthetic pyrethroids (Wheelock et al. 2008, Sanchez Hernandez et al. 2011, 2015). In fact, in some pest insects and soil organisms an enhancement of carboxylesterase activity has been associated with high resistance to pesticides such as organophosphorus and synthetic pyrethroids (Sogorb and Vilanova 2002, Wheelock et al. 2008, Colovic et al. 2013). However, some specific carboxylesterase activities could be also inhibited almost completely due to the exposure to OPs such as chlorpyrifos (Collange et al. 2010) and parathion (Jouni et al. 2018). The concomitant evaluation of both esterase activities and the use of different species represents probably the best approach to a broader mechanistic assessment of pesticide effects in soil organisms.

The multidrug resistance (MDR) proteins are effectors of phase-III xenobiotic biotransformation process. These proteins are a superfamily of translocating ATPase pumps found in the eukaryotic plasma-membranes as well as vacuole membranes (Bambeke et al. 2003). They have a protective function by eliminating a variety of harmful and toxic molecules and metabolites from the cell. In prokaryotes, the efflux pumps mainly confer the extrusion of cytotoxic compounds including drugs, however after a conjugation step (phase II). In this process the extrusion of drugs is coupled to the inflow of protons. These pumps are, in fact, H^+ antiporters, and are classified into a different number of families (Paulsen et al. 1996, Saier et al. 1999, Pao et al. 1998, Ward et al. 2001, Chung and Saier 2001, Tseng et al. 2003, Jack et al. 2001). In contrast to prokaryotes, the major efflux mechanism in eukaryotes depends on proteins that exploit the energy obtained from ATP hydrolysis. This class of protein have been widely studied in marine organisms (Bard 2000, Kurelec 1992, Loncar et al. 2010), but have been poorly investigated in terrestrial invertebrates. More recently, a study by Hackenberger et al. (2012) described the presence of these proteins in earthworms. Many of these studies have suggested the MDR activity as a potential biomarker of exposure, since the transport activity increases in proportion to the level of environmental pollution (Smital and Kurelec 1998, Minier et al. 1999, Albertus and Laine 2001).

Among various type of used biomarkers in environmental assessment there are the antioxidant molecules and enzymatic activities such as catalase, superoxide dismutase, glutathione transferase, glutathione reductase and glutathione peroxidase (Regoli and Principato 1995, Livingstone et al. 1992, Winston et al. 1990). The antioxidants molecules (such as glutathione) and the aforementioned enzymes protect cells and tissues from the toxicity of reactive oxygen species (ROS). Glutathione plays a key role in the defence of cells against ROS and xenobiotics; glutathione peroxidase neutralizes noxious compounds formed in the cell due to pollutant exposure and glutathione reductase plays an important role in the conversion of the oxidized form to the reduced and active one. The relationships between metal poisoning, free radicals, free radical scavenging, and glutathione metabolism are considered critical for cell survival (Sunderman 1987, Halliwell and Gutteridge 1990, Laszczyca 2004). In the same way, antioxidant activities are regarded as fast and prognostic markers of individual response to the environmental stress (Cortet et al. 1999, Lagadic et al. 1994, Walker et al. 1998, Scott-Fordsmand and Weeks 2000). Several studies have also shown that the exposure to pollutants such as metal ions leads to the generation of ROS followed by an increase in lipid peroxidation. These effects have been mainly described in mammals exposed to cadmium, copper, iron, mercury and others (Christie and Costa 1984, Sunderman 1985, Halliwell and Gutteridge 1990). However, similar results have been also documented in some groups of soil invertebrates (Lionetto et al. 2012, Leomanni et al. 2015, Ribeiro et al. 2015, Chaitanya et al. 2016).

Alterations of DNA

Many mutagenic and carcinogenic environmental pollutants, such as PHAs and dioxins, can damage DNA in various ways, causing single and double strand breakages, crosslinks, fragmentation of chromosomes or mutations onset. Each of these changes is worth of investigation and can be quantitatively determined as a biomarker (McCarthy and Shugart 1990).

The single cell gel electrophoresis, known as the comet assay, and micronucleus test are the two most extensively used methods in the detection of pollutant effect related to DNA alteration. The comet assay provides a simple and effective method for evaluating nuclear damage virtually in all eukaryotic cells, including vegetables cells and invertebrate haemocytes (Singh et al. 1988, Speit and Hartmann 1999). This technique is capable to detect several types of DNA injures such as single and double strand breaks, incomplete excision repair sites, cross-linking and apoptotic cells (Cotelle and Ferard 1999, Burlinson 2012). However, the most common use of this method is based on the hydrolysis of DNA alkali-labile sites originated by exposure to oxidant compounds or other types of oxidative stress. In cells that have accrued damage to DNA, the alkali treatment unwinds DNA, releasing fragments that are detected when subjected to an electric field. During electrophoresis of cells/ nuclei embedded in a high-resolution agarose micro-gel, the negatively charged DNA migrates toward the anode and the forming comet length/shape reflects the DNA damage (Fig. 2). The characteristics of the comet tail including length, width and DNA content are useful in assessing qualitative differences in the type of DNA damage. This alkali version of the comet assay represents one of the more sensitive

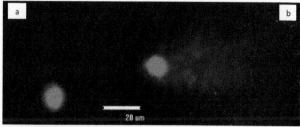

Fig. 2. Image of a whole nucleus (a), and a damaged nucleus (b).

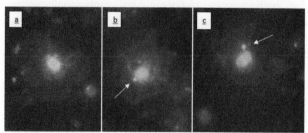

Fig. 3. Example of micronuclei formation. Whole cell (a), cells with micronuclei (b,c).

method to detect DNA damage by different pollutants and environmental stressors (Lee and Steinert 2003, Kilemade et al. 2004, Galloway et al. 2010, Bernardeschi et al. 2010, Almeida et al. 2013), however, it does not allow a clear prediction of genotoxicity because single strand damage can be usually repaired.

The micronucleous test is, therefore, the method of choice to detect DNA alterations. Indeed, it detects only double strand brakes and chromosomes anomalies formed during mitotic (or meiotic) processes. Micronuclei may arise from different abnormalities of cell division processes such as an acentric chromosome fragment detaching from a chromosome after breakage which does not integrate in the daughter cell (Fig. 3). The micronuclei test as well as the comet assay can be used to evaluate pollutant DNA damage on different soil organisms such as bacteria, protozoan, mussels, earthworm and plant (Ma et al. 1983, Majer et al. 2002, Lionetto et al. 2012, De Lapuente et al. 2015, Dondero and Calisi 2015, Imanikia et al. 2016, Dawood et al. 2017).

Cytological and Histological Biomarkers

Cytology and histology are two branches of concern to understand and evaluate the effects of pollutants as well as the compensatory mechanisms coping with toxic substances. In fact, the cell is the site where toxins are accumulated, metabolized and detoxified. Yet, cell biology has given considerable impetus to the development and application of cellular biomarkers (Kammenga et al. 2000, Sanchez Hernandez 2006, Fontanetti et al. 2011, Dondero and Calisi 2015). In cells exposed to pollutants, it can be found functional and structural alterations in response to chemical stress. Such parameters consist of alterations in the transmembrane transport mechanisms, intracellular signal transduction, structure of biological membranes, destabilization of the lysosomal membrane, cytoskeleton alterations. The destabilization of lysosomal membranes represents one of the most studied biomarkers in the scientific literature (Weeks and Svendsen 1996, Kammenga et al. 2000, Svendsen et al. 2004, Sanchez Hernandez 2006, Dondero et al. 2006, Lionetto et al. 2012, Dondero and Calisi 2015, Leomanni et al. 2015, 2016, Calisi et al. 2009, 2011b, 2013, 2016). This is an index related to a general stress syndrome and it is often linked with contamination (Svendsen et al. 1996, Weeks and Svendsen 1996, Svendsen et al. 2004, Koukozika and Dimitriadis 2005). Lysosomal membrane is one of the first biological target of toxicants. In fact, both inorganic and organic pollutants usually accumulate therein, leading to destabilization of the lysosomal membranes (Kammenga et al. 2000, Moore et al. 2006a,b, Dondero et al. 2006, Dondero and Calisi 2015). This effect can lead to the formation of leaky lysosomes and the release of hydrolytic enzymes in cytoplasm with onset of cell and tissue pathology, and eventually cell death (Stine and Brown 1996, Moore et al. 2006a,b). Furthermore, pollutants can interact with lysosomes producing other toxic effects such as abnormal accumulation of lipids, or lipofuscins that are lipoperoxidation final products (Regoli 2000, Moore et al. 2006, 2007, Fontanetti et al. 2011, Dondero and Calisi 2015).

The immune system is the most important natural defence system against external invasions from the environment. The immune system of invertebrate is composed by a series of circulating cells usually referred as to haemocytes or coelomocytes. These types of cells may also represent the first internal barrier against toxicants. As a consequence, all measurable alterations of such cells may represent a suitable biomarker of either effect or general stress (Calisi et al. 2008, 2009, 2011b). In particular, some authors demonstrated a general response of immune cells consisting in the enlargement of cells size accompanied by a rounding shape and the loss of pseudopods (Calisi et al. 2008, 2009, 2011b, 2013, 2016, Leomanni et al. 2016).

The validity of cytological markers has been confirmed in several soil organisms such as isopods (Novak et al. 2012), earthworms (Lionetto et al. 2012, Calisi et al. 2013), and snails (Leomanni et al. 2016). In some studies with earthworms Calisi et al. (2009, 2011b) demonstrated the validity of pollutant-induced morphometric alterations in both *E. fetida* and *L. terrestris* granulocytes with possible applications as sensitive, simple, and quick cellular biomarker for monitoring and assessment applications. These cellular biomarkers are useful to study the effect of pollutant on the cytoskeleton that has been demonstrated to be a target of different contaminants such as heavy metals and xenobiotics (Gomez-Mendikute and Cajaraville 2003). Similar results were reported by Leomanni et al. (2016) in terrestrial snails exposed to Hg. These organisms showed a reduction of the survival capability of haemocytes.

Behavioral Biomarkers

Behavior is defined as the integrated response of an organism in which both physiological and environmental factors are considered (Dell'Omo 2002). Past studies already indicated that animal behavior is a promising tool for ecological risk assessment and environmental toxicology (Drummond and Russom 1990, Scherrer

1992, Cohn and MacPhail 1996, Amiard-Triquet and Amiard 2013). Nowadays, behavioral studies are becoming more common in the assessment of soil toxicity (Loureiro et al. 2005, 2009, Capowiez et al. 2010, Dittbrenner et al. 2010, Matos Moreira et al. 2011, Drahokoupilová and Tuf 2012, Zizek and Zidar 2013, Martinez Morcillo et al. 2013). The main advantage in gathering at animal behavior is that behavioral differences can be displayed easily. Behavioral changes can be indirectly caused by pollutant exposure since these may have direct effects on the nervous system, or animal sensory organs (Dell'Omo 2002, Sanchez Hernandez 2006, Conti 2008, Hellou 2011). Following pollutant exposure, organisms may implement a defence strategy in order to protect themselves from the adverse effects of pollutants, e.g., avoidance. Behavioral studies can be considered a non-invasive tool, in fact, measurements and evaluations can be performed with minimal physical damage to the organism. On the other hand, there are some drawbacks in using behavior in biomonitoring due to the lack of standardized methods and the possible influence of biotic and abiotic factors that could lead to biased assessment. The following behavior responses are worth of investigation: changes in locomotion and movement capacity; excavation and burrow formation; alterations in tactile sensitivity and chemoceptor activity; alterations in predatory and escape activities; aggressiveness. The avoidance test is the most widely employed test for the evaluation of soil contamination (Loureiro 2005, 2009, De Silva and van Gestel 2009, Matos Moreira et al. 2011, Zizek and Zidar 2013, Martinez Morcillo et al. 2013). This assay, in fact, has been standardized by International Standard Organization (ISO 2008).

"Omics" Biomarkers

Omics approaches such as transcriptomics, proteomics and metabolomics, have the great advantage of allowing a comprehensive mechanistic assessment of the exposure to uncharacterized or neglected environmental toxins, such as emerging contaminants or (complex) mixtures.

The nematode *Caenorhabditis elegans* is one the best characterized model organism, both biologically and genetically. It was the first metazoan genome sequenced in 1998 by the The *C. elegans* Sequencing Consortium. It has been frequently used in biomedicine, but indeed it represents a soil organism and a valid model also in environmental toxicology (Brulle et al. 2010). More recently, high-throughput post-genomic molecular analyses have become available also to other soil organisms, in particular earthworms and collembola due to their long-term use as bioindicators of soil pollution (Sturzembaum et al. 2003, Nota et al. 2008, Pirooznia et al. 2008).

The pioneer study by Bundy et al. (2008) demonstrated that coordinated transcriptomic and metabolomic analyses could be a powerful tool for identifying changes in energetic metabolism, i.e., carbohydrate use and oxidative phosphorylation, predicting high order level outcomes, i.e., reproduction and mortality.

In another study with *E. fetida*, Zhang et al. (2014) used Illumina RNA sequencing to analyze the effects of Dechlorane Plus (DDC-CO), a polychlorinated flame retardant. They showed that despite the acute toxicity was very low, some molecular biomarkers of oxidative stress (e.g., catalase, peroxiredoxin), DNA repair (endonuclease FEN-1) and neuronal damage were readily activated disclosing an ecological risk otherwise underestimated. Moreover, high throughput sequencing allowed the identification of pathways related to antioxidant enzymes, stress responses, neurological dysfunctions, calcium binding, and signal-transduction.

In a study with the soil arthropod *Folsomia candida*, Chen et al. (2014) tested the application of global transcriptomics in a triad-based quality assessment of sediments from a heavy metal polluted river. The authors using a 180 K feature high-resolution microarray found 32 different genes significantly correlating with a combined index of toxic pressure, accounting for the environmental concentration of each contaminant and its corresponding cumulative species sensitivity distribution profile. This kind of results clearly suggests that gene expression and other omics approaches are nowadays ready to be integrated in complex ecological risk assessment procedures and that technical and interpretative bottlenecks have been exceeded.

Concluding Remarks

In this chapter, several methodological and conceptual aspects relayed to the use of biomarkers for monitoring of soil quality have been discussed. We have summarized the most common classifications of biomarkers for a better understanding of the advantages and limitations in the use of these biological tools. It is easy to understand how these tools can also be useful for studying and evaluating the effects that toxicants promote in the terrestrial ecosystems. It should be noted, however, that for soil monitoring it would be advisable to use different indicator species or sentinels and biomarkers to achieve a complete assessment of the impact of contaminants on soil system. Compared to the use of toxicity tests, the biomarker approach is a tool for early identification of the effects of chemicals on terrestrial environments. Moreover, the use of a multi-biomarker approach would allow the identification of the major classes of contaminant implied in a soil pollution scenario.

Finally, biomarkers described in this chapter are suitable toxicity endpoints to assess the efficacy of remediation program. It is well-known that the main goal of any remediation procedure is the removal of contaminants up to meet the regulatory standards. However, long-term exposure to low levels of contaminants still present in soil after remediation is rarely assessed. In this context, the use of selected biomarkers according with the chemical nature of contamination is a recommended strategy to ensure inactivation of remaining contaminants during remediation. Occasionally, bioremediation of soils contaminated by organic contaminants such as pesticides may lead to production of highly toxic metabolites. This is the case of organophosphorus pesticides that during biodegradation originate oxidized metabolites (named "oxon") which display a higher toxicity than the parent compounds. The measurement of esterase inhibition, for instance, could be an excellent candidate to assess side-effects coming from bioremediation of organophosphorus-contaminated soils. In conclusion, we can state that the measurement of biomarkers in multiple indicator species can be a workable strategy in the monitoring of bioremediation programs, using the multiple strategies that the ecotoxicology provides (e.g., laboratory toxicity testing, mescocosm or field survey).

References

Abbasi, A., A. Sajid, N. Haq, S. Rahman, Z.L. Misbah, G. Sanober et al. 2013. Agricultural pollution: An emerging issue. pp. 347–387. *In*: Ahmad, P., M.R. Wani, M.M. Azooz and L.P. Tran (eds.). Improvement of Crops in the Era of Climatic Changes. Springer, New York, NY, U.S.A.

Achazi, R.K., C. Flenner, D.R. Livingstone, L.D. Peters, K. Schaub and E. Scheiwe. 1998. Cytochrome P450 and dependent activities in unexposed and PAH-exposed terrestrial annelids. Comp. Biochem. Physiol. C 121: 339–350.

Aelion, C.M. 2004. Soil contamination monitoring. pp. 148–175. *In*: Inyang, H.I. and J.L. Daniels (eds.). Environmental Monitoring. EOLSS Publishers Co. Ltd., Oxford, UK.

Albertus, J.A. and R.O. Laine. 2001. Enhanced xenobiotic transporter expression in normal teleost hepatocytes: Response to environmental and chemotherapeutic toxins. J. Exp. Biol. 204: 217–227.

Almeida, C., C.G. Pereira, T. Gomes, C. Cardoso, M.J. Bebianno and A. Cravo. 2013. Genotoxicity in two bivalve species from a coastal lagoon in the south of Portugal. Mar. Environ. Res. 89: 29–38.

Amiard Triquet, C., J.C. Amiard and P.S. Rainbow. 2013. Ecological Biomarkers: Indicators of Ecotoxicological Effects. CRC Press, Boca Raton, FL, U.S.A.

Amiard-Triquet, C. and J.C. Amiard. 2013. Behavioral ecotoxicology. pp. 253–278. *In*: Amiard Triquet, C., J.C. Amiard and P.S. Rainbow (eds.). Ecological Biomarkers: Indicators of Ecotoxicological Effects. CRC Press, Boca Raton, FL, U.S.A.

APAT. 2004. Versione aggiornata sulla base delle indicazioni contenute nella strategia tematica del suolo dell'Unione Europea, CTN-TES 2003-TES-T-MAN-03-02, APAT, Rome, Italy.

Ashraf, M.A., M.J. Maah and I. Yusoff. 2014. Soil contamination, risk assessment and remediation. pp. 3–56. *In*: Hernandez-Soriano, M.C. (ed.). Environmental Risk Assessment of Soil Contamination. InTech-Open Access Publisher in Science, Technology and Medicine, Rijeka, Croatia.

Bambeke, V.F., Y. Glupczynski, J.C. Plesiat Pechere and P.M. Tulkens. 2003. Antibiotic efflux pumps in prokaryotic cells: Occurrence, impact on resistance and strategies for the future of antimicrobial therapy. J. Antimicrob. Chemoth. 51: 1055–1065.

Banni, M., F. Dondero, J. Jebali, H. Guerbej, H. Boussetta and A. Viarengo. 2007. Assessment of heavy metal contamination using real-time PCR analysis of mussel metallothionein mt10 and mt20 expression: a validation along the Tunisian coast. Biomarkers 12: 369–383.

Bard, S.M. 2000. Multixenobiotic resistance as a cellular defence mechanism in aquatic organisms. Aquat. Toxicol. 48: 357–389.

Bernardeschi, M., P. Guidi, V. Scarcelli, G. Frenzilli and M. Nigro. 2010. Genotoxic potential of TiO_2 on bottlenose dolphin leukocytes. Anal. Bioanal. Chem. 396: 619–623.

Bradham, K.D., E.A. Dayton, N.T. Basta, J. Schroder, M. Payton and R.P. Lanno. 2006. Effect of soil properties on lead bioavailability and toxicity to earthworms. Environ. Toxicol. Chem. 25: 769–775.

Brown, P.J., S.M. Long, D.J. Spurgeon, C. Svendsen and P.K. Hankard. 2004. Toxicological and biochemical responses of the earthworm *Lumbricus rubellus* to pyrene, a non-carcinogenic polycyclic aromatic hydrocarbon. Chemosphere 57: 1675–1681.

Brulle, F., A.J. Morgan, C. Cocquerelle and F. Vandenbulcke. 2010. Transcriptomic underpinning of toxicant-mediated physiological function alterations in three terrestrial invertebrate taxa: A review. Environ. Poll. 158: 2793–2808.

Bundy, J.G, J.K. Sidhu, F. Rana, D.J. Spurgeon, C. Svendsen, J.F. Wren et al. 2008. Systems toxicology approach identifies coordinated metabolic responses to copper in a terrestrial non-model invertebrate, the earthworm *Lumbricus rubellus*. BMC Biol. 6: 25.

Burlinson, B. 2012. The *in vitro* and *in vivo* Comet assays. Methods Mol. Biol. 817: 143–163.

Calabrese, V., R. Lodi, C. Tonon, V.D. Agata, M. Sapienza, G. Scapagnini et al. 2005. Oxidative stress, mitochondrial dysfunction and cellular stress response in Friedreich's ataxia. J. Neurol. Sci. 233: 145–162.

Calisi, A., M.G. Lionetto, R. Caricato, M.E. Giordano and T. Schettino. 2008. Morphometric alteration in *Mytilus galloprovincialis*: A new biomarker. Environ. Toxicol. Chem. 27: 1435–1441.

Calisi, A., M.G. Lionetto and T. Schettino. 2009. Pollutant-induced alterations of granulocyte morphology in the earthworm *Eisenia foetida*. Ecotoxicol. Environ. Saf. 72: 1369–1377.

Calisi, A., M.G. Lionetto, J.C. Sanchez-Hernandez and T. Schettino. 2011a. Effect of heavy metal exposure on blood haemoglobin concentration and methemoglobin percentage in *Lumbricus terrestris*. Ecotoxicology 20: 847–854.

Calisi, A., M.G. Lionetto and T. Schettino. 2011b. Biomarker response in the earthworm *Lumbricus terrestris* exposed to chemical pollutants. Sci. Tot. Environ. 409: 4456–4464.

Calisi, A., N. Zaccarelli, M.G. Lionetto and T. Schettino. 2013. Integrated biomarker analysis in the earthworm *Lumbricus terrestris*: application to the monitoring of soil heavy metal pollution. Chemosphere 90: 2637–2644.

Calisi, A., A. Grimaldi, A. Leomanni, M.G. Lionetto, F. Dondero and T. Schettino. 2016. Multibiomarker response in the earthworm *Eisenia fetida* as tool for assessing multi-walled carbon nanotube ecotoxicity. Ecotoxicology 25: 677–687.

Cao, X., R. Bi and Y. Song. 2017. Toxic responses of cytochrome P450 sub-enzyme activities to heavy metals exposure in soil and correlation with their bioaccumulation in *Eisenia fetida*. Ecotox. Env. Saf. 144: 158–165.

Capowiez, Y., N. Dittbrenner, M. Rault, M. Hedde, R. Triebskorn and C. Mazzia. 2010. Earthworm cast production as a new behavioural biomarker for toxicity testing. Environ. Pollut. 158: 388–393.

Chaffai, A.H. 2014. Usefulness of bioindicators and biomarkers in pollution biomonitoring. Int. J. Biotech. Well. Industr. 3: 19–26.

Chaitanya, R.K., K. Shashank and P. Sridevi. 2016. Oxidative stress in invertebrate systems. pp. 51–68. *In*: Ahmad, R. (ed.). Free Radicals and Diseases. InTech-Open Access Publisher in Science, Technology and Medicine. Rijeka, Croatia.

Chapman, P.M. 1990. The sediment quality triad approach to determining pollution-induced degradation. Sci. Tot. Environ. 97–98: 815–825.

Chen, G., T.E. de Boer, M. Wagelmans, C.A. van Gestel, N.M. van Straalen and D. Roelofs. 2014. Integrating transcriptomics into triad-based soil-quality assessment. Environ. Toxicol. Chem. 33: 900–909.

Christie, N.T. and M. Costa. 1984. *In vitro* assessment of the toxicity of metal compounds. IV. Disposition of metals in cells: Interactions with membranes, glutathione, metallothionein, and DNA. Biol. Trace El. Res. 6: 139–158.

Chung, Y.J. and H.H. Saier. 2001. SMR-type multi drug resistance pumps. Curr. Opin. Drug. Disc. 4: 237–245.

Coeurdassier, M., M. Saint-Denis, A. Gomot-de Vaufleury, D. Ribera and P.M. Badot. 2001. The garden snail (*Helix aspersa*) as a bioindicator of organophosphorus exposure: effects of dimethoate on survival, growth, and acetylcholinesterase activity. Environ. Toxicol. Chem. 20: 1951–1957.

Cohn J. and R.C. Mac Phail. 1996. Ethological and experimental approaches to behavior analysis: Implications for ecotoxicology. Environ. Health Persp. 104: 299–305.

Collange, B., C.E. Wheelock, M. Rault, C. Mazzia, Y. Capowiez and J.C. Sanchez-Hernandez. 2010. Inhibition, recovery and oxime-induced reactivation of muscle esterases following chlorpyrifos exposure in the earthworm *Lumbricus terrestris*. Environ. Poll. 158: 2266–2272.

Colovic, M.B., D.Z. Krstic, T.D. Lazarevic-Pasti, A.M. Bondzic and V.M. Vasic. 2013. Acetylcholinesterase inhibitors: Pharmacology and toxicology. Curr. Neuropharmacol. 11: 315–335.

Conti, M.E. 2008. Biomarkers for environmental monitoring. pp. 25–46. *In*: Conti, M.E. (ed.). Biological Monitoring: Theory and Applications. Wit Press Ashurts Lodge, Ashurst Southempton, U.K.

Cortet, J., A. Gomot-De Vauflery, N. Poinsot-Balaguer, L. Gomot, C. Texier and D. Cluzeau. 1999. The use of invertebrate soil fauna in monitoring pollutant effects. Eur. J. Soil Biol. 35: 115–134.

Cotelle, S. and J.F. Ferard. 1999. Comet assay in genetic ecotoxicology: A review. Environ. Mol. Mutagen. 34: 246–255.

Criel, P., K. Lock, H. Van Eeckhout, K. Oorts, E. Smolders and C.R. Janssen. 2008. Influence of soil properties on copper toxicity for two soil invertebrates. Environ. Toxicol. Chem. 27: 1748–1755.

Dallinger, R., B. Berger, R. Triebskorn-Köhler and H. Köhler. 2001. Soil biology and ecotoxicology. pp. 489–525. *In*: Baker, G.M. (ed.). The Biology of Terrestrial Molluscs. CABI Publishing, London, U.K.

Dawood, M., A. Wahid, B. Zakariya, M.Z. Hashmi, S. Mukhtar and Z. Malik. 2017. Use of earthworms in biomonitoring of soil xenobiotics. pp. 73–88. *In*: Hashmi, M.Z., V. Kumar and A. Varma (eds.). Xenobiotics in the Soil Environment. Springer International Publishing, New York, NY, U.S.A.

De Lapuente, J., J. Lourenco, S.A. Mendo, M. Borras, M.G. Martins, P.M. Costa et al. 2015. The comet assay and its applications in the field of ecotoxicology: a mature tool that continues to expand its perspectives. Front. Genet. 6: 180.

De Silva, P.M.C.S. and C.A.M. van Gestel. 2009. Comparative sensitivity of *Eisenia andrei* and *Perionyx excavatus* in earthworm avoidance tests using two soil types in the tropics. Chemosphere 77: 1609–1613.

Dell'Omo, G. 2002. Behavioral Ecotoxicology. John Wiley & Sons, Chichester, West Sussex, U.K.

Dittbrenner, N., R. Triebskorn, I. Moser and Y. Capowiez. 2010. Physiological and behavioural effects of imidacloprid on two ecologically relevant earthworm species (*Lumbricus terrestris* and *Aporrectodea caliginosa*) Ecotoxicology 19: 1567–1573.

Dondero, F., L. Piacentini, M. Banni, M. Rebelo, B. Burlando and A. Viarengo. 2005. Quantitative PCR analysis of two molluscan metallothionein genes unveils differential expression and regulation. GENE 345: 259–270.

Dondero, F., A. Dagnino, H. Jonsson, F. Caprì, L. Gastaldi and A. Viarengo. 2006. Assessing the occurrence of a stress syndrome in mussels (*Mytilus edulis*) using a combined biomarker/gene expression approach. Aquat. Toxicol. 78: 13–24.

Dondero, F. and A. Calisi. 2015. Evaluation of pollution effects in marine organisms: "Old" and "New Generation" biomarkers. pp. 143–192. *In:* Sebastià, M.T. (ed.). Coastal Ecosystems: Experiences and Recommendations for Environmental Monitoring Programs. Nova Science Publishers, New York, NY, U.S.A.

Doran, J.W. and T.B. Parkin. 1996. Quantitative indicators of soil quality: A minimum data set. pp. 25–37. *In*: Doran, J.W. and A.J. Jones (eds.). Methods for Assessing Soil Quality. Soil Science Society of America, Special Publication 49, Madison, WI, U.S.A.

Drahokoupilová, T. and I.H. Tuf. 2012. The effect of external marking on the behaviour of the common pill woodlouse *Armadillidium vulgare*. ZooKeys 176: 145–154.

Drummond, R.A. and C.L. Russom. 1990. Behavioral toxicity syndromes: A promising tool for assessing toxicity mechanisms in juvenile fathead minnows. Environ. Toxicol. Chem. 9: 37–46.

Eckwert, H., G. Alberti and H.R. Köhler. 1997. The induction of stress proteins (hsp) in *Oniscus asellus* (Isopoda) as a molecular marker of multiple heavy metal exposure. I. Principles and Toxicological Assessment Ecotoxicology 6: 249–262.

Edwards, C.A., S. Subler, S.K. Chen and D.M. Bogomolov. 1995. Essential criteria for selecting bioindicator species, processes, or systems to assess the environmental impact of chemicals on soil ecosystems. pp. 67–84. *In*: Van Straalen, N.M. and D.A. Krivolutskii (eds.) New Approaches to the Development of Bioindicator Systems for Soil Pollution. Kluwer Academic Publishers, Amsterdam, The Netherlands.

Feder, M.E. and G.E. Hoffmann. 1999. Heat shock proteins, molecular chaperones and the stress response: evolutionary and ecological physiology. Annu. Rev. Ph. Physiol. 61: 243–282.

Ferreira, N.G.C., R.F. Domingues, C.F. Calhôa, A.M.V.M. Soares and S. Loureiro. 2010. Acetylcholinesterase characterization in the terrestrial isopod *Porcellionides pruinosus*. pp. 227–236. *In*: Hamamura, N., S. Suzuki, S. Mendo, C.M. Barroso, H. Iwata and S. Tanabe (eds.). Interdisciplinary Studies on Environmental Chemistry—Biological Responses to Contaminants. TERRAPUB, Okusawa, Setagaya-ku, Tokyo, Japan.

Filser, J., S. Wiegmann and B. Schröder. 2014. Collembola in ecotoxicology—Any news or just boring routine? Appl. Soil Ecol. 83: 193–199.

Fontanetti, S.C., R.L. Nogarol, B.R. de Souza, G.D. Perez and T.G. Maziviero. 2011. Bioindicators and biomarkers in the assessment of soil toxicity. pp 143–168. *In:* Pascucci, S. (ed.). Soil Contamination. InTech-Open Access Publisher in Science, Technology and Medicine, Rijeka, Croatia.

Forbes, V. E., A. Palmqvist and L. Bach. 2006. The use and misuse of biomarkers in ecotoxicology. Environ. Toxicol. Chem. 25: 272–280.

Fountain, M.T. and S.P. Hopkin. 2005. *Folsomia candida* (Collembola): A "Standard" soil arthropod. Annu. Rev. Entomol. 50: 201–222.

Galloway, T.S., C. Lewis, I. Dolciotti, B.D. Johnstone, J. Moger and F. Regoli. 2010. Sublethal toxicity of nano-titanium dioxide and carbon nanotubes in a sediment dwelling marine polychaete. Environ. Pollut. 158: 1748–1755.

Gerhardt, A. 2002. Bioindicator species and their use in biomonitoring. pp. 1–50. *In:* UNESCO (ed.). Encyclopedia of Life Support Systems. UNESCO, EOLSS. Oxford U.K.

Gomez Mendikute, A. and M.P. Cajaraville. 2003. Comparative effects of cadmium, copper, paraquat and benzo(a)pyrene on the actin cytoskeleton and production of reactive oxygen species (ROS) in mussel haemocytes. Toxicology *in vitro* 17: 539–546.

Hackenberger, B.K., M. Velki, S. Stepic and D.K. Hackenberger. 2012. First evidence for the presence of efflux pump in the earthworm *Eisenia andrei.* Ecotox. Environ. Saf. 75: 40–45.

Hagger, J.A., M.B. Jones, D.R. Leonard, R. Owen and T.S. Galloway. 2006. Biomarkers and integrated environmental risk assessment: are there more questions than answers? Integr. Environ. Assess. Manag. 2: 312–329.

Halliwell, B. and M.C. Gutteridge. 1990. Role of free radicals and catalytic metal ions in human disease: An overview. pp. 1–85. *In:* Packer, L. and A.N. Glatzer (eds.). Oxygen Radicals in Biological Systems. Part B. Meth. Enzymol. Academic Press, New York, NY, U.S.A.

Hellou, J. 2011. Behavioural ecotoxicology, an "early warning" signal to assess environmental quality. Environ. Sci. Pollut. Res. Int. 18: 1–11.

Hopkin, S.P. 1993. *In situ* biological monitoring of pollution in terrestrial and aquatic ecosystems. pp. 397–427. *In:* Calow, P. (ed.). Handbook of Ecotoxicology. Blackwell Scientific Publications, Oxford, U.K.

Hyne, R.V. and W.A. Maher. 2003. Invertebrate biomarkers: Links to toxicosis that predict population decline. Ecotox. Environ. Saf. 54: 366–374.

Imanikia, S., F. Galea, E.R. Nagy, D.H. Phillips, S.R. Stürzenbaum and V.M. Arlta. 2016. The application of the comet assay to assess the genotoxicity of environmental pollutants in the nematode *Caenorhabditis elegans.* Environ. Toxicol. Pharmacol. 45: 356–361.

Iserentant, R.E. and J. De Sloover. 1976. Le concept de bioindicateur. Mémoires de la Société Royale de Botanique de Belgique 7: 15–24.

ISO (International Standard Organization). 1996. Soil quality. Vocabulary. Part 1: Terms and definitions relating to the protection and pollution of the soil. International Organization for Standardization. Geneva, Switzerland.

ISO (International Standard Organization). 2008. Soil Quality: Avoidance Test for Testing the Quality of Soils and Effects of Chemicals on Behavior—Part 1: Test with Earthworms (*Eisenia fetida* and *Eisenia andrei*). ISO/DIS 17512-1.2. International Organization for Standardization. Geneva, Switzerland.

Jack, D.L., N.M. Yang and M.H. Saier. 2001. The drug/metabolite transporter super-family. Eur. J. Biochem. 68: 3620–3639.

Jouni F., J.C. Sanchez-Hernandez, C. Mazzia, M. Jobin, Y. Capowiez and M. Rault. 2018. Interspecific differences in biochemical and behavioral biomarkers in endogeic earthworms exposed to ethyl-parathion. Chemosphere 202: 85–93.

Kammenga, J.E., R. Dallinger, M.H. Donker, H.R. Köhler, V. Simonsen, R. Triebskorn et al. 2000. Biomarkers in terrestrial invertebrates for ecotoxicological soil risk assessment. Rev. Environ. Contamin. Toxicol. 164: 93–147.

Khati, W., K. Oualia, C. Mouneyrac and A. Banaoui. 2012. Metallothioneins in aquatic invertebrates: Their role in metal detoxification and their use in biomonitoring. Energy Procedia 18: 784–794.

Kilemade, M.F., M.G.J. Harti, D. Sheehan, C. Mothersill, F.N.A.M. van Pelt, J. O'Halloran et al. 2004. Genotoxicity of field-collected inter-tidal sediments from Cork Harbor, to juvenile turbot (*Scophthalmus maximus* L.) as measured by the Comet assay. Environ. Mol. Mutagen. 44: 54–64.

Kohler, H.R., R. Triebskorn, W. Stocker, P.M. Kloetzel and G. Alberti. 1992. The 70 kD heat shock protein (hsp70) in soil invertebrates: a possible tool for monitoring environmental toxicants. Arch. Environ. Contam. Toxicol. 22: 334–338.

Kohler, H.R., C. Knodler and M. Zanger. 1999. Divergent kinetics of hsp70 induction in *Oniscus asellus* (Isopoda) in response to four environmentally relevant organic chemicals (B[a]P, PCB52, gamma-HCH, PCP): suitability and limits of a biomarker. Arch. Environ. Contam. Toxicol. 36: 179–185.

Koukouzika, N. and V.K. Dimitriadis. 2005. Multiple biomarker comparison in *Mytilus galloprovincialis* from the Greece coast: Lysosomal membrane stability, neutral red retention, micronucleus frequency and stress on stress. Ecotoxicology 14: 449–463.

Kurelec, B. 1992. The multixenobiotic resistance mechanism in aquatic organisms. Crit. Rev. Toxicol. 22: 23–43.

Lagadic, L., T. Caquet and F. Ramade. 1994. The role of biomarkers in environmental assessment (5). Invertebrate populations and communities. Ecotoxicology 3: 193–208.

Lagadic, L., T. Caquet, J.C. Amiard and F. Ramade. 2000. Use of Biomarkers for Environmental Quality Assessment. Science Publishers Inc., Enfield, CT, U.S.A.

Lagzi, I., R. Meszaros, G. Gelybo and A. Leelossy. 2014. Atmospheric Chemistry. Eotvos Lorand University, Budapest, Hungary.

Lal, R. 2005. Encyclopedia of Soil Science—2nd Edition. CRC Press Taylor and Francis Group, Boca Raton FL, U.S.A.

Laszczyca, P., M. Augustyniak, A. Babczynska, K. Bednarska, A. Kafel, P. Migula et al. 2004. Profiles of enzymatic activity in earthworms from zinc, lead and cadmium polluted areas near Olkusz (Poland). Environ. Int. 30: 901–910.

Le Roux, S., P. Baker and A. Crouch. 2016. Bioaccumulation of total mercury in the earthworm *Eisenia Andrei*. Springerplus 5: 681.

Lee, R.F. and S. Steinert. 2003. Use of the single cell gel electrophoresis/comet assay for detecting DNA damage in aquatic (marine and freshwater) animals. Mutat. Res. 544: 43–64.

Leomanni, A., T. Schettino, A. Calisi, S. Gorbi, M. Mezzelani, F. Regoli et al. 2015. Antioxidant and oxidative stress related responses in the mediterranean land snail *Cantareus apertus* exposed to the carbammate pesticide carbaryl. Comp. Biochem. Physiol. C 168: 20–27.

Leomanni, A., T. Schettino, A. Calisi and M.G. Lionetto. 2016. Mercury induced haemocyte alterations in the terrestrial snail *Cantareus apertus* as novel biomarker. Comp. Biochem. Physiol. C 183: 20–27.

Lionetto, M.G., R. Caricato, A. Calisi and T. Schettino. 2011. Acetylcholinesterase inhibition as relevant biomarker in environmental biomonitoring: New insights and perspectives. pp. 87–116. *In*: Visser, J.E. (ed.). Ecotoxicology Around the Globe. Nova Science Publishers, New York, NY, U.S.A.

Lionetto, M.G., A. Calisi and T. Schettino. 2012. Earthworms biomarkers as tools for soil pollution assessment. pp. 305–332. *In*: Hernandez-Soriano, M.C. (ed.). Soil Health and Land use Management. InTech-Open Access Publisher in Science, Technology and Medicine, Rijeka, Croatia.

Lionetto, M.G., R. Caricato, A. Calisi, M.E. Giordano and T. Schettino. 2013. Acetylcholinesterase as biomarkers in environmental and occupational medicine: New insights and future perspectives. Biomed. Res. Int. Epub: 1–8.

Livingstone, D.R., F. Lips, P. Garcia Martinez and R.K. Pipe. 1992. Antioxidant enzymes in the digestive gland of the common mussel, *Mytilus edulis*. Mar. Biol. 112: 265–276.

Loncar, J., M. Popović, R. Zaja and T. Smital. 2010. Gene expression analysis of the ABC efflux transporters in rainbow trout (*Oncorhynchus mykiss*). Comp. Biochem. Physiol. C 151: 209–215.

Loureiro, S., A.M.V.M. Soares and A.J.A. Nogueira. 2005. Terrestrial avoidance behavior tests as screening tool to assess soil contamination. Environ. Pollut. 138: 121–131.

Loureiro, S., M.J.B. Amorim, B. Campos, S.M.G. Rodrigues and A.M.V.M. Soares. 2009. Assessing joint toxicity of chemicals in *Enchytraeus albidus* (Enchytraeidae) and *Porcellionides pruinosus* (Isopoda) using avoidance behaviour as an endpoint. Environ. Poll. 157: 625–636.

Lowe, W.R. and R.J. Kendal. 1990. Sentinel species and sentinel bioassay. pp. 309–331. *In*: McCarthy, J.F. and L.R. Shugart (eds.). Biomarkers of Environmental Contamination. Lewis Publisher, Boca Raton, FL., U.S.A.

Ma, T.H., V.A. Anderson, M.M. Harris and J.L. Bare. 1983. Tradescantia-Micronucleus (Trad-MCN) test on the genotoxicity of malathion. Environ. Mutagen. 5: 127–137.

Majer, B.J., D. Tscherko, A. Paschke, R. Wennrich, M. Kundi, E. Kandeler et al. 2002. Effects of heavy metal contamination of soils on micronucleus induction in Tradescantia and on microbial enzyme activities: a comparative investigation. Mutat. Res. 515: 111–24.

Martínez Morcillo, S., J.L. Yela, Y. Capowiez, C. Mazzia, M. Rault and J.C. Sanchez-Hernandez. 2013. Avoidance behaviour response and esterase inhibition in the earthworm, *Lumbricus terrestris*, after exposure to chlorpyrifos. Ecotoxicology 22: 597–607.

Matos-Moreira, M., J.C. Niemeyer, J.P. Sousa, M. Cunha and E. Carral. 2011. Behavioral avoidance tests to evaluate effects of cattle slurry and dairy sludge application to soil. Revista Brasileira de Ciência do Solo 35: 1471–1477.

Mazzia, C., Y. Capowiez, J.C. Sanchez-Hernandez, H.R. Köhler, R. Triebskorn and M. Rault. 2011. Acetylcholinesterase activity in the terrestrial snail *Xeropicta derbentina* transplanted in apple orchards with different pesticide management strategies. Environ. Pollut. 159: 319–323.

McCarthy, F. and L.R. Shugart. 1990. Biomarkers of environmental contamination. Lewis Publisher, Chelsea, MA, U.S.A.

Minier, C., N. Eufemia and D. Epel. 1999. The multi-xenobiotic resistance phenotype as a tool to biomonitor the environment. Biomarkers 4: 442–454.

Moore, M.N. 1985. Cellular responses to pollutants. Mar. Poll. Bull. 16: 134–139.

Moore, M.N., M.H. Depledge, J.W. Readman and P. Leonard. 2004. An integrated biomarker based strategy for ecotoxicological evaluation of risk in environmental management. Mut. Res. 552: 247–268.

Moore, M.N., J.I. Allen and A. McVeigh. 2006. Environmental prognostics: An integrated model supporting lysosomal stress responses as predictive biomarkers of animal health status. Mar. Environ. Res. 61: 278–304.

Moore, M.N., A. Viarengo, P. Donkin and A.J.S. Hawkins. 2007. Autophagic and lysosomal reactions to stress in the hepatopancreas of blue mussels. Aquat. Toxicol. 84: 80–91.

Mouneyrac, C. and C. Amiard-Triquet. 2013. Biomarkers of ecological relevance. pp. 221–236. *In:* Firard, J. and C. Claise (eds.). Encyclopedia of Aquatic Ecotoxicology. Springer, Berlin, Germany.

Nadeau, D., S. Corneau, I. Plante, G. Morrow and R.M. Tanguay. 2001. Evaluation for Hsp70 as a biomarkers of effect of pollutants on the earthworm *Lumbricus terrestris*. Cell. Stress. Chap. 6: 153–163.

Nesheim, M.C. 2015. Environmental effects of the U.S. food system. pp. 127–166. *In*: Nesheim, M.C., M. Oria and P.T. Yih (eds.). A Framework for Assessing Effects of the Food System. The National Academic Press, Washington, DC, U.S.A.

Niemi, G.J. and M.E. Mc Donald. 2004. Application of ecological indicators. Ann. Rev. Ecol. Evol. System 35: 89–111.

Nota, B., M.J. Timmermans, O. Franken, K. Montagne-Wajer, J. Mariën, M.E.D. Boer et al. 2008. Gene expression analysis of collembola in cadmium containing soil. Environ. Sci. Techn. 42: 8152–8157.

Novak, S., D. Drobne, J. Valant, Z. Pipan-Tkalec, P. Pelicon, P. Vavpetič et al. 2012. Cell membrane integrity and internalization of ingested TiO_2 nanoparticles by digestive gland cells of a terrestrial isopod. Environ. Toxicol. Chem. 31: 1083–1090.

Pao, S.S., I.T. Paulsen and M.H. Saier, Jr. 1998. Major facilitator superfamily. Microbiol. Mol. Biol. Rev. 62: 1–34.

Paulsen, I.T., M.H. Brown and R.A. Skurray. 1996. Proton-dependent multidrug efflux systems. Microbiol. Rev. 60: 575–608.

Peakall, D.B. and L.R. Shugart. 1993. Biomarkers: Research and Application in Assessment of Environmental Health. Springer-Verlag, Berlin, Germany.

Pérès, G., F. Vandenbulcke, M. Guernion, M. Hedde, T. Beguiristain, F. Douay et al. 2011. Earthworm indicators as tools for soil monitoring, characterization and risk assessment. An example from the national Bioindicator programme (France). Pedobiologia 54: 77–87.

Pirooznia, M., P. Gong, X. Guan, L.S. Inouye, K. Yang, E.J. Perkins et al. 2007. Cloning, analysis and functional annotation of expressed sequence tags from the earthworm Eisenia fetida. BMC Bioinformatics 8: 7.

Rault, M., C. Mazzia and Y. Capowiez. 2007. Tissue distribution and characterization of cholinesterase activity in six earthworm species. Comp. Biochem. Physiol. B 147: 340–346.

Regoli, F. and G. Principato. 1995. Glutathione, glutathione-dependent and antioxidant enzymes in mussels, *Mytilus galloprovincialis*, exposed to metals in different field and laboratory conditions: Implications for a proper use as biochemical biomarkers. Aquat. Toxicol. 31: 143–164.

Regoli, F. 2000. Total oxyradical scavenging capacity (TOSC) in polluted and translocated mussels: a predictive biomarker of oxidative stress. Aquatic Toxicol. 50: 351–361.

Regoli, F. 2001. Monitoraggio della contaminazione chimica: ecotossicologia e biomarkers. pp. 189–222. *In*: Danovaro, R. (ed.). Recupero ambientale. UTET, Torino, Italy.

Ribeiro, M.J., V.L. Maria, J.J. Scott-Fordsmand and M.J.B. Amorim. 2015. Oxidative stress mechanisms caused by Ag nanoparticles (NM300K) are different from those of $AgNO_3$: effects in the soil invertebrate *Enchytraeus crypticus*. Int. J. Environ. Res. 12: 9589–9602.

Ribeiro, S., L. Guilhermino, J.P. Sousa and A.M.V.M. Soares. 1999. Novel bioassay based on acetylcholinesterase and lactate dehydrogenase activities to evaluate the toxicity of chemicals to soil isopods. Ecotoxicol. Environ. Saf. 44: 287–293.

Saier, M.H., J.T. Beatty, A. Goffeau, K.T. Harley, W.H. Heijne, S.C. Huang et al. 1999. The major facilitator superfamily. J. Mol. Microbiol. Biotechnol. 1: 257–279.

Saint-Denis, M., J.F. Narbonne, C. Arnaud, E. Thybaud and D. Ribera. 1999. Biochemical responses of the earthworm *Eisenia fetida andrei* exposed to contaminated artificial soil: Effects of benzo (a)pyrene. Soil Biol. Biochem. 31: 1837–1846.

Sánchez-Bayo, F., P.J. Van den Brink and R.M. Mann. 2011. Ecological impacts of toxic chemicals. Bentham Science Publishers Ltd, U.S.A.

Sanchez-Hernandez, J.C. 2006. Earthworms biomarkers in ecological risk assessment. Rev. Environ. Contam. Toxicol. 188: 85–126.

Sanchez-Hernandez, J.C. 2010. Environmental applications of earthworm esterases in the agroecosystem. J. Pest. Sci. 35: 290–301.

Sanchez-Hernandez, J.C. 2011. Pesticide Biomarkers in terrestrial invertebrates. pp. 213–240. *In*: Stoytcheva, M. (ed.). Pesticides in the Modern World—Pests Control and Pesticides Exposure and Toxicity, Assessment. InTech- Open Access Publisher in Science, Technology and Medicine, Rijeka, Croatia.

Sanchez-Hernandez, J.C., J. Notario del Pino and J. Domínguez. 2015. Earthworm-induced carboxylesterase activity in soil: Assessing the potential for detoxification and monitoring organophosphorus pesticides. Ecotoxicol. Environ. Saf. 122: 303–12.

Sanchez, W. and J.M. Porcher. 2009. Fish biomarkers for environmental monitoring within the water framework directive of the European Union. Trend. Anal. Chem. 28: 150–158.

Santos, M.J.G., N.G.C. Ferreira, A.M.V.M. Soares and S. Loureiro. 2010. Toxic effects of molluscicidal baits to the terrestrial isopod *Porcellionides pruinosus* (Brandt, 1833). J. Soil. Sed. 10: 1335–1343.

Scherrer, E. 1992. Behavioural responses as indicators of environmental alterations: Approaches, results, developments J. Appl. Ichthyol. 8: 122–131.

Scott-Fordsmand, J.J. and J.M. Weeks. 2000. Biomarkers in earthworms. Rev. Environ. Contam. Toxicol. 165: 117–159.

Singh, N.P., M.T. McCoy, R.R. Tice and E.L. Schneider. 1988. A simple technique for quantification of low levels of DNA damage in individual cells. Exp. Cell. Res. 175: 184–191.

Smina, A.H., B. Samira, D. Mohamed and B. Houria. 2016. Evaluation of acetylcholinesterase, glutathione S-transferase and catalase activities in the land snail *Helix aspersa* exposed to thiamethoxam. J. Ent. Zool. 4: 369–374.

Smital, T. and B. Kurelec. 1998. The chemosensitizers of multixenobiotic resistance mechanism in aquatic invertebrates: A new class of pollutants. Mutat. Res. 399: 43–53.

Sogorb, M.A. and E. Vilanova. 2002. Enzymes involved in the detoxification of organophosphorus, carbamate and pyrethroid insecticides through hydrolysis. Toxicol. Lett. 128: 215–228.

Speit, G. and A. Hartmann. 1999. The comet assay (single-cell gel test)—a sensitive genotoxicity test for the detection of DNA damage and repair. pp. 203–212. *In*: Henderson, D.S. (ed.). Methods in Molecular Biology, Vol. 113, DNA Repair Protocols: Eucaryotic Systems. Humana Press, Totowa, NJ, U.S.A.

Spurgeon, D.J., S. Lofts, P.K. Hankard, M. Toal, D. McLellan, S. Fishwick et al. 2006. Effect of pH on metal speciation and resulting metal uptake and toxicity for earthworms. Environ. Toxicol. Chem. 25: 788–796.

Stanek, K., D. Drobne and P. Trebše. 2006. Linkage of biomarkers along levels of biological complexity in juvenile and adult diazinon fed terrestrial isopod (*Porcellio scaber*, Isopoda, crustacea). Chemosphere 64: 1745–1752.

Stine, K.E. and T.M. Brown. 1996. Principles of toxicology. Lewis Publishers, Boca Raton, FL, U.S.A.

Sturzenbaum, S.R., J. Parkinson, M. Blaxter, A.J. Morgan, P. Kille and O. Georgiev. 2003. The earthworm expressed sequence tag project. Pedobiologia 47: 447–451.

Sunderman, F.W. 1985. Metals and lipid peroxidation. Acta Pharmacol. Toxicolo. 59: 248–255.

Sunderman, F.W. 1987. Biochemical indices of lipid peroxidation in occupational and environmental medicine. pp. 151–158. *In*: Foa, V., E.A. Emmett, B. Maroni and A. Colombi (eds.). Occupational and Environmental Chemical Hazards. Cellular and Biochemical Indices for Monitoring Toxicity. Ellis Horwood, Chicehester, West Sussex, U.K.

Svendsen, C., A.A. Meharg, P. Freestone and J.M. Weeks. 1996. Use of an earthworm lysosomal biomarker for the ecological assessment of pollution from an industrial plastics fire. Appl. Soil. Ecol. 3: 99–107.

Svendsen, C., D.J. Spurgeon, P.K. Hankard and J.M. Weeks. 2004. A review of lysosomal membrane stability measured by neutral red retention: is it a workable earthworm biomarker? Ecotox. Environ. Saf. 57: 20–29.

The *C. elegans* Sequencing Consortium. 1998. Genome Sequence of the Nematode *C. elegans*: A platform for investigating biology. Science 282: 2012–2018.

Tseng, T.T., K.S. Gratwick, J. Kollman, D. Park, D.H. Nies, A. Goffeau et al. 2003. The RND permease family: An ancient, ubiquitous and diverse family that includes human disease and development proteins. J. Mol. Microbiol. Biotechnol. 1: 107–125.

Van der Oost, R., J. Beyer and N.P.E. Vermeulen. 2003. Fish bioaccumulation and biomarkers in environmental risk assessment: A review. Environ. Toxicol. Pharmacol. 13: 57–149.

Viarengo, A., B. Burlando, F. Dondero, A. Marro and R. Fabbri. 1999. Metallothionein as a tool in biomonitoring programmes. Biomarkers 4: 455–466.

Vijver, M.G, J.P.M. Vink, C.J.H. Miermans and C.A.M. Van Gestel. 2003. Oral sealing using glue: a new method to distinguish between intestinal and dermal uptake of metals in earthworms. Soil Biol. Biochem. 35: 125–132

Walker, C.H. 1998. Biomarker strategies to evaluate the environmental effects of chemicals. Environ. Health Persp. 106: 613–620.

Ward, A., C. Hoyle, S. Palmer, J. O'Reilly, J. Griffith, M. Pos et al. 2001. Prokaryote multidrug efflux proteins of the major facilitator superfamily: amplified expression, purification and characterisation. J. Mol. Microbiol. Biotechnol. 3: 193–200.

Weeks, J.M. and C. Svendsen. 1996. Neutral red retention by lysosome from earthworm (*Lumbricus rubellus*) coelomocytes: A simple biomarker of exposure to soil copper. Environ. Toxicol. Chem. 15: 1801–1805.

Weeks, J.M., D.J. Spurgeon, C. Svendsen, P. Hankard, J.E. Kammenga, R. Dallinger et al. 2004. Critical analysis of soil invertebrate biomarkers: A field case study in avonmouth, UK. Ecotoxicology 13: 817–822.

Wheelock, C.E., B.M. Phillips, B.S. Anderson, J.L. Miller, M.J. Miller and B.D. Hammock. 2008. Applications of carboxylesterase activity in environmental monitoring and toxicity identification evaluations (TIEs). Rev. Environ. Contam. Toxicol. 195: 117–178.

Winston, G.W., D.R. Livingstone and F. Lips. 1990. Oxygen reduction metabolism by the digestive gland of the common marine mussel *Mytilus edulis* L. J. Exp. Zool. 255: 296–308.

Zhang, L., F. Ji, M. Li, Y. Cui and B. Wu. 2014. Short-term effects of Dechlorane Plus on the earthworm *Eisenia fetida* determined by a systems biology approach. J. Hazard Mat. 273: 239–246.

Zizek, S. and P. Zidar. 2013. Toxicity of the ionophore antibiotic lasalocid to soil-dwelling invertebrates: Avoidance tests in comparison to classic sublethal tests. Chemosphere 92: 570–575.

Index

Editor's Biography

Dr. Juan C. Sanchez-Hernandez (ORCID: 0000-0002-8295-0979) received a PhD degree in Environmental Sciences from the University of Siena (Italy) in 1999. He is a professor in Environmental Toxicology and Animal Physiology at the University of Castilla-La Mancha (Toledo, Spain). His research activity began in terrestrial ecotoxicology, particularly studying the bioaccumulation and toxicity of metals and persistent organic pollutants in organisms inhabiting remote areas (Antarctic continent and the Andean range). Recently, his research interests include the ecotoxicity of agrochemicals on non-target terrestrial organisms, and vermiremediation or the use of earthworms to degrade pesticides in soil. He has published more than 70 peer-reviewed articles (Researcher ID: E-8928-2011) on these topics. Besides, Dr. Sanchez-Hernandez has supervised more than 10 postgraduate and postdoctoral fellows and visiting scientists from many countries, including Portugal, Italy, France, Argentina, Brazil, and Chile. He is a frequent reviewer both for several of top journals in the field of environmental sciences (publons.com/a/730765) as well as for a number of international research institutions (e.g., the Spanish National Agency, Chilean Commission of Science and Technology or CONICYT, Argentinian Minister for Science, Technology and Innovation or FONCyT, the Georgian Shota Rustaveli National Science Foundation, and Croatian Science Foundation).

Editor's Biography

Printed and bound by CPI Group (UK) Ltd, Croydon, CR0 4YY

24/10/2024

01778307-0009